Mathematics: From Theory to Applications

Mathematics: From Theory to Applications

Editor: Lucas Lincoln

NY RESEARCH PRESS
New York

Published by NY Research Press
118-35 Queens Blvd., Suite 400,
Forest Hills, NY 11375, USA
www.nyresearchpress.com

Mathematics: From Theory to Applications
Edited by Lucas Lincoln

International Standard Book Number: 978-1-63238-554-3 (Hardback)

Cataloging-in-Publication Data

Mathematics : from theory to applications / edited by Lucas Lincoln.
p. cm.
Includes bibliographical references and index.
ISBN 978-1-63238-554-3
1. Mathematics. I. Lincoln, Lucas.
QA36 .M38 2017
510--dc23

Printed in the United States of America.

Contents

Preface VII

Chapter 1 **Barrier Option Under Lévy Model : A PIDE and Mellin Transform Approach** 1
Sudip Ratan Chandra and Diganta Mukherjee

Chapter 2 **Fractional Schrödinger Equation in the Presence of the Linear Potential** 19
André Liemert and Alwin Kienle

Chapter 3 **Tight State-Independent Uncertainty Relations for Qubits** 33
Alastair A. Abbott, Pierre-Louis Alzieu, Michael J. W. Hall and Cyril Branciard

Chapter 4 **Dynamics and the Cohomology of Measured Laminations** 50
Carlos Meniño Cotón

Chapter 5 **New Approach for Fractional Order Derivatives: Fundamentals and Analytic Properties** 70
Ali Karcı

Chapter 6 **Microtubules Nonlinear Models Dynamics Investigations through the exp(-Φ(ξ))-Expansion Method Implementation** 85
Nur Alam and Fethi Bin Muhammad Belgacem

Chapter 7 **Chaos Control in Three Dimensional Cancer Model by State Space Exact Linearization Based on Lie Algebra** 98
Mohammad Shahzad

Chapter 8 **Birkhoff Normal Forms, KAM Theory and Time Reversal Symmetry for Certain Rational Map** 109
Erin Denette, Mustafa R. S. Kulenović and Esmir Pilav

Chapter 9 **Coefficient Inequalities of Second Hankel Determinants for Some Classes of Bi-Univalent Functions** 121
Rayaprolu Bharavi Sharma and Kalikota Rajya Laxmi

Chapter 10 **A Note on Burg's Modified Entropy in Statistical Mechanics** 132
Amritansu Ray and S. K. Majumder

Chapter 11 **Multiplicative Expression for the Coefficient in Fermionic 3-3 Relation** 149
Igor Korepanov

Chapter 12 **Qualitative Properties of Difference Equation of Order Six** 165
Abdul Khaliq and E.M. Elsayed

Chapter 13 **Skew Continuous Morphisms of Ordered Lattice Ringoids** 179
Sergey Victor Ludkowski

Permissions

List of Contributors

Index

Preface

This book aims to highlight the current researches and provides a platform to further the scope of innovations in this area. This book is a product of the combined efforts of many researchers and scientists from different parts of the world. The objective of this book is to provide the readers with the latest information in the field.

This book on mathematics discusses the innovative theories and applications of the field. This book elucidates the concepts and innovative models around prospective developments with respect to mathematics. Principles and theorems that are developed in this field are of great importance to various allied fields. The various advancements in this field are glanced at and their applications as well as ramifications are looked at in detail. The chapters covered in this extensive book deal with the core subject of mathematics. For someone with an interest and eye for detail, this book covers the most significant topics in the field of mathematics.

I would like to express my sincere thanks to the authors for their dedicated efforts in the completion of this book. I acknowledge the efforts of the publisher for providing constant support. Lastly, I would like to thank my family for their support in all academic endeavors.

Editor

Barrier Option Under Lévy Model : A PIDE and Mellin Transform Approach

Sudip Ratan Chandra [1,*] and Diganta Mukherjee [2]

Academic Editor: Anatoliy Swishchuk

1 Department of Mathematics, Jadavpur University, West Bengal 700032, India
2 Indian Statistical Institute, Kolkata, West Bengal 700108, India; diganta@isical.ac.in
* Correspondence: sudipratan@gmail.com

Abstract: We propose a stochastic model to develop a partial integro-differential equation (PIDE) for pricing and pricing expression for fixed type single Barrier options based on the Itô-Lévy calculus with the help of Mellin transform. The stock price is driven by a class of infinite activity Lévy processes leading to the market inherently incomplete, and dynamic hedging is no longer risk free. We first develop a PIDE for fixed type Barrier options, and apply the Mellin transform to derive a pricing expression. Our main contribution is to develop a PIDE with its closed form pricing expression for the contract. The procedure is easy to implement for all class of Lévy processes numerically. Finally, the algorithm for computing numerically is presented with results for a set of Lévy processes.

Keywords: Barrier option pricing; Lévy process; numerical inverse Mellin transform; simulation

1. Introduction

Barrier options are derivatives with a pay-off that depends on whether a reference entity has crossed a certain boundary. Common examples are the knock-in and knock-out call and put options that are activated or deactivated when the underlying crosses a specified Barrier-level. Barrier and Barrier-type options belong to the most widely traded exotic options in the financial markets.

A class of models that has been shown to be capable of generating a good fit of observed call and put option price data is formed by the infinite activity Lévy models, such as normal inverse Gaussian, CGMY and Meixner. This class of models has been extensively studied and we refer for background and further references to the book by [1]. In this paper, we consider Barrier options driven by Lévy processes with infinite activity. This class contains many of the Lévy models used in financial modelling as the fore-mentioned ones.

Several approaches have been proposed during the last few years. The calculation of first-passage distributions and Barrier option prices in (specific) Lévy models has been investigated in a number of papers. In [2], the authors proposed a Laplace transformed based approach to compute the prices and greeks of Barrier options for a class of Lévy process with Wiener-Hopf factorisation. The authors of [3] calculated prices and deltas of double Barrier options under the Black-Scholes model. For spectrally one-sided Lévy processes with a Gaussian component [4] derived a method to evaluate first-passage distributions. The authors of [5–7] followed a transform approach to obtain Barrier prices for a jump-diffusion with exponential jumps. In the setting of infinite activity Lévy processes with jumps in two directions Cont and [8] investigated discretisation of the associated integro-differential equations. In [9], the author employed Fourier methods to investigate Barrier option prices for Lévy processes of regular exponential type. These approaches are based on exponential Lévy process with a risk neutral measure considering a complete market, involving extremely complex techniques and applicable for a specific class of Lévy process.

Summarizing all the issues in the previous work, we find a few challenges in pricing the Barrier option under Lévy processes. First of all, the Lévy market is incomplete and more than one measure exists leading to multiple prices for a single contract and hedging is not possible. Therefore, the pricing model requires the selection of the correct measure from the market and finding market price of risk with the help of market price available by calibration method with better goodness of fit. Secondly, as the distribution of the underlying stock prices is unknown, in general no explicit analytical expression is available. Finally, it is also difficult to derive a closed form expression of the contract. Our model is proposed to take care of all the challenges. The approach first developed a PIDE for pricing and solved it using Mellin transform and its inverse. In [10], the author proposed a similar method for Asian options of arithmetic type but used Fourier transform instead of Mellin transform. The advantage of our model is that it has a closed form expression of the Mellin transform applicable for any class of Lévy processes and the standard inverse Mellin transform can be applied to construct prices. The Mellin transform based method for option pricing was proposed earlier by [11–13] for pricing American options.

The organization of different sections in this paper is as follows. Section 2 recalls some basic facts about exponential Lévy processes and provides a model used in this paper. Section 3 derives the partial integro-differential equation (PIDE) for the option pricing of Barrier options. It also provides a pricing formula in terms of the inverse Mellin transform. Numerical results are provided in Section 4 and a brief conclusion is provided in Section 5.

2. Model with Lévy Processes

We denote the stock-price of the underlying asset at a given time t by $S(t)$. It is well known that contrary to the Brownian process the log-return of stock-price (that is, $\log(S(t))$) is neither Gaussian, nor homogeneous and it does *not* have independent increments (see, e.g., [14]). Thus, we study the return considering the stock price as the exponential Lévy process described by the following equations:

$$\begin{aligned} S(t) &= S(0)e^{Z(t)}, \\ dZ(t) &= \mu dt + \sigma dW(t) + \int_{\mathbb{R}} x\tilde{N}(dt, dx) \end{aligned} \tag{1}$$

with $\tilde{N}(dt, dx) = N(dt, dx) - \nu(dx)dt$, where N is the jump measure of Z and $W(t)$ is the Brownian motion. The Lévy triplet for Z is (μ, σ^2, ν) with respect to some measure $\mathbb{P}$.

For convenience, we assume $S(0) = 1$ for the rest of the paper. The parameters σ, and μ are called the *volatility* and *drift* of stock price respectively. We assume that $Z(t)$ has finite moments $\int_{|x|\geq 1} |x|^p \nu(dx) < \infty$, for all positive integer p (see [15]). The examples of such a class of Lévy processes are the infinite activity processes like VG, NIG, CGMY, Meixner processes. Some of these processes are described in Appendix B. Details of these processes are also described in [1].

We briefly describe the procedure of finding the equivalent martingale measure. All the details are provided in the Appendix A. To find an equivalent martingale measure Q for the stock-price process $S(t)$, let Y be a Lévy type stochastic integral of the form

$$dY(t) = G(t)dt + F(t)W(t) + \int_{\mathbb{R}-\{0\}} H(t,x)\tilde{N}(ds, dx)$$

where $\sqrt{G(t)}, F(t) \in \mathcal{P}_2(t)$ and $H \in \mathcal{P}_2(t, \mathbb{R} - \{0\})$ for each $t \geq 0$ (where $\mathcal{P}_2$ is defined in the Appendix A). The equivalent martingale measure Q, on a fixed time interval $[0, T]$, satisfies $\frac{dQ}{d\mathbb{P}} = e^{Y(T)}$, for $0 \leq t \leq T$.

Clearly, the Lévy triplet of Z with respect to $\mathbb{Q}$ in terms of the Lévy triplet with respect to $\mathbb{P}$ is given by

$$\left(\mu_Q, \sigma^2, e^{H(t,x)}\nu(dx)\right), \quad \mu_Q = \mu + \sigma F(t) + \int_{\mathbb{R}} x(e^{H(t,x)} - 1)\nu(dx) \tag{2}$$

We make the following assumption related to the nature of the function $H(t,x)$.

Assumption 1. $\int_{|x|\geq 1} e^x \nu_Q(dx) = \int_{|x|\geq 1} e^{x+H(t,x)}\nu(dx) < \infty$.

Therefore, with respect to the equivalent martingale measure $\mathbb{Q}$, the dynamics of $S(t)$ is given by

$$\frac{dS(t)}{S(t-)} = r\,dt + \sigma dW_Q(t) + \int_{\mathbb{R}} (e^x - 1)\tilde{N}_Q(dt, dx) \tag{3}$$

It is clear from Equations (2) and (A3) that there are non-unique ways (depending on various choices of $F(t)$ and $H(t,x)$) of selecting density function Y. The choice for the equivalent martingale measure $\mathbb{Q}$ in this paper will be the *Föllmer-Schweizer minimal measure* which minimizes the quadratic risk of the associated cost function. In this procedure there is an unique measure $\mathbb{Q}$ for which $\frac{d\mathbb{Q}_t}{d\mathbb{P}_t} = e^{Y(t)}$, so that

$$d(e^{Y(t)}) = e^{Y(t)}\, P(t)\left(\sigma dW(t) + \int_{\mathbb{R}} x\tilde{N}(dt, dx)\right)$$

for some adapted process $P(t)$ which satisfies

$$\sigma P(t) = F(t), \quad xP(t) = e^{H(t,x)} - 1$$

for $t \geq 0$ and $x \in \mathbb{R}$. We define

$$\rho_1 = \int_{\mathbb{R}} x^2\nu(dx), \quad \rho_2 = \int_{\mathbb{R}} x(e^x - 1 - x1_{|x|<1})\nu(dx), \quad \rho_3 = \int_{\mathbb{R}} (e^x - 1 - x1_{|x|<1})\nu(dx)$$

Then we obtain the following expression from Equation (A3).

$$P(t) = \frac{r - \mu - \frac{\sigma^2}{2}}{1 + \rho_1 + \rho_2 + \rho_3} = \rho \tag{4}$$

We note that given r and the Lévy triplet of Z with respect to measure $\mathbb{P}$, *i.e.*, (μ, σ^2, ν), Equation (4) gives a constant function , for $P(t) = \rho$. Thus, $F(t) = \sigma\rho$ is also constant. On the other hand, $H(t,x)$ is a function of x alone and it is given by $H(t,x) = \log(1 + \rho x)$. Consequently, the Lévy density $\nu_Q(dx) = (1 + \rho x)\nu(dx)$. The derived parameter ρ is also known as the *market price of risk* for the Lévy market.

In [16] it is shown that this method coincides with the general procedure described by Föllmer and Schweitzer (see [17]) which works by constructing a replicating portfolio of value $V(t) = \alpha(t)\ S(t) + \beta(t)\ W(t)$ and discounting it to obtain $\tilde{V}(t) = \alpha(t)\tilde{S}(t) + \beta(t)W(0)$. If we now define the cumulative cost $C(t) = \tilde{V}(t) - \int_0^t \alpha(t)d\tilde{S}(s)$, then $\mathbb{Q}$ minimizes the risk $E\left[(C(T) - C(t))^2|\mathcal{F}_t\right]$.

3. Pricing Barrier Options

In this section, we present two main theorems related to single Barrier options. Let S be the stock price and B is a fixed single Barrier. In general, there are four different types of Barrier options according to the payoff functions. Let T be the time of expiry of the option. For *fixed strike* (K) call and put Up-And-Out Barrier options payoffs are given by $(S-K)^+, 0 \leq S \leq B$ and $(K-S)^+, 0 \leq S \leq B$ respectively. For *fixed strike* call and put Down-And-Out Barrier options the payoffs are given by $(S-K)^+, B \leq S$ and $(K-S)^+, B \leq S$ respectively. In this section, we develop a technique for pricing fixed strike call for both Up-And-Out and Down-And-Out options. Option pricing for other type options can be done by a very similar procedure. We first show that the price of the both Up-And-Out and Down-And-Out Barrier option is given by a PIDE.

For the convenience of notation, in this section, we write simply W and $\tilde{N}$ in lieu of W_Q and $\tilde{N}_Q$ respectively. Since in this section we mostly work with the equivalent martingale measure $\mathbb{Q}$ this abuse of notation will not create any confusion. However, we will keep the notation for the Lévy density with respect to $\mathbb{P}$ and $\mathbb{Q}$ as the same as in the previous section, viz. ν and ν_Q respectively. For the Föllmer Schweizer minimal equivalent martingale measure $\mathbb{Q}$,

$$\nu_Q(dx) = (1+\rho x)\nu(dx)$$

where ρ is given by Equation (4). Also, assume the Lévy density corresponding to Lévy measures ν_Q and ν are denoted as $w_Q(x)$ and $w(x)$ respectively. Thus for the Föllmer Schweizer case

$$w_Q(x) = (1+\rho x)w(x) \tag{5}$$

Theorem 1. *The price of Up-And-Out and Down-And-Out Barrier call option* $C(t, S(t))$, *where the stock-price dynamics is described by Equation* (1), *is given by*

$$\begin{aligned}\frac{\partial C(t,S)}{\partial t} &+ rS\frac{\partial C}{\partial S}(t,S) + \frac{1}{2}\sigma^2 S^2\frac{\partial^2 C}{\partial S^2}(t,S) - rC(t,S)\\ &+ \int_{\mathbb{R}} \nu_Q(dx)\left[C(t,Se^x) - C(t,S) - S(e^x-1)\frac{\partial C}{\partial S}(t,S)\right] = 0\end{aligned} \tag{6}$$

with final condition

$$C(T,S) = (S-K)^+, 0 \le S \le B \text{ for Up-And-Out option} \tag{7}$$

$$= (S-K)^+, B \le S < \infty \text{ for Down-And-Out option} \tag{8}$$

Proof. Under an equivalent martingale measure $\mathbb{Q}$, the Up-And-Out and Down-And-Out Barrier call option can be written as

$$C(t,S(t)) = e^{-r(T-t)}E_Q\left[H(S_T)|\mathcal{F}_t\right]$$

where

$$\begin{aligned} H(S_t) &= \left(S(t)-K\right)^+ \mathbb{1}_{S(t)\le B} \text{ for Up-And-Out option}\\ &= \left(S(t)-K\right)^+ \mathbb{1}_{S(t)\ge B} \text{ for Down-And-Out option}\end{aligned}$$

From the dynamics of the stock price under $\mathbf{Q}$ is given by Equation (3). We define the continuous part and jump of $S(t)$ by

$$dS^c(t) = S(t-)rdt + \sigma S(t-)dW(t)$$

and

$$\Delta S = S(t) - S(t-)$$

respectively.
The continuous part of S(t) is defined to be

$$dS^c(t) \quad = \quad rS(t)dt + \sigma S(t)dW(t)$$

Now $S(t)$ has a smooth C^2 density with derivative vanishing at infinity and so $C(t, S(t))$ is a smooth function of S and we can apply Itô formula.Let us consider $S(t) = S$ and $\widetilde{C}(t, S(t)) = e^{r(T-t)}C(t, S(t))$ and if we can apply Itô's formula to this function,

$$\begin{aligned} d\widetilde{C}(t, S(t)) &= e^{r(T-t)}\left[\frac{\partial C}{\partial t} + rS\frac{\partial C}{\partial S} + \frac{1}{2}\sigma^2 S^2\frac{\partial^2 C}{\partial S^2} - rC\right. \\ &\quad \left. + \int_{\mathbb{R}}\left(C(t, Se^x) - C(t, S) - (e^x - 1)S\frac{\partial C}{\partial S}\right)\nu_Q(dx)\right] dt \\ &\quad + e^{r(T-t)}\frac{\partial C}{\partial S}\sigma S dW(t) \\ &\quad + e^{r(T-t)}\int_{\mathbb{R}}\left\{C(t, Se^x) - C(t, S)\right\}\widetilde{N}(dt, dx) \\ &= a(t)dt + dM(t) \end{aligned}$$

where

$$\begin{aligned} a(t) &= e^{r(T-t)}\left[\frac{\partial C}{\partial t} + rS\frac{\partial C}{\partial S} + \frac{1}{2}S^2\sigma^2\frac{\partial^2 C}{\partial S^2} - rC\right. \\ &+ \left. \int_{\mathbb{R}}\left(C(t, Se^x) - C(t, S) - S(e^x - 1)\frac{\partial C}{\partial S}\right)\nu_Q(dx)\right] \end{aligned}$$

and

$$dM(t) = e^{r(T-t)}\frac{\partial C}{\partial S}\sigma S dW(t) + e^{r(T-t)}\int_{\mathbb{R}}\left\{C(t, Se^x) - C(t, S)\right\}\widetilde{N}(dt, dx)$$

Clearly, $M(t)$ is a Martingale. By construction $\widetilde{C}(t, S(t)) = E[H(S(t))|\mathcal{F}_t]$ and $M(t)$ both are martingales, then $\widetilde{C}(t, S(t)) - M(t)$ is also a martingale. But $\widetilde{C}(t, S(t)) - M(t) = \int_0^t a(s)ds$ is a continuous process with finite variation. Thgerefore, we must have $a(t) = 0$ almost surely. Thus, we obtain the partial integro-differential equation (PIDE),

$$\begin{aligned} \frac{\partial C(t, S)}{\partial t} &+ rS\frac{\partial C}{\partial S}(t, S) + \frac{1}{2}\sigma^2 S^2\frac{\partial^2 C}{\partial S^2}(t, S) - rC(t, S) \\ &+ \int_{\mathbb{R}}\nu_Q(dx)\left[C(t, Se^x) - C(t, S) - S(e^x - 1)\frac{\partial C}{\partial S}(t, S)\right] = 0 \end{aligned} \tag{9}$$

for $0 \le t \le T$ and $0 < S < \infty$ and $C(t, S) \to \infty$ as $S \to \infty$ with the boundary conditions are

Up and Out Barrier Option

$$C(t, 0) = 0, 0 \le t \le T,$$
$$C(t, B) = 0, 0 \le t < T$$
$$C(T, S) = (S - K)^+, 0 \le S \le B$$

Down and Out Barrier Option

$$C(t, 0) = 0, 0 \le t \le T$$
$$C(t, B) = 0, 0 \le t < T$$
$$C(T, S) = (S - K)^+, B \le S < \infty$$

□

Theorem 2. *The Mellin transform of the price of Barrier option* $C(t, S(t))$ *is given by*

$$C(t, S(t)) = S\frac{1}{2\pi i}\int_{c-i\infty}^{c+i\infty}\left(\frac{K}{S}\right)^{-\eta}\left[H(\eta)e^{\psi(\eta)(T-t)}\right]d\eta \tag{10}$$

with

$$H(\eta) = \begin{cases} \frac{1}{\eta(\eta+1)} - \left[\frac{(K/B)^{\eta}}{\eta} - \frac{(K/B)^{\eta+1}}{\eta+1}\right] & \textit{for Up-And-Out option} \\ \frac{(\frac{K}{B})^{\eta}}{\eta} - \frac{(\frac{K}{B})^{\eta+1}}{\eta+1}, \text{ if } \frac{K}{B} \leq 1 & \textit{for Down-And-Out option} \\ \frac{1}{\eta(\eta+1)} \text{ if } \frac{K}{B} \geq 1 & \textit{for Down-And-Out option} \end{cases}$$

and

$$\psi(\eta) = -\frac{1}{2}\sigma^2\eta(\eta+1) + r\eta + I(\eta) \tag{11}$$

with

$$I(\eta) = \int_{\mathbb{R}} \nu_Q(dx)\left[e^{(\eta+1)x} - (1+\eta)e^x + \eta\right] \tag{12}$$

Proof. Let us assume that $y = \frac{K}{S(t)}$, then

$$\begin{aligned} C(t,S) &= e^{-r(T-t)}E_Q\left[H(S_t)|\mathcal{F}_t\right] \\ &= S(t)f(t,y) \end{aligned}$$

where

$$\begin{aligned} f(t,y) &= E_Q\left[(1-y)^+|\mathcal{F}_t\right].\mathbb{1}_{y\geq\frac{K}{B}}, \text{ for Up-And-Out} \\ &= E_Q\left[(1-y)^+|\mathcal{F}_t\right].\mathbb{1}_{y\leq\frac{K}{B}}, \text{ for Down-And-Out} \end{aligned}$$

Using above we have as follows,

$$\begin{aligned} \frac{\partial f}{\partial t} &- ryf_y - \frac{1}{2}\sigma^2 y^2 f_{yy} \\ &+ \int_{\mathbb{R}} \nu_Q(dx)\left[e^x\left\{f(t,ye^{-x}) - f(t,y)\right\} + (e^x-1)yf_y\right] = 0 \end{aligned} \tag{13}$$

with the following boundary conditions

(1) Up and Out Barrier Option

$$\begin{aligned} f(T,y) &= (1-y)^+, \text{when } \infty > y \geq \frac{K}{B} \\ &= 0 \text{ else} \end{aligned}$$

(2) Down and Out Barrier Option

$$\begin{aligned} f(T,y) &= (1-y)^+, \text{when } 0 \leq y \leq \frac{K}{B} \leq 1 \\ &= 0 \text{ else} \end{aligned}$$

Now, the Mellin transform of the PIDE, gives us,

$$\begin{aligned} \frac{d\hat{f}(t,\eta)}{dt} &+ r\eta\hat{f}(t,\eta) - \frac{1}{2}\sigma^2\eta(\eta+1)\hat{f}(t,\eta) \\ &+ \int_{\mathbb{R}} \nu_Q(dx)\left[e^{(\eta+1)x} - (\eta+1)e^x + \eta\right]\hat{f}(t,\eta) = 0 \end{aligned}$$

At boundary condition $t = T$, $\hat{f}(T,\eta) = \hat{H}(\eta)$, and we can write

$$\hat{f}(t,\eta) = \hat{H}(\eta)e^{\psi(\eta)(T-t)} \tag{14}$$

where

$$\psi(\eta) = -\frac{1}{2}\sigma^2\eta(\eta+1) + r\eta + I(\eta)$$

and

$$I(\eta) = \int_{\mathbb{R}} \nu_Q(dx)\left[e^{(\eta+1)x} - (1+\eta)e^x + \eta\right]$$

Mellin Tranform of the boundary condition $\hat{H}(\eta)$ Up-and-Out Barrier option

$$\begin{aligned} \hat{H}(\eta)=\hat{f}(T,\eta) &= \int_{K/B}^{1}(1-y)y^{\eta-1}dy \\ &= \frac{1}{\eta(\eta+1)}-\left[\frac{(K/B)^{\eta}}{\eta}-\frac{(K/B)^{\eta+1}}{\eta+1}\right] \end{aligned} \tag{15}$$

and for Down-and-Out Barrier option is

$$\begin{aligned} \hat{H}(\eta)=\hat{f}(T,\eta) &= \int_{0}^{\frac{K}{B}}(1-y)y^{\eta-1}dy \\ &= \frac{\left(\frac{K}{B}\right)^{\eta}}{\eta}-\frac{\left(\frac{K}{B}\right)^{\eta+1}}{\eta+1}, \text{ if } \frac{K}{B}\leq 1 \\ &= \int_{0}^{1}(1-y)y^{\eta-1}dy=\frac{1}{\eta(\eta+1)} \text{ if } \frac{K}{B}\geq 1 \end{aligned} \tag{16}$$

Hence, we can derive the expression for Call price for the both type of options described in Equation (10) .

□

Theorem 3. *The Mellin transform of the sensitivities of Barrier option is given by*

$$\Delta(t,S(t))=\frac{1}{2\pi i}\int_{c-i\infty}^{c+i\infty}(\eta+1)\left(\frac{K}{S}\right)^{-\eta}\left[H(\eta)e^{\psi(\eta)(T-t)}\right]d\eta \tag{17}$$

$$\Gamma(t,S(t))=\frac{1}{S}\frac{1}{2\pi i}\int_{c-i\infty}^{c+i\infty}\eta(\eta+1)\left(\frac{K}{S}\right)^{-\eta}\left[H(\eta)e^{\psi(\eta)(T-t)}\right]d\eta \tag{18}$$

$$\Theta(t,S(t))=\frac{1}{2\pi i}\int_{c-i\infty}^{c+i\infty}\psi(\eta)\left(\frac{K}{S}\right)^{-\eta}\left[H(\eta)e^{\psi(\eta)(T-t)}\right]d\eta \tag{19}$$

with

$$H(\eta)=\begin{cases}\frac{1}{\eta(\eta+1)}-\left[\frac{(K/B)^{\eta}}{\eta}-\frac{(K/B)^{\eta+1}}{\eta+1}\right] & \textit{for Up-And-Out option}\\ \frac{\left(\frac{K}{B}\right)^{\eta}}{\eta}-\frac{\left(\frac{K}{B}\right)^{\eta+1}}{\eta+1}, \text{ if } \frac{K}{B}\leq 1 & \textit{for Down-And-Out option}\\ \frac{1}{\eta(\eta+1)} \text{ if } \frac{K}{B}\geq 1 & \textit{for Down-And-Out option}\end{cases}$$

and

$$\psi(\eta)=-\frac{1}{2}\sigma^{2}\eta(\eta+1)+r\eta+I(\eta) \tag{20}$$

with

$$I(\eta)=\int_{\mathbb{R}}\nu_{Q}(dx)\left[e^{(\eta+1)x}-(1+\eta)e^{x}+\eta\right] \tag{21}$$

Proof. Since

$$C(t,S(t))=S\frac{1}{2\pi i}\int_{c-i\infty}^{c+i\infty}\left(\frac{K}{S}\right)^{-\eta}\left[H(\eta)e^{\psi(\eta)(T-t)}\right]d\eta$$

and

$$\Delta(t,S(t))=\frac{\partial C}{\partial S};\Gamma(t,S(t))=\frac{\partial^{2}C}{\partial S^{2}};\Theta(t,S(t))=\frac{\partial C}{\partial t}$$

By differentiating, we will have the desired result. □

4. Numerical Results

As the Lévy market is incomplete, there exists more than one or mathematically infinite number of equivalent martingale measures. We describe a method to determine an unique Lévy measure ν from the market data by using *non-parametric calibration*. Given observed market prices of options, we follow the non-parametric approach for identification of the Lévy measure.

Let us consider the (observed) market prices $C^*(T_i, S_i, B)$, $i = 1, \dots, n$, for a set of liquid put options. The objective is to find constants ν such that

$$C^{\nu}(T_i, S_i, B) = C^*(T_i, S_i, B), \tag{22}$$

where C^{ν} is the option price computed for parameters ν. The popular approach to non-linear least squares is

$$(\nu^*) = \arg\inf_{\nu} \sum_{i=1}^{N} \{C^{\nu}(T_i, S_i, B) - C^*(T_i, S_i, B)\}^2$$

The usual formulations of the inverse problems via nonlinear least squares are ill-posed and in [18] a regularization method is proposed on relative entropy. In [18] the calibration problem is reformulated into problem of finding a risk-neutral jump-diffusion model that reproduces the observed option prices and has the smallest possible relative entropy with respect to a chosen prior model. In the calibration for the present paper we use this technique. The following parameters estimated by calibration of S&P 500 options (1970 to 2001) in [1], has been considered for computing the prices

Algorithm 1 Algorithms for computing the Barrier Call option

Require: Initial time t and stock price $S(t)$, Maturity time T, Stock growth r and Volatility σ , Lévy triplet (m, k, ν) and put price available from Market.
Ensure: $C(t, S(t))$

{**Step 1**}
Estimate the Lévy triplet (m, k, ν)

$$H(\eta) = \begin{cases} \frac{1}{\eta(\eta+1)} - \left[\frac{(K/B)^{\eta}}{\eta} - \frac{(K/B)^{\eta+1}}{\eta+1}\right] & \text{for Up-And-Out option} \\ \frac{(\frac{K}{B})^{\eta}}{\eta} - \frac{(\frac{K}{B})^{\eta+1}}{\eta+1}. \text{ if } \frac{K}{B} \leq 1 & \text{for Down-And-Out option} \\ \frac{1}{\eta(\eta+1)} \text{ if } \frac{K}{B} \geq 1 & \text{for Down-And-Out option} \end{cases}$$

{**Step 2**}
for $n \leftarrow 1, L$ **do**
  Evaluate $I(n) = \int_{\mathbb{R}} \nu_Q(dx) \left[e^{(n+1)x} - (1+n)e^x + n\right]$ using Clenshaw Curtis quadrature rule in the Appendix C taking examples of Levy Process from Appendix B
  $\psi(n) = -\frac{1}{2}\sigma^2 n(n+1) + m\eta + I(n)$
  $C(t, n) = H(n)e^{\psi(n)(T-t)}$
  $fVal(n) = \frac{P(t,n)}{2^n\Gamma(n)}$
end for
for $k \leftarrow 1, L$ **do**
  temp=0
  for $n \leftarrow 1, k$ **do**
    $temp = temp + (-1)^{n-1}\binom{k-1}{n-1} fVal(n)$
  end for$C(k) = temp$;
end for
for $k \leftarrow 1, L$ **do**
  $C(t, S(t)) = C(t, S(t)) + C(k) * e^{-\frac{S}{2}} L_{k-1}\left(\frac{S}{2}\right)$;
end for
$C(t, S(t))$

Algorithm 1 describes the procedure for computing the call price of the both Down-And-Out and Up-And-Out Barrier options. We have used above calibrated parameters to plot the call price plot against the Time-to-Maturity and Initial stock price for NIG, CGMY and Meixner processes in Figures 1–6. This help us to understand how the call price changes with the change in stock price and maturity. The change of call price and sensitivities are also computed with the change of parameters such as volatility σ, Interest rate r, initial stock price S_0 and Barrier B.

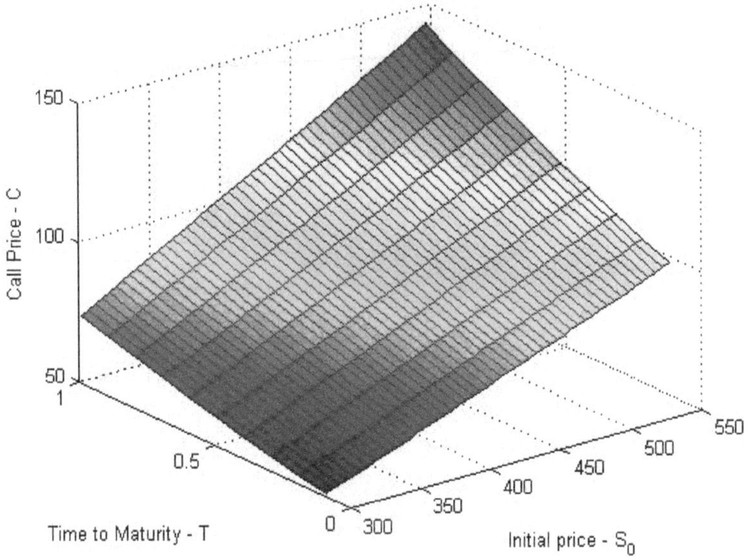

Figure 1. Down-And-Out call with NIG process with Stock Price $S_0 = 450$, Strike price K = 150, Barrier B = 350, $\sigma = 0.1812$, r = 0.167 and Time to maturity T = 1.1.

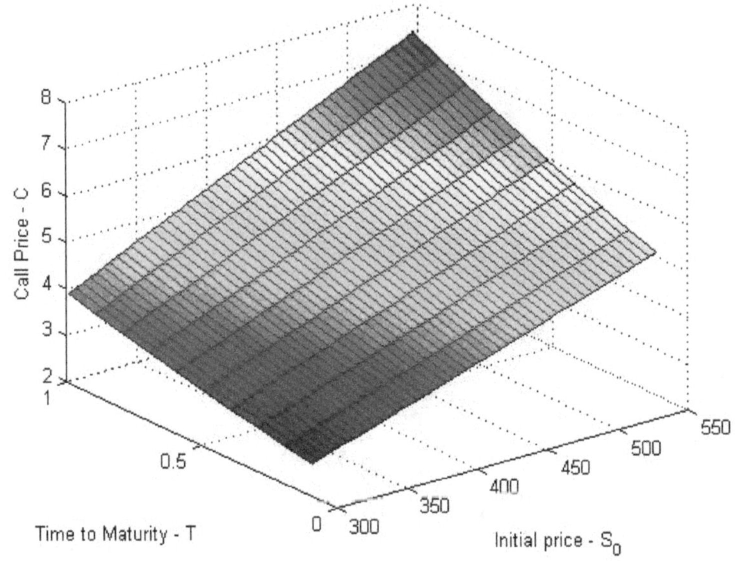

Figure 2. Up-And-Out call with NIG ($\alpha = 6.1882$, $\beta = -3.8941$, $\delta = 0.1622$) with Stock Price $S_0 = 450$, Strike price K = 150, Barrier B = 350, $\sigma = 0.1812$, r = 0.167 and Time to maturity T = 1.1.

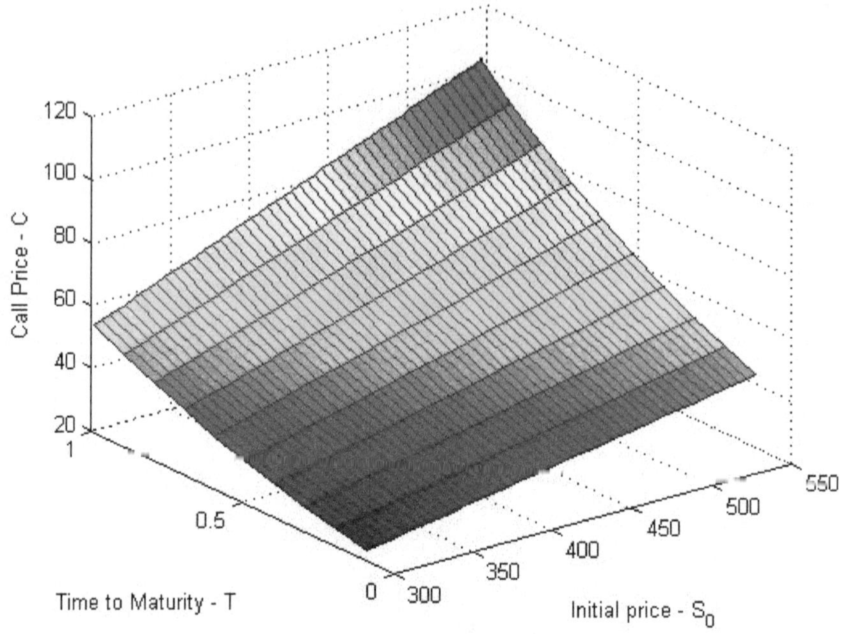

Figure 3. Down-And-Out call with CGMY(C = 0.0244, G = 0.0765, M = 7.5515, Y = 1.2945) with Stock Price S_0 = 450, Strike price K = 150, Barrier B = 350, σ = 0.1812, r = 0.167 and Time to maturity T = 1.1.

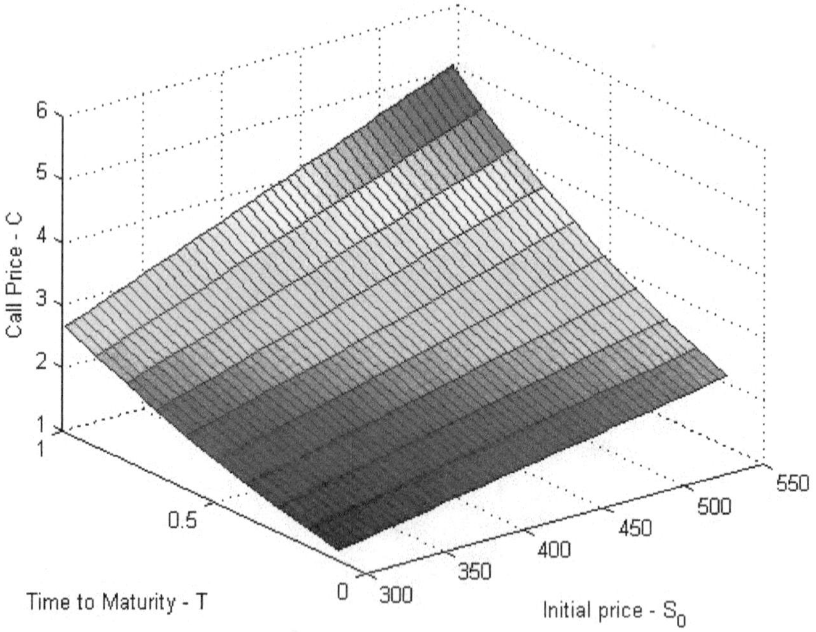

Figure 4. Up-And-Out call with CGMY(C = 0.0244, G = 0.0765, M = 7.5515, Y = 1.2945) with Stock Price S_0 = 450, Strike price K = 150, Barrier B = 350, σ = 0.1812, r = 0.167 and Time to maturity T = 1.1.

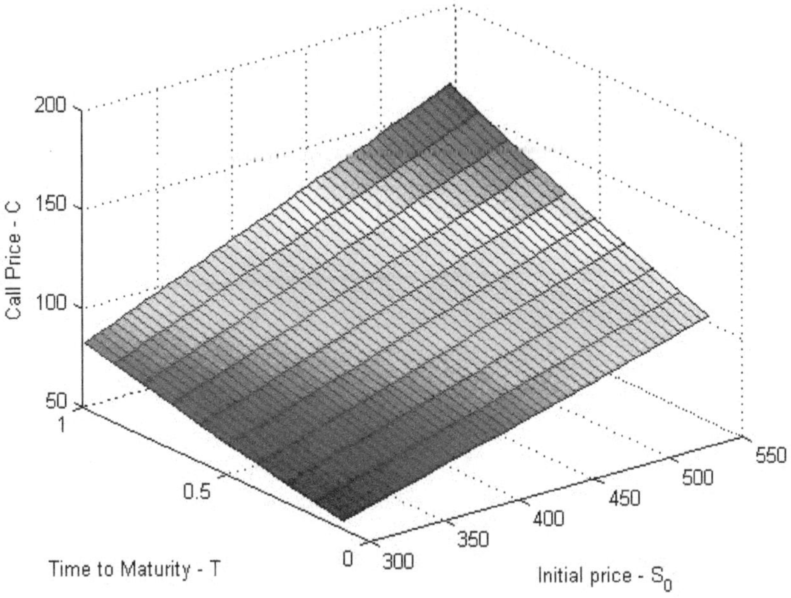

Figure 5. Down-And-Out call with Meixner($\alpha = 0.3977$, $\beta = -1.494$, $\delta = 0.3462$) with Stock Price $S_0 = 450$, Strike price K = 150, Barrier B = 350, $\sigma = 0.1812$, $r = 0.167$ and Time to maturity T = 1.1.

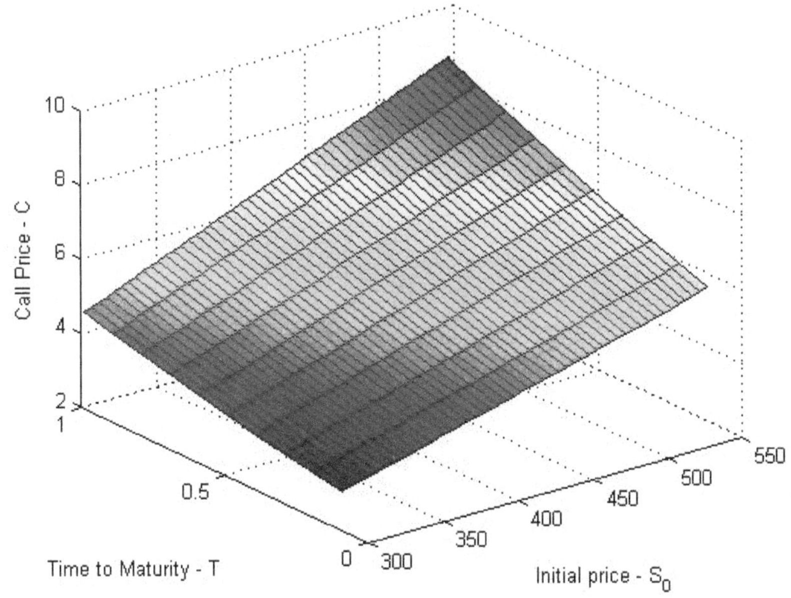

Figure 6. Up-And-Out call with Meixner($\alpha = 0.3977$, $\beta = -1.494$, $\delta = 0.3462$) with Stock Price $S_0 = 450$, Strike price K = 150, Barrier B = 350, $\sigma = 0.1812$, $r = 0.167$ and Time to maturity T = 1.1.

In Table 1 we provide the calibration results for the given data set with three different processes (as Lévy density)- NIG, CGMY and Meixner. The Algorithm 1 used to compute the call price and sensitivities and result listed in Tables 2–5. This result is also generated with the change of time-to-maturity, growth and volatility of the stock for different types of Lévy process.

Table 1. Estimated parameters for Levy processes.

Model	Parameters			
NIG	$\alpha = 6.1882$	$\beta = -3.8941$	$\delta = 0.1622$	
CGMY	$C = 0.0244$	$G = 0.0765$	$M = 7.5515$	$Y = 1.2945$
Meixner	$\alpha = 0.3977$	$\beta = -1.494$	$\delta = 0.3462$	

Table 2. Change in Call Price with different types of Lévy Process.

t	r	σ	NIG (α, β, δ) Call		CGMY (C, G, M, Y) Call		Meixner (α, β, δ) Call	
			Down-Out	Up-Out	Down-Out	Up-Out	Down-Out	Up-Out
1	0.167	0.5	8.8249	9.5132	8.4641	9.1207	8.8331	9.5222
	0.167	0.2	8.9626	9.6631	8.8112	9.4983	8.9689	9.6699
0.8	0.167	0.5	8.6620	9.3362	7.6423	8.2272	8.6063	9.3626
	0.167	0.2	9.0740	9.7843	8.6218	9.2924	9.0932	9.8052
0.5	0.167	0.5	8.4232	9.0767	6.5559	7.0477	8.4706	9.1282
	0.167	0.2	9.2436	9.9691	8.3454	8.9918	9.2827	10.0117

Call option with stock Initial value S = 300, Strike price K = 150, Barrier B = 450 and Time to maturity T = 1.1.

Table 3. Call Price & Sensitivities change with Barrier.

Barrier (B)	Call		Delta		Gamma		Theta	
	Down-Out	Up-Out	Down-Out	Up-Out	Down-Out	Up-Out	Down-Out	Up-Out
250	15.8960	3.4417	0.0410	0.0058	8.7909	1.9033	−0.6604	−0.1417
300	14.0080	5.3296	0.0374	0.0094	7.7468	2.9474	−0.5825	−0.2196
350	12.4266	6.9111	0.0342	0.0126	3.8220	6.8722	−0.5172	−0.2850

Option with Stock price S = 350, K = 150, σ = 0.1812, r = 0.167, Time to maturity T = 1.1 and NIG $(\alpha = 6.1882, \beta = -3.8941, \delta = 0.1622)$ as Lévy Process.

Table 4. Change of Delta and Gamma over Stock price change.

S_0	NIG (α, β, δ) Delta		NIG Gamma		CGMY (C, G, M, Y) Delta		CGMY Gamma		Meixner (α, β, δ) Delta		Meixner Gamma	
	Down-Out	Up-Out	Down-Out	Up-Out	Down-Out	Up-Out	Down-Out	Up-Out	Down-Out	Up-Out	Down-Out	Up-Out
350	0.03	0.01	6.87	3.82	0.02	0.01	4.97	4.61	0.02	0.01	5.53	5.13
400	0.03	0.01	9.09	5.05	0.02	0.01	6.58	6.09	0.03	0.01	7.32	6.78
450	0.04	0.01	11.64	6.45	0.03	0.01	8.42	7.80	0.03	0.02	9.38	8.67

Barrier Call option with Strike price K = 150, Barrier B = 350, σ = 0.1812, r = 0.167 and Time to maturity T = 1.1.

Table 5. Change of Theta over Time to expire.

t	NIG (α, β, γ) Theta		CGMY (C, G, M, Y) Theta		Meixner (α, β, δ) Theta	
	Down-Out	Up-Out	Down-Out	Up-Out	Down-Out	Up-Out
0.4	−0.6851	−0.3763	0.7374	0.6763	−0.5246	−0.4801
0.6	−0.6753	−0.3710	0.7536	0.6910	−0.5175	−0.4736
0.8	−0.6655	−0.3657	0.7702	0.7061	−0.5104	−0.4672
1.0	−0.6560	−0.3605	0.7871	0.7215	−0.5034	−0.4608

Barrier Call option with Stock Price S_0 = 450, Strike price K = 150, Barrier B = 350, σ = 0.1812, r = 0.167 and Time to maturity T = 1.1.

The Sensitivities like Delta, Gamma and Theta of the option with respect to initial stock price S_t and t will be denoted by

$$\Delta = \frac{\partial}{\partial S}C(S,B,t); \Gamma = \frac{\partial^2}{\partial S^2}C(S,B,t); \Theta = \frac{\partial}{\partial t}C(S,B,t)$$

Using the above equations for sensitivities, we will check how the Call, Delta, Gamma & Theta changes with the change of Barrier for a specific type of Lévy process (in this case NIG) in the Table 3.

The Call Price and Sensitivities (Delta, Gamma and Theta) computed (Tables 4 and 5) for different types of Lévy process with its parameters.

5. Conclusions

In this paper, we have focused on three types of Lévy process with infinite activity but finite moments to option pricing and compared the results. We developed alternative techniques to compute prices and sensitives of the Barrier options. Here, we first determined the modified Lévy process under measure for incomplete market followed by development of a Partially Integro-Differential Equation and subsequently used the Mellin transform technique to get an expression for options. The expression computed numerically with a class of Lévy process with infinite activity where the distribution of the process is unknown.

Acknowledgments: We are thankful to Mrinal Ghosh at Indian Institute of Science (India), Gopal Basak at Indian Statistical Institute (India) and Indranil SenGupta, Department of Mathematics, North Dakota State University, Fargo, North Dakota, USA for providing valuable ideas.

Author Contributions: Both authors have contributed equally to this work and they agree to the final version.

Conflicts of Interest: The authors declare no conflict of interest.

Appendix

A. Derivation of the Stock Dynamics under the Equivalent Martingale Measure

Define $\mathcal{P}_2(t, E)$ to be the set of all equivalence class of mappings $f : [0,t] \times \mathcal{B}(\mathbb{R}) \times \Omega \to \mathbb{R}$ which coincide everywhere with respect to $\rho_\Sigma \times \mathbb{P}$, and satisfy the following conditions

1. f is predictable.
2. $\mathbb{P}\left(\int_0^t \int_E |f(s,x)|^2 \rho_\Sigma(ds,dx) < \infty\right) = 1,$

where ρ_Σ is a σ-finite measure on $\mathbb{R}^+ \times E$. Analogously it is possible to define $\mathcal{P}_2(t)$. Let Y be a Lévy type stochastic integral of the form

$$dY(t) = G(t)dt + F(t)W(t) + \int_{\mathbb{R}-\{0\}} H(t,x)\widetilde{N}(ds,dx)$$

where $\sqrt{G(t)}, F(t) \in \mathcal{P}_2(t)$ and $H \in \mathcal{P}_2(t, \mathbb{R}-\{0\})$ for each $t \geq 0$.

The goal is to find the equivalent martingale measure Q, on a fixed time interval $[0,T]$, for which $\frac{dQ}{d\mathbb{P}} = e^{Y(T)}$, for $0 \leq t \leq T$. We consider the associated process e^Y is a martingale and hence $G(t)$ is determined by $F(t)$ and $H(t)$. With respect to the new measure Q, $W_Q(t) = W(t) - \int_0^t F(s)\, ds$ is a Brownian motion and

$$\widetilde{N}_Q(t,E) = \tilde{N}(t,E) - \int_0^t \int_E (e^{H(s,x)} - 1)\nu(dx)\, ds$$

is a martingale (see [19], Section 5.6.3). Thus with respect to the new measure Q the dynamics of Z is given by

$$dZ(t) = \left(\mu + \sigma F(t) + \int_{\mathbb{R}} x(e^{H(s,x)} - 1)\nu(dx)\right) dt + \sigma dW_Q(t) + \int_{\mathbb{R}} x\widetilde{N}_Q(dt,dx) \tag{A1}$$

Also,

$$\int_{\mathbb{R}} x\widetilde{N}_Q(dt,dx) = \int_{\mathbb{R}} x(N(dt,dx) - \nu_Q(dx)dt)$$

where

$$\nu_Q(dx) = e^{H(t,x)}\nu(dx) \tag{A2}$$

is the Lévy measure with respect to Q. Thus from Equations (A1) and (A2) it is clear that the Lévy triplet of Z with respect to Q in terms of Lévy triplet with respect to $\mathbb{P}$ is given by

$$\left(\mu_Q, \sigma^2, e^{H(t,x)}\nu(dx)\right), \quad \mu_Q = \mu + \sigma F(t) + \int_{\mathbb{R}} x(e^{H(t,x)} - 1)\nu(dx)$$

Remark A1. *Using Girsanov's theorem (see [20]), there exist a deterministic process $\beta(t)$ and a measurable non negative deterministic process $Y(t,x)$ such that*

$$\mu_Q = \mu + \sigma^2\beta(t) + \int_{\mathbb{R}} x(Y-1)\nu(dx), \quad {\sigma_Q}^2 = \sigma^2, \quad \nu_Q(dx) = Y\nu(dx)$$

Comparing with Equation (2) *we obtain* $\beta(t) = \frac{F(t)}{\sigma}$ *and* $Y(t,x) = e^{H(t,x)}$.

With respect to Q, $e^{-rt}S(t) = e^{-rt+Z(t)}$ is a martingale. By Proposition 3.18(2), [21] and Equation (2), we thus obtain (since we have Assumption 1)

$$\frac{\sigma^2}{2} + \mu_Q + \int_{\mathbb{R}} (e^x - 1 - x1_{|x|\leq 1})\nu_Q(dx) = r \tag{A3}$$

Therefore the dynamics of stock-price is given by the following theorem.

Theorem A1. *With respect to the equivalent martingale measure Q, the dynamics of $S(t)$ is given by*

$$\frac{dS(t)}{S(t-)} = r\,dt + \sigma dW_Q(t) + \int_{\mathbb{R}} (e^x - 1)\widetilde{N}_Q(dt,dx)$$

Proof. Using results for exponential of a Lévy process (see Proposition 8.20, [21]) we obtain,

$$\frac{dS(t)}{S(t-)} = \left(\mu_Q + \frac{\sigma^2}{2} + \int_{\mathbb{R}} (e^x - 1 - x1_{|x|\leq 1})\nu_Q(dx)\right) dt + \sigma dW_Q(t) + \int_{\mathbb{R}} (e^x - 1)\widetilde{N}_Q(dt,dx)$$

Thus the proof follows from Equation (A3). □

B. Examples of Lévy Process

B.1. Lévy Process with Infinite Activity

We have considered the following Lévy processes with infinite activity but $\int_{\mathbb{R}} x^2\nu(dx)\mathbb{1}_{\{x<1\}} < \infty$.

1. The Normal Inverse Gaussian

   The NIG distribution with parameters $\alpha > 0, \alpha < \beta < \alpha$ and $\delta > 0, \mathrm{NIG}(\alpha,\beta,\delta)$, has a characteristic function

   $$E\left[e^{iuX}\right] = \exp\left(-\delta\left(\sqrt{\alpha^2 - (\beta + iu)^2} - \sqrt{\alpha^2 - \beta^2}\right)\right)$$

   The Lévy measure is given by

   $$\nu_{NIG}(dx) = \frac{\delta\alpha}{\pi}\frac{\exp(\beta x)\,K_1(\alpha|x|)}{|x|}dx \tag{B1}$$

where $K_\lambda(x)$ is the modified Bessel function of third kind with index λ.

An NIG process has no Brownian component and its Lévy triplet is

$$[\gamma, 0, \nu_{NIG}(dx)], \text{ where}$$

$$\gamma = \frac{2\delta\alpha}{\pi}\int_0^1 \sinh(\beta x)\ K_1(\alpha x)dx$$

2. The CGMY Process

The CGMY(C, G, M, Y) distribution is four parameter distribution with characteristic function

$$E[e^{iuX}] = \exp\Big(C\Gamma(-Y)\big((M-iu)^Y - M^Y + (G+iu)^Y - G^Y\big)\Big)$$

The Lévy measure of this process admits the representation

$$\nu_{CGMY}(dx) = C\Big(\frac{e^{-Mx}}{x^{1+y}}\mathbb{1}_{x>0} + C\frac{e^{Gx}}{|x|^{1+y}}.\mathbb{1}_{x<0}\Big)dx \text{ when } C, G, M > 0 \text{ and } y < 2 \quad \text{(B2)}$$

The CGMY process is a pure jump Lévy process with Lévy triplet

$$[\gamma, 0, \nu_{CGMY}(dx)]$$

where

$$\gamma = C\Big(\int_0^1 x^{-Y}e^{-Mx}dx - \int_{-1}^0 |x|^{-Y}e^{Gx}dx\Big)$$

The characteristic function of the pure jump KoBol process of order $\nu \in (0,2), \nu \neq 1$ is given by

$$\phi_{KoBol}(u) = \exp\Big(-i\mu u + c\Gamma(-\nu)\big[\lambda_+^\nu - (\lambda_+ + iu)^\nu - (-\lambda_-)^\nu - (-\lambda_- - iu)^\nu\big]\Big) \quad \text{(B3)}$$
$$\text{where } c > 0, \mu \in \mathbb{R}, \lambda_- < -1 < 0 < \lambda_+$$

An ordinary KoBoL process is obtained from this definition by specializing to the case where $\nu_+ = \nu_- = \nu$ and $c_+ = c_- = c$ The relation between these parameters and the parameters C, G, M, Y is as follows: $C = c, G = \lambda_+, M = -\lambda_-, Y = \nu$

3. The Meixner Process

The Meixner process is defined by Meixner$(\alpha, \beta, \delta), \alpha > 0, -\pi < \beta < \pi, \delta > 0$ then Lévy measure is defined by

$$\nu_{Meixner}(dx) = \delta\frac{\exp(\beta x/\alpha)}{x\ \sinh(\pi x/\alpha)}dx \quad \text{(B4)}$$

Since $\int_{-1}^{+1} |x|\nu(dx) = \infty$, the process is of infinite variation but moments of all order exists. The first parameter of Lévy triplet

$$\gamma = \alpha\delta\ \tan(\beta/2) - 2\delta\int_1^\infty \frac{\sinh(\beta x/\alpha)}{\sinh(\pi x/\alpha)}dx$$

It has no Brownian part and a pure jump part governed by the Lévy measure.

The Lévy triplet is given by

$$[\gamma, 0, \nu_{Meixner}(dx)]$$

C. Numerical Techniques

C.1. Computing $I(\eta)$ *by Clenshaw Curtis Quadrature Rule*

In this section, we will use Clenshaw-Curtis rule for integration [22] to calculate the integral $I(\eta)$ because of its high accuracy level and low computational time. According to Clenshaw-Curtis rule for integration, any integral in $[-1,1]$ can be written with the help of interpolation polynomial $L_n(x)$ as

$$I_n(f) = \int_{-1}^{1} f(x)dx = \int_{-1}^{1} L_n(x)dx = \int_{-1}^{1} \sum_{k=0}^{M} c_k T_k(x)dx = \sum_{k=0}^{M} c_k \mu_k \tag{C1}$$

where $\mu_k = \int_{-1}^{1} T_k(x)dx$ are the moments of the Chaebyshev polynomials, $c_k = \frac{2}{M}\sum_{j=0}^{M} f(x_j)\cos\left(\frac{kj\pi}{M}\right)$ which is the real part of an FFT, and $x_j = \cos(j\pi/M)$.
The μ_k can be written,

$$\mu_k = \int_{-1}^{1} T_k(x)dx = \begin{cases} 0 & \text{if k odd} \\ 2/(1-k^2) & \text{if k even} \end{cases}$$

A fast and accurate algorithm for computing the weights in the Fejér and Clenshaw-Curtis rules in $O(M\log M)$ computation has been given by [22]. The weights are obtained as the inverse FFT of certain vectors given by explicit rational expression.
Converting the any integration from interval $[a,b]$ to $[-1,1]$, we have

$$\int_{a}^{b} f(x)dx = \frac{b-a}{2}\int_{-1}^{1} f\left(\frac{b-a}{2}x + \frac{a+b}{2}\right)dx$$

C.2. Properties of Mellin Transform

The Mellin transform of real valued function $\phi(z)$ defined on $(0,\infty)$ where Mellin transform with respect to s which is a real number, is definded as

$$\mathcal{M}\{\phi(z)\} = \Phi(s) = \int_{0}^{\infty} z^{s-1}\phi(z)dz, \quad s \in \mathbb{R}$$

where its inverse is

$$\mathcal{M}^{-1}\{\Phi(s)\} = \phi(z) = \frac{1}{2\pi i}\int_{c-i\infty}^{c+i\infty} z^{-s}\Phi(s)ds, \quad c > 0$$

There are some interesting properties of Mellin Transform on scaling and derivaties of first and second order available as follows (See [23],

$$\mathcal{M}\{\phi(az)\} = a^{-s}\Phi(s)$$

$$\mathcal{M}\{z\frac{\partial\phi(z)}{\partial z}\} = -s\Phi(s)$$

$$\mathcal{M}\{z^2\frac{\partial^2\phi(z)}{\partial z^2}\} = (-1)^2 s(s+1)\Phi(s)$$

C.3. Numerical Mellin Inversion

The Mellin transform is defined by the formulae [19]:

$$\Phi(s) = \int_0^\infty z^{s-1}\phi(z)dz, \quad Re(s) > 0 \tag{C2}$$

and its inverse is

$$\phi(z) = \frac{1}{2\pi i}\int_{c-i\infty}^{c+i\infty} z^{-s}\Phi(s)ds, \quad c > 0$$

where one-to-one correspondence is denoted as follows, if the inverse $\Phi(s)$ function exists:

$$\phi(z) \leftrightarrow \Phi(s) \text{ or } \Phi(s) = \mathcal{M}\{\phi(z)\}$$

The numerical Mellin inverse is first presented by [24] and later by [25].We have followed the approach proposed by [24] and can write the numerical inverse of Mellin as,

$$\phi(t) \simeq \sum_{s=1}^{N} c_s e^{-\frac{t}{2}} L_{s-1}\left(\frac{t}{2}\right) \tag{C3}$$

where

$$c_s = \sum_{n=1}^{s} (-1)^{n-1}\binom{s-1}{n-1} H_n, \quad s = 1(1)N \tag{C4}$$

and

$$H_s \equiv H(s) \equiv \frac{\Phi(s)}{2^s\Gamma(s)} \tag{C5}$$

Now, we have observed that H_s is defined in integer domain and so $\Phi(s)$. But, in real case it is quite likely that the Mellin transform $\Phi(s^*) = \mathcal{M}\{f(t)\}$ will have a strip of existence for $s^* \in (a^*, b^*)$ where s^* is not an integer rather real.In such case, we will apply a linear transform as to keep H_s defined in integer domain as follows,

$$s^* = As + B, \quad s \in [1, N] \tag{C6}$$

with

$$A = \frac{b^* - a^*}{N-1}, \quad B = \frac{a^*N - b^*}{N-1} \text{which maps the interval } [1, N] \text{ onto } [a^*, b^*]$$

Since the function exists in interval $[a^*, b^*]$ we can invert $\Phi(As + B)$ to recover the function $g(t)$ with the following

$$\mathcal{M}\{g(t)\} = G(s) \equiv \Phi(s^*) = \Phi(As + B), \quad s \in [1, N] \tag{C7}$$

and thereafter original function $f(t) = \mathcal{M}^{-1}\Phi(s)$ can be derived by the following transformation:

$$f(t) = A\frac{g(t^A)}{t^B}$$

References

1. Shoutens, W. Chapter 5. In *Lévy Processes in Finance: Pricing Financial Derivatives*; Wiley: Hoboken, NJ, USA, 2003; p. 82.
2. Jeannin, M.; Pistorius, M. A transform approach to compute prices and Greeks of Barrier options driven by a class of Lévy processes. *Quantative Financ.* **2010**, *10*, 629–644.
3. Geman, H.; Yor, M. Pricing and hedging double Barrier options: A probabilistic approach. *Math. Financ.* **1996**, *6*, 365–378.
4. Rogers, L. Evaluating first-passage probabilities for spectrally one-sided Lévy processes. *J. Appl. Probab.* **2000**, *37*, 1173–1180.
5. Kou, S.; Petrella, G.; Wang, H. Pricing path-dependent options with jump risk via Laplace transforms. *Kyoto Econ. Rev.* **2005**, *74*, 1–23.
6. Lipton, A. *Mathematical Methods for Foreign Exchange: A Financial Engineer's Approach*; World Scientific: Singapore, Singapore, 2001.
7. Sopp, A. Pricing Barrier options under Local Volatility. Available online: http://kodu.ut.ee/spartak/papers/locvols.pdf (accessed on 24 December 2015).
8. Cont, R.; Voltchkova, E. A finite difference scheme for option pricing in jump diffusion and exponentiel Lévy models. *SIAM J. Numer. Anal.* **2005**, *43*, 1596–1626.
9. Boyarchenko, S.; Levendorskii, S. Barrier options and touch-and-out options under regular Lévy processes of exponential type. *Ann. Appl. Probab.* **2002**, *12*, 1261–1298.
10. Chandra, S.R.; Mukherjee, D.; Indranil, S. PIDE and Solution Related to Pricing of Lévy Driven Arithmetic Type Floating Asian Options. *Stoch. Anal. Appl.* **2015**, *33*, 630–652.
11. Panini, R.; Srivastav, R.P. Option pricing with Mellin Transform. *Math. Comput. Model.* **2004**, *40*, 43–56.
12. Frontczak, R.; Schöbel, R. *Pricing American options with Mellin Transforms*, No. 319; School of Business and Economics, University of TÃijbingen: Tübinger, Germany, 2010.
13. Kamden, S.J. Option pricing with Lévy process using Mellin Transform. Available online: https://hal.archives-ouvertes.fr/file/index/docid/58139/filename/sadefo_resume_Mellin_levy_optionpricing.pdf (accessed on 29 December 2015).
14. Amaral, L.; Plerou, V.; Gopikeishnan, P.; Meyer, M.; Stanly, H. The distribution of returns of stock prices. *Int. J. Theor. Appl. Financ.* **2000**, *3*, 365–369.
15. Applebaum, D. *Lévy Processes and Stochastic Calculus*; Cambridge University Press: Cambridge, UK, 2004.
16. Chan, T. Pricing Contingent Claims on Stocks Derived by Lévy Processes. *Ann. Appl. Probab.* **1999**, *9*, 504–528.
17. Föllmer, H.; Schweizer, M. Hedging of Contingent Claims under Incomplete Information. In *Applied Stochastic Analysis, Stochastics Monographs*; Davis, M.H.A., Elliott, R.J., Eds.; Gordon and Breach: New York, NY, USA, 1991; Volume 5, pp. 389–414.
18. Cont, R.; Tankov, P. Calibration of Jump-Diffusion Option Pricing Models: A Robust Non-Parametric Approach. *SSRN Electron. J.* **2002**, doi:10.2139/ssrn.332400.
19. Doetsch, G. *Handbuch der Laplace—Transformtion*; Verlag Birkhauser: Basel, Switzerland, 1950.
20. Jacod, J.; Shiryaev, A.N. *Limit Theorems for Stochastic Processes*, 2nd ed.; Springer: New York, NY, USA, 2003.
21. Rama, C.; Peter, T. Chapter 2.6.3. In *Financial Modelling with Jump Processes*; Chapman & Hall/CRC Financial Mathematics Series; Chapman & Hall/CRC: London, UK, 2004.
22. Waldvogel, J. Fast construction of the Fejér and Clenshaw-Curtis quadrature rules. *BIT Numer. Math.* **2004**, *43*, 1–18.
23. Debnath, L.; Bhatta, D. Chapter 8. In *Integral Transforms and Their Applications*, 2nd ed.; Chapman and Hall/CRC: London, UK, 2010.
24. Theocaris, P.; Chrysakis, A.C. Numerical inversion of the Mellin transform. *J. Math. Appl.* **1977**, *20*, 73–83.
25. Iqbal, M. Spline regularization of Numerical Inversion of MellinTransform. *Approx. Theory Appl.* **2000**, *16*, 1–16.

2

Fractional Schrödinger Equation in the Presence of the Linear Potential

André Liemert * and Alwin Kienle

Institut für Lasertechnologien in der Medizin und Meßtechnik an der Universität Ulm, Helmholtzstr. 12, D-89081 Ulm, Germany; alwin.kienle@ilm-ulm.de
* Correspondence: andre.liemert@ilm-ulm.de

Academic Editor: Rui A. C. Ferreira

Abstract: In this paper, we consider the time-dependent Schrödinger equation:

$$i\frac{\partial\psi(x,t)}{\partial t} = \frac{1}{2}(-\Delta)^{\frac{\alpha}{2}}\psi(x,t) + V(x)\psi(x,t), \quad x \in \mathbb{R}, \quad t > 0$$

with the Riesz space-fractional derivative of order $0 < \alpha \leq 2$ in the presence of the linear potential $V(x) = \beta x$. The wave function to the one-dimensional Schrödinger equation in momentum space is given in closed form allowing the determination of other measurable quantities such as the mean square displacement. Analytical solutions are derived for the relevant case of $\alpha = 1$, which are useable for studying the propagation of wave packets that undergo spreading and splitting. We furthermore address the two-dimensional space-fractional Schrödinger equation under consideration of the potential $V(\boldsymbol{\rho}) = \mathbf{F} \cdot \boldsymbol{\rho}$ including the free particle case. The derived equations are illustrated in different ways and verified by comparisons with a recently proposed numerical approach.

Keywords: Riesz fractional derivative; Caputo fractional derivative; Mittag-Leffler matrix function; fractional Schrödinger equation

1. Introduction

The time-dependent Schrödinger equation is a fundamental equation in quantum mechanics for studying the dynamics and evolution of wave packets over time. Several years ago, the classical Schrödinger equation has been generalized to a fractional partial differential equation that takes into account the Riesz space-fractional derivative instead of the conventional Laplacian [1,2]. Apart from quantum mechanics, there are many other equations occurring in science that have been reconsidered in terms of fractional derivatives such as the diffusion-wave equation [3–6], the Langevin equation [7] or the radiative transport equation [8]. Analytical solutions to fractional differential equations are, in general, not available in terms of elementary functions. As an example, the fundamental solution to the Schrödinger equation for a free particle can only be written in closed form under consideration of the Fox H-function [9]. Very recently, the space-fractional Schrödinger equation has been employed for studying the propagation dynamics of wave packets in the presence of the harmonic potential [10] as well as of the free particle [11]. In addition, the fractional Schrödinger equation subject to a periodic $\mathcal{PT}$-symmetric potential has been used to investigate the conical diffraction of a light beam [12]. An optical realization of the space-fractional Schrödinger equation, based on transverse light dynamics in aspherical optical cavities, has been recently proposed in the study [13]. Besides the free particle and the harmonic potential, the case of a linear potential is also a fundamental problem in quantum mechanics that can be treated and solved analytically [14,15]. In this context, we refer to the paper [16] dealing with the nonlinear Schrödinger equation in the presence of uniform acceleration.

We additionally note on the time-dependent Schrödinger equation with a nonlocal term that has been analyzed in the publication [17].

In this article, we consider the time-dependent Schrödinger equation with the Riesz space-fractional derivative of order $0 < \alpha \leq 2$ in the presence of the linear potential. In the one-dimensional (1D) case, the wave function in momentum space is derived for an arbitrary fractional order α in closed form, which allows the determination of other important quantities such as the mean square displacement (MSD). For the case of $\alpha = 1$, we provide some explicit analytical solutions useable for studying the time evolution of wave packets that undergo both spreading and splitting. We additionally address the space-fractional Schrödinger equation in two spatial dimensions. The obtained equations are illustrated in different ways and partly verified by comparisons with a recently proposed numerical approach.

2. 1D Space-Fractional Schrödinger Equation

The 1D space-fractional Schrödinger equation in the presence of the linear potential $V(x) = \beta x$ is given by

$$i\frac{\partial\psi(x,t)}{\partial t} = \frac{1}{2}(-\Delta)^{\frac{\alpha}{2}}\psi(x,t) + \beta x\psi(x,t), \quad x \in \mathbb{R}, \quad t > 0 \tag{1}$$

where $\psi(x,t)$ is the wave function, $0 < \alpha \leq 2$ is the order of the Riesz space-fractional derivative and $\beta \in \mathbb{R}$. The nonlocal operator occurring in Equation (1) is defined in the sense of [18]

$$\mathcal{F}\{(-\Delta)^{\frac{\alpha}{2}}f(x)\}(\omega) = |\omega|^{\alpha}F(\omega) \tag{2}$$

with $F(\omega) = \int_{-\infty}^{\infty} f(x)\exp(-i\omega x)\,dx$ being the Fourier transform of $f(x)$. Alternatively, it can also be defined directly in real space in terms of the left- and right-sided Riemann-Liouville derivatives [19]. In this section, Equation (1) is considered under the initial and boundary conditions

$$\psi(x,0^{+}) = \psi_0(x), \qquad \lim_{|x|\to\infty}\psi(x,t) = 0 \tag{3}$$

where $\psi_0(x)$ is the initial wave packet that is normalized according to $\int_{-\infty}^{\infty}|\psi_0(x)|^2\,dx = 1$. The corresponding Schrödinger equation in momentum space is given by

$$i\frac{\partial\psi(\omega,t)}{\partial t} = \frac{1}{2}|\omega|^{\alpha}\psi(\omega,t) + i\beta\frac{\partial\psi(\omega,t)}{\partial\omega}, \quad \omega \in \mathbb{R}, \quad t > 0 \tag{4}$$

where $\psi(\omega,0^{+}) = \psi_0(\omega)$. In order to solve this equation, we introduce a wave function of the form $\psi(\omega,t) = e^{i\varphi(\omega)}u(\omega+\beta t)$, where φ and u are unknown. Inserting this function into Equation (4) results in

$$i\beta e^{i\varphi}u'(\omega+\beta t) = \frac{1}{2}|\omega|^{\alpha}e^{i\varphi}u(\omega+\beta t) + i\beta e^{i\varphi}u'(\omega+\beta t) - \beta e^{i\varphi}u(\omega+\beta t)\frac{d\varphi}{d\omega} \tag{5}$$

where u' denotes the first derivative of the unknown function u. Comparing both sides of Equation (5) leads to the equation

$$\frac{d\varphi(\omega)}{d\omega} = \frac{|\omega|^{\alpha}}{2\beta} \tag{6}$$

which is solved by the function

$$\varphi(\omega) = \frac{\omega|\omega|^{\alpha}}{2\beta(1+\alpha)} + C, \quad \omega \in \mathbb{R} \tag{7}$$

with C being an arbitrary constant. The unknown quantity u can be found by employing the initial condition according to

$$\psi(\omega,0) = \psi_0(\omega) = e^{i\varphi(\omega)}u(\omega) \quad \Rightarrow \quad u(\omega) = \psi_0(\omega)e^{-i\varphi(\omega)} \tag{8}$$

Therefore, the general solution of the Schrödinger Equation (4) in momentum space is found to be

$$\psi(\omega,t) = \psi_0(\omega+\beta t)\exp\left(i\frac{\omega|\omega|^\alpha - (\omega+\beta t)|\omega+\beta t|^\alpha}{2\beta(1+\alpha)}\right) \tag{9}$$

for all $0 < \alpha \leq 2$. It can be seen that $|\psi(\omega,t)|^2 = |\psi_0(\omega+\beta t)|^2$, which is a measure for the probability density in momentum space, is independent of the fractional order. The wave function in real space is formally given by

$$\psi(x,t) = \frac{1}{2\pi}\int_{-\infty}^{\infty} \psi(\omega,t)\exp(i\omega x)\,d\omega \tag{10}$$

In general, this integral must be carried out numerically using one of the established quadrature rules. For the classical case of $\alpha = 2$, the wave function can also be given in terms of the convolution

$$\psi(x,t) = \frac{e^{-i\beta x t}e^{-i\beta^2 t^3/6}}{\sqrt{2\pi i t}}\int_{-\infty}^{\infty}\psi_0(y)\exp\left(i\frac{(x-y+\beta t^2/2)^2}{2t}\right)dy \tag{11}$$

The probability density function (pdf) associated with the position of the particle is defined as $\rho(x,t) = |\psi(x,t)|^2$. Furthermore, in view of experimental activities, the MSD is known to be an important quantity that can be computed by means of the pdf as $\langle x^2(t)\rangle = \int_{-\infty}^{\infty} x^2\rho(x,t)\,dx$. In our case, when $\rho(x,t)$ is not given, the MSD can be determined from the wave function in momentum space according to

$$\langle x^2(t)\rangle = \frac{1}{2\pi}\int_{-\infty}^{\infty}\left|\frac{\partial\psi(\omega,t)}{\partial\omega}\right|^2 d\omega \tag{12}$$

Using Equation (9), one obtains the MSD for an arbitrary initial stage as

$$\langle x^2(t)\rangle = \frac{1}{2\pi}\int_{-\infty}^{\infty}\left|\frac{d\psi_0(\omega)}{d\omega} + i\psi_0(\omega)\frac{|\omega-\beta t|^\alpha - |\omega|^\alpha}{2\beta}\right|^2 d\omega \tag{13}$$

In the following, we also want to address the case of the free particle which is obtained for $\beta = 0$. The corresponding solution in momentum space Equation (9) reduces to

$$\psi(\omega,t) = \psi_0(\omega)\exp(-it|\omega|^\alpha/2), \quad 0 < \alpha \leq 2 \tag{14}$$

However, also in this case, the wave function in real space is not available in elementary form. Instead, we have

$$\psi(x,t) = \int_{-\infty}^{\infty} G(x-y,t)\psi_0(y)\,dy \tag{15}$$

with $G(x,t)$ being Green's function, which is given for $1 < \alpha \leq 2$ by

$$G(x,t) = \frac{1}{\sqrt{\pi}|x|}H_{1,2}^{1,1}\left[\frac{2}{it}\frac{|x|^\alpha}{2^\alpha}\,\middle|\,\begin{matrix}(1,1)\\(1/2,\alpha/2),(1,\alpha/2)\end{matrix}\right] \tag{16}$$

The Fox H-function occurring above is defined in terms of the Mellin-Barnes integral [9]

$$H_{p,q}^{m,n}\left[z\,\middle|\,\begin{matrix}(a_p,A_p)\\(b_q,B_q)\end{matrix}\right] = \frac{1}{2\pi i}\int_L \frac{\prod_{j=1}^m\Gamma(b_j+B_js)\prod_{j=1}^n\Gamma(1-a_j-A_js)}{\prod_{j=n+1}^p\Gamma(a_j+A_js)\prod_{j=m+1}^q\Gamma(1-b_j-B_js)}z^{-s}\,ds \tag{17}$$

where $\Gamma(s)$ is the Gamma function and L denotes an appropriate integration path in the complex plane. The fundamental solution Equation (16) can be evaluated by means of the series

$$G(x,t) = \frac{(-2i/t)^{1/\alpha}}{\alpha\pi} \sum_{k=0}^{\infty} \frac{(-(-2i/t)^{2/\alpha}x^2)^k}{(2k)!}\Gamma\left(\frac{2k+1}{\alpha}\right) \tag{18}$$

where $G(0,t) = (-2i/t)^{1/\alpha}\Gamma(1/\alpha)/(\alpha\pi)$. The corresponding MSD for $\beta = 0$ becomes

$$\langle x^2(t)\rangle = \frac{1}{2\pi}\int_{-\infty}^{\infty} \left| \frac{d\psi_0(\omega)}{d\omega} - i\psi_0(\omega)\frac{\alpha t}{2}\,\mathrm{sign}(\omega)|\omega|^{\alpha-1} \right|^2 d\omega \tag{19}$$

where $\mathrm{sign}(x)$ is the sign function. It can be verified that the MSD Equation (19) as function of time describes for all fractional orders a parabola. As an example, for the commonly applied Gaussian wave packet

$$\psi_0(x) = \frac{1}{\sqrt{\sigma\sqrt{\pi}}}\exp\left(-\frac{x^2}{2\sigma^2}\right) \tag{20}$$

one obtains in the free particle case the MSD

$$\langle x^2(t)\rangle = \frac{(\alpha\sigma t)^2}{(2\sigma)^{2\alpha}}\frac{\Gamma(2\alpha-1)}{\Gamma(\alpha)} + \frac{\sigma^2}{2} \tag{21}$$

which exists for all $1/2 < \alpha \leq 2$. Figure 1 displays the MSD Equation (21) for different values of the fractional order.

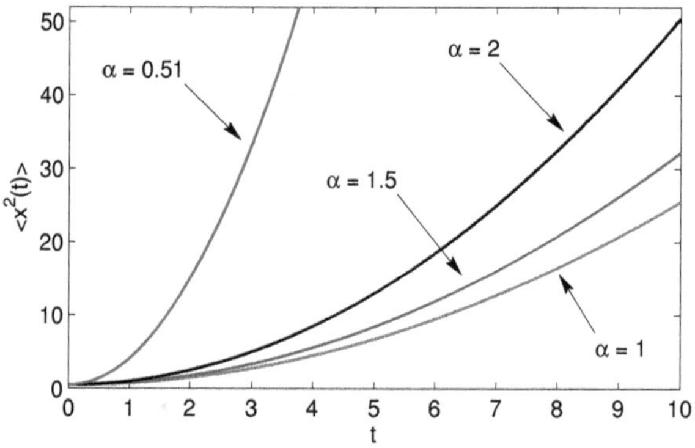

Figure 1. Illustration of the mean square displacement (MSD) Equation (21) for different values of the fractional order α under consideration of the Gaussian initial wave packet Equation (20) with $\sigma = 1$.

Explicit Solutions for the Square Root of the Laplacian $(-\Delta)$

In this subsection, we provide some explicit analytical solutions of Equation (1) for the order $\alpha = 1$. In this case, the fractional Laplacian can be represented as [20]

$$(-\Delta)^{\frac{1}{2}}f(x) = \mathcal{F}^{-1}\{|\omega|F(\omega)\}(x) = \frac{d}{dx}\mathcal{H}\{f(x)\} \tag{22}$$

where $\mathcal{H}$ is the Hilbert transform that is defined by

$$\mathcal{H}\{f(x)\} = \frac{1}{\pi}\mathcal{P}\int_{-\infty}^{\infty}\frac{f(y)}{x-y}dy \tag{23}$$

and $\mathcal{P}$ denotes the Cauchy principal value. It should be noted that, in recent time, the case of $\alpha = 1$ has attracted attention in the context of the propagation dynamics of wave packets under the use of the space-fractional Schrödinger equation [10,11,13]. In this study, we investigate the time evolution of the Cauchy initial wave packet

$$\psi_0(x) = \sqrt{\frac{2\varepsilon}{\pi}} \frac{\varepsilon}{x^2 + \varepsilon^2} \tag{24}$$

where $\varepsilon > 0$ is a measure for the beam width. We afore have to note that the time evolution of the Gaussian wave packet and the Cauchy wave packet shows a similar behavior. However, in the latter case, we are able to derive simple analytical solutions that are easy to implement and reproduce. Our starting point is the evaluation of the Fourier integral Equation (10) under consideration of the wave function Equation (9). For this, we also need the Cauchy wave packet Equation (24) in momentum space, that is $\psi_0(\omega) = \sqrt{2\pi\varepsilon}\exp(-\varepsilon|\omega|)$. Splitting the integrand into three sections and assuming that $\beta \geq 0$ enables the derivation of the following closed-form expression

$$\begin{aligned} \psi(x,t) =& \sqrt{\frac{\varepsilon}{2\pi}} e^{-i\beta t^2/4} \left[\frac{e^{-\beta\varepsilon t}}{\varepsilon - i(x - t/2)} + \frac{e^{-i\beta x t}}{\varepsilon + i(x + t/2)} \right] + \frac{1}{2}\sqrt{\frac{\beta\varepsilon}{i}} e^{-\beta t(\varepsilon + it/4)} \\ & \times e^{i\beta(x - t/2 + i\varepsilon)^2/2} \left[\operatorname{erf}\sqrt{\frac{i\beta}{2}}(x + t/2 + i\varepsilon) - \operatorname{erf}\sqrt{\frac{i\beta}{2}}(x - t/2 + i\varepsilon) \right] \end{aligned} \tag{25}$$

where $\operatorname{erf}(z)$ is the error function. Note that the case $\beta < 0$ can be treated in the same manner. The associated MSD belonging to this process can be obtained from Equation (13) and summarized as

$$\langle x^2(t) \rangle = \frac{t^2}{4} - \frac{1 + e^{-2\beta\varepsilon t}}{4\beta\varepsilon} t + \frac{1 + 4\beta^2\varepsilon^4 - e^{-2\beta\varepsilon t}}{(2\beta\varepsilon)^2} \tag{26}$$

In this context, we note on the limits

$$\lim_{\beta \to 0} \langle x^2(t) \rangle = \lim_{\beta \to \infty} \langle x^2(t) \rangle = \frac{t^2}{4} + \varepsilon^2 \tag{27}$$

In the case of the free particle, there is the possibility to provide the general solution for an arbitrary initial stage. In doing so, we need the inverse Fourier transform of the impulse response $G(\omega, t) = \exp(-it|\omega|/2)$, which can be written as

$$G(x,t) = \frac{1}{2\pi} \int_{-\infty}^{\infty} \exp(-it|\omega|/2) \exp(i\omega x)\, d\omega = \frac{t}{2}\delta(x^2 - t^2/4) + \frac{i}{\pi} \frac{2t}{4x^2 - t^2} \tag{28}$$

Inserting this finding into Equation (15) leads to the general solution of the Schrödinger equation

$$\psi(x,t) = \frac{\psi_0(x - t/2) + \psi_0(x + t/2)}{2} + \frac{it}{2\pi} \mathcal{P} \int_{-\infty}^{\infty} \frac{\psi_0(y)}{(x - y)^2 - t^2/4}\, dy \tag{29}$$

In the actual case of the Cauchy initial wave packet Equation (24), we find the time evolution

$$\psi(x,t) = \sqrt{\frac{\varepsilon}{2\pi}} \frac{2\varepsilon + it}{x^2 + (\varepsilon + it/2)^2} \tag{30}$$

which can be confirmed via Equation (25) by setting $\beta = 0$. The resulting pdf belonging to this process is

$$\rho(x,t) = \frac{2\varepsilon}{\pi} \frac{\varepsilon^2 + t^2/4}{(x^2 + \varepsilon^2 - t^2/4)^2 + (\varepsilon t)^2} \tag{31}$$

In addition, the cumulative distribution function $F(x,t) = \int_{-\infty}^{x} \rho(y,t)\, dy$ is found to be

$$F(x,t) = \frac{1}{2} + \frac{t^2 + 4\varepsilon^2}{2\pi t} \operatorname{Im}\left(\frac{\arctan[x/(\varepsilon - it/2)]}{\varepsilon - it/2} \right) \tag{32}$$

Within the classical quantum mechanics, it is known that a pdf belonging to the free particle solution typically spreads out but it does not split. However, as has been very recently shown, the pdf belonging to the space-fractional Schrödinger equation additionally undergoes splitting [11]. For the Cauchy wave packet, there is the possibility to provide some more details concerning this feature. Taking the first derivative of the pdf Equation (31) with respect to the space and setting $\rho_x(x,t) = 0$, yields three critical positions, namely $x = 0$ or $x = \pm\sqrt{t^2/4 - \varepsilon^2}$. From the behavior of $\rho(x,t)$, it follows that $x = 0$ is an absolute maximum as long as $0 \leq t \leq 2\varepsilon$. On the other side, for time values $t > 2\varepsilon$, the position $x = 0$ becomes a relative minimum. At the same time, we obtain two absolute maxima located at $x = \pm\sqrt{t^2/4 - \varepsilon^2}$. Consequently, the probability density spreads out for $t > 0$, and additionally, starts to split at $t = 2\varepsilon$. For illustration purposes, Figure 2 shows the time evolution of a Cauchy initial wave packet according to Equation (31).

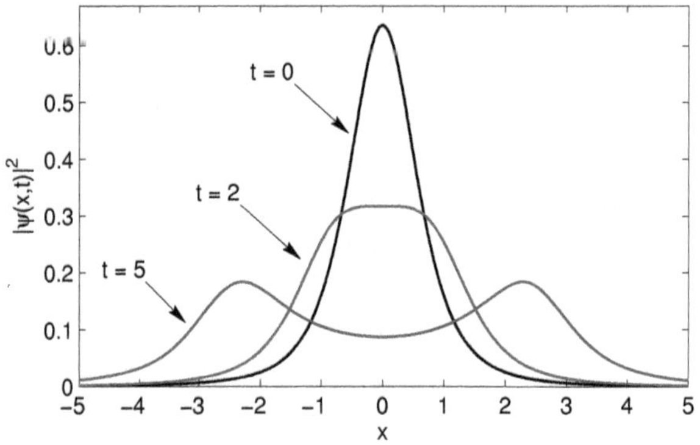

Figure 2. Time evolution of the Cauchy initial wave packet Equation (24) for the parameter $\varepsilon = 1$.

A similar behavior can be observed in the case of the Gaussian initial wave packet Equation (20), where the critical time regarding the splitting process must be determined numerically from the wave function

$$\psi(x,t) = \frac{\psi_0(x - t/2)}{2}\left[1 + i\,\mathrm{erfi}\left(\frac{x - t/2}{\sqrt{2}\sigma}\right)\right] + \frac{\psi_0(x + t/2)}{2}\left[1 - i\,\mathrm{erfi}\left(\frac{x + t/2}{\sqrt{2}\sigma}\right)\right] \quad (33)$$

where $\mathrm{erfi}(x)$ is the imaginary error function and $\psi_0(x)$ is given by Equation (20). Figure 3 displays the same situation as shown in Figure 2 for the the Gaussian initial wave packet Equation (20).

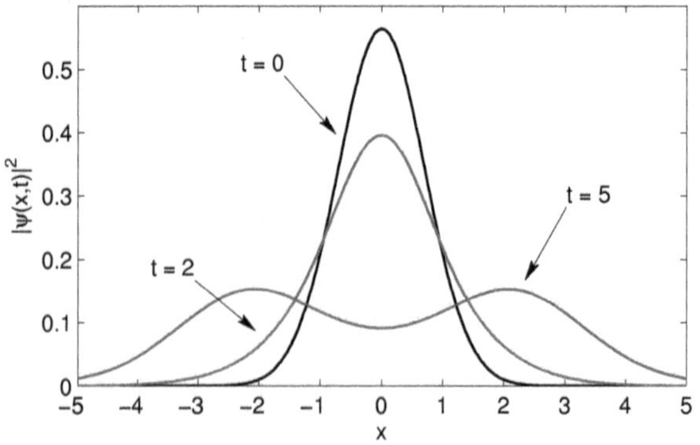

Figure 3. Time evolution of the Gaussian initial wave packet Equation (20) for the parameter $\sigma = 1$.

Concerning an experimental verification/realization of the derived theoretical results we refer to the recent publication [11], which proposes an optical system composed of two convex lenses and a phase mask. We furthermore note on the free particle solution to the classical Schrödinger equation. In the case of the Cauchy wave packet Equation (24), we have

$$\psi(x,t) = \frac{1}{2}\sqrt{\frac{\varepsilon}{it}}\left[w\left(\frac{i\varepsilon + x}{\sqrt{2it}}\right) + w\left(\frac{i\varepsilon - x}{\sqrt{2it}}\right)\right] \tag{34}$$

where $w(z) = \exp(-z^2)\,\mathrm{erfc}(-iz)$ is the Faddeeva function. The corresponding MSD belonging to this process is $\langle x^2(t)\rangle = t^2/(2\varepsilon^2) + \varepsilon^2$. Figure 4 displays the time evolution of the wave function Equation (34) for the same time values considered in Figure 2. It can can be seen that the probability density indeed spreads out but it does not split.

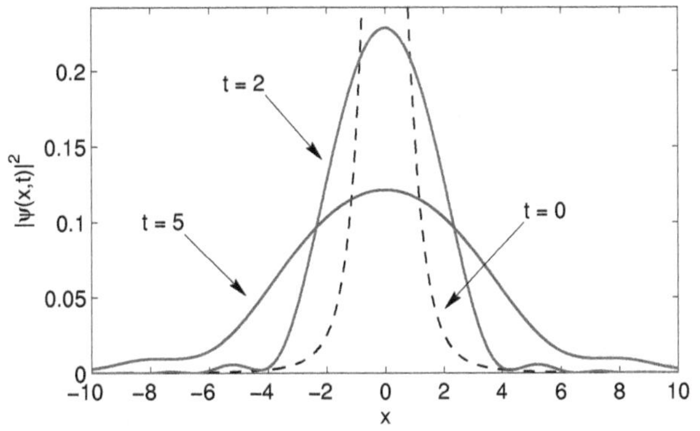

Figure 4. Time evolution of the Cauchy initial wave packet Equation (24) for $\varepsilon = 1$ within the classical quantum mechanics.

3. 2D Space-Fractional Schrödinger Equation

The 2D space-fractional Schrödinger equation in the presence of the potential $V(\rho) = \mathbf{F}\cdot\rho$ is given by

$$i\frac{\partial\psi(\rho,t)}{\partial t} = \frac{1}{2}(-\Delta)^{\frac{\alpha}{2}}\psi(\rho,t) + V(\rho)\psi(\rho,t), \quad \rho\in\mathbb{R}^2, \quad t>0 \tag{35}$$

where $\mathbf{F} = (F_1, F_2)$ and $F = |\mathbf{F}|$. Again, the 2D fractional Laplacian is most convenient defined by

$$\mathcal{F}\{(-\Delta)^{\frac{\alpha}{2}}f(\rho)\}(\mathbf{q}) = |\mathbf{q}|^{\alpha}F(\mathbf{q}) \tag{36}$$

where $|\mathbf{q}| = q$ denotes the length of the wave vector $\mathbf{q} = (q_1, q_2)$ that belongs to the Fourier transform $F(\mathbf{q}) = \int f(\rho)\exp(-i\mathbf{q}\cdot\rho)\,d\rho$. In addition, for $\alpha = 1$, the 2D fractional Laplacian can be represented as

$$(-\Delta)^{\frac{1}{2}}f(\rho) = \nabla\cdot\mathcal{R}\{f(\rho)\} \tag{37}$$

where $\mathcal{R}$ denotes the Riesz transform [21], that maps a scalar function $f(\rho)$ into a vector field according to

$$\mathcal{R}\{f(\rho)\} = \frac{1}{2\pi}\int\frac{f(\mathbf{s})(\rho-\mathbf{s})}{|\rho-\mathbf{s}|^3}\,d\mathbf{s} \tag{38}$$

We note that relation Equation (37) is a generalization of the 1D differential operator Equation (22). Concerning the initial and boundary conditions in 2D, we have

$$\psi(\rho,0^+) = \psi_0(\rho), \qquad \lim_{\rho\to\infty}\psi(\rho,t) = 0 \tag{39}$$

where $\rho = |\boldsymbol{\rho}|$. The Schrödinger Equation (35) in momentum space becomes

$$\frac{\partial \psi(\mathbf{q},t)}{\partial t} = \frac{q^\alpha}{2i}\psi(\mathbf{q},t) + \mathbf{F}\cdot\nabla\psi(\mathbf{q},t), \quad \mathbf{q}\in\mathbb{R}^2, \quad t>0 \tag{40}$$

where the Del operator has to be applied with regard to the Fourier variables. Similar as in 1D, we seek a wave function of the form $\psi(\mathbf{q},t) = \exp[i\varphi(\mathbf{q})]u(\mathbf{q}+\mathbf{F}t)$. Inserting this ansatz into Equation (40), leads to the first order partial differential equation

$$F_1\frac{\partial\varphi(\mathbf{q})}{\partial q_1} + F_2\frac{\partial\varphi(\mathbf{q})}{\partial q_2} = \frac{q^\alpha}{2} \tag{41}$$

In contrast to the 1D case, we are not able to solve this equation for all fractional orders in elementary form. However, for the relevant case of $\alpha = 1$, an exact solution can be derived and summarized as

$$\varphi(\mathbf{q}) = \frac{q}{4}\frac{F_1q_1+F_2q_2}{F_1^2+F_2^2} + \frac{1}{4}\frac{(F_1q_2-F_2q_1)^2}{(F_1^2+F_2^2)^{3/2}}\ln\left(2q+2\frac{F_1q_1+F_2q_2}{\sqrt{F_1^2+F_2^2}}\right) \tag{42}$$

The solution belonging to the classical case with $\alpha = 2$ is less complicated and given by $\varphi(\mathbf{q}) = q_1^3/(3F_1) + q_2^3/(3F_2)$. Again, the unknown function u can be recovered under consideration of the initial condition such as

$$\psi(\mathbf{q},0) = \psi_0(\mathbf{q}) = \exp[i\varphi(\mathbf{q})]u(\mathbf{q}) \quad \Rightarrow \quad u(\mathbf{q}) = \psi_0(\mathbf{q})\exp[-i\varphi(\mathbf{q})] \tag{43}$$

Therefore, we obtain the wave function in momentum space according to

$$\psi(\mathbf{q},t) = \psi_0(\mathbf{q}+\mathbf{F}t)\exp[i(\varphi(\mathbf{q})-\varphi(\mathbf{q}+\mathbf{F}t))] \tag{44}$$

In the case of the free particle with $\mathbf{F} = \mathbf{0}$, the solution in momentum space can be given for all orders as

$$\psi(\mathbf{q},t) = \psi_0(\mathbf{q})\exp(-itq^\alpha/2), \quad 0<\alpha\le 2 \tag{45}$$

Moreover, if the initial wave packet exhibits rotational symmetry which implies that $\psi_0(\mathbf{q}) = \psi_0(q)$, the solution in real space can be evaluated by means of the inverse Hankel transform

$$\psi(\rho,t) = \frac{1}{2\pi}\int_0^\infty \psi_0(q)\exp(-itq^\alpha/2)J_0(q\rho)q\,dq, \quad 0<\alpha\le 2 \tag{46}$$

with $J_0(x)$ being the zero order Bessel function. For $\psi_0(q) = 1$ and $1<\alpha\le 2$, Equation (46) yields the 2D Green's function

$$G(\rho,t) = \frac{1}{\pi\rho^2}H_{1,2}^{1,1}\left[\frac{2}{it}\frac{\rho^\alpha}{2^\alpha}\,\middle|\,\begin{matrix}(1,1)\\(1,\alpha/2),(1,\alpha/2)\end{matrix}\right] \tag{47}$$

which can be evaluated by means of the series

$$G(\rho,t) = \frac{(-2i/t)^{2/\alpha}}{2\alpha\pi}\sum_{k=0}^{\infty}\frac{(-(-2i/t)^{2/\alpha}\rho^2/4)^k}{(k!)^2}\Gamma\left(\frac{2k+2}{\alpha}\right) \tag{48}$$

where $G(0,t) = (-2i/t)^{2/\alpha}\Gamma(2/\alpha)/(2\alpha\pi)$. In the case of $\alpha = 2$, this relation coincides with the known result $G(\rho,t) = \exp(i\rho^2/(2t))/(2\pi it)$. On the other side, Green's function for $\alpha = 1$ and $\rho \ne t/2$ is found to be

$$G(\rho,t) = \frac{2i}{\pi}\frac{\partial}{\partial t}\frac{1}{\sqrt{4\rho^2-t^2}} = \frac{2}{\pi}\frac{it}{(4\rho^2-t^2)^{3/2}} \tag{49}$$

Based on this relation, we obtain the wave function

$$\psi(\rho, t) = \frac{2i}{\pi}\frac{\partial}{\partial t}\int \frac{\psi_0(\rho - \mathbf{s})}{\sqrt{4s^2 - t^2}}\, d\mathbf{s} \tag{50}$$

which is the general solution of Equation (35) for $\alpha = 1$ and $\mathbf{F} = 0$. In the following, we investigate the time evolution of the rotationally symmetric Cauchy wave packet

$$\psi_0(\rho) = \sqrt{\frac{2}{\pi}}\frac{\varepsilon^2}{(\rho^2 + \varepsilon^2)^{3/2}} \tag{51}$$

which is normalized according to $\int |\psi_0(\boldsymbol{\rho})|^2\, d\boldsymbol{\rho} = 1$. Similar as in 1D, we consider the case of $\alpha = 1$ in detail. The desired wave function can be obtained either via the convolution Equation (50) or by means of the inverse Hankel transform Equation (46). In the latter case, we need the initial wave packet Equation (51) in momentum space, that is $\psi_0(q) = 2\varepsilon\sqrt{2\pi}\exp(-\varepsilon q)$. The required inverse Hankel transform can be taken exactly and summarized as

$$\psi(\rho, t) = \sqrt{\frac{2}{\pi}}\frac{\varepsilon(\varepsilon + it/2)}{[\rho^2 + (\varepsilon + it/2)^2]^{3/2}} \tag{52}$$

The associated probability density to this wave function is given by

$$|\psi(\rho, t)|^2 = \frac{2\varepsilon^2}{\pi}\frac{\varepsilon^2 + t^2/4}{[(\rho^2 + \varepsilon^2 - t^2/4)^2 + (\varepsilon t)^2]^{3/2}} \tag{53}$$

In accordance with the 1D case, the probability density in 2D also undergoes spreading and splitting. Remarkable, the probability density Equation (53) starts to split at $t = 2\varepsilon$, which is exactly the same time value as for the 1D pendant Equation (31). Figure 5 displays the time evolution of a rotationally symmetric Cauchy initial wave packet for different time values.

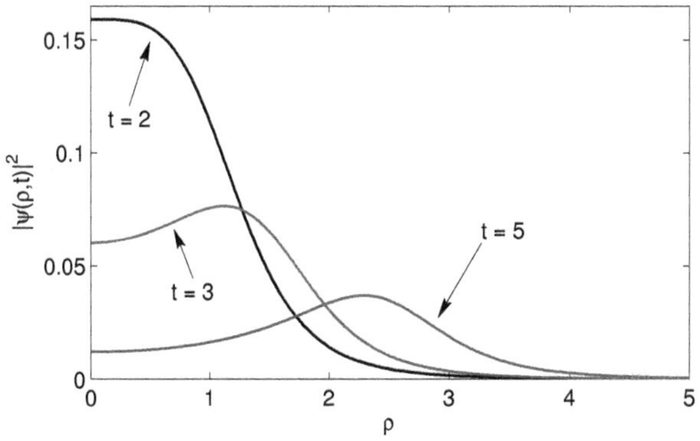

Figure 5. Time evolution of the rotationally symmetric Cauchy initial wave packet Equation (51) for the parameter $\varepsilon = 1$.

We note that the solution for the rotationally symmetric Gaussian wave packet $\psi_0(\rho) = 2\exp(-2\rho^2/\sigma^2)/(\sqrt{\pi}\sigma)$ can be derived in a similar manner. The required Hankel transform for evaluation of the wave function Equation (46) is given by $\psi_0(q) = \sqrt{\pi}\sigma\exp(-\sigma^2 q^2/8)$.

4. Numerical Solution of the Fractional Schrödinger Equation

In this section, we describe a recently proposed matrix approach [22] that has been used for solving numerically the 1D space-fractional diffusion equation. In this context, we additionally refer to the recent publication [23], which takes into account the so-called Adomian decomposition method.

We adopt the matrix approach in order to solve the space-fractional Schrödinger Equation (1) subject to an arbitrary potential function $V(x)$. Within this framework, we consider the computational domain $|x| \leq L/2$ together with the uniform grid $x_j = -L/2 + jh$ for $j = 0, \ldots, N$ as well as the node spacing $h = L/N$. Furthermore, the fractional Laplacian has to be restricted to a bounded domain. If $0 < \alpha < 1$, we have the representation

$$(-\Delta)^{\frac{\alpha}{2}} f(x) = \frac{1}{2\cos(a\pi/2)\Gamma(1-a)} \frac{d}{dx} \int_{-L/2}^{L/2} \frac{\text{sign}(x-y) f(y)}{|x-y|^{\alpha}} dy \tag{54}$$

whereas for $1 < \alpha < 2$ it is given by

$$(-\Delta)^{\frac{\alpha}{2}} f(x) = \frac{1}{2\cos(a\pi/2)\Gamma(2-a)} \frac{d^2}{dx^2} \int_{-L/2}^{L/2} \frac{f(y)}{|x-y|^{\alpha-1}} dy \tag{55}$$

In particular, the square root of the negative Laplacian on a bounded domain becomes

$$(-\Delta)^{\frac{1}{2}} f(x) = \frac{1}{\pi} \frac{d}{dx} \int_{-L/2}^{L/2} \frac{f(y)}{x-y} dy \tag{56}$$

For our applications, the quantity L has to be chosen relatively large in order to obtain a reasonable model for the infinite domain. On the other side, the matrix approach is designed for solving the fractional Schrödinger on a bounded domain, where analytical solutions are typically not available. The key point here is the approximation of the fractional Laplacian by means of the fractional centered difference [22]

$$(-\Delta)^{\frac{\alpha}{2}} f(x_j) \approx \frac{1}{h^{\alpha}} \sum_{k=1}^{N-1} g_{|j-k|} f(x_k) \tag{57}$$

which can be applied for all $0 < \alpha \leq 2$. The corresponding weights are given by [22]

$$g_k = \frac{(-1)^k \Gamma(1+\alpha)}{\Gamma(1+k+\alpha/2)\Gamma(1-k+\alpha/2)} \tag{58}$$

We note that within the definition Equation (57), we have taken into account the boundary data $f(x_0) = f(x_N) = 0$. The weights given in Equation (58) can be computed efficiently by means of the recursion [22]

$$g_{k+1} = \frac{2k-\alpha}{2k+\alpha+2} g_k \tag{59}$$

Applying the fractional centered difference scheme Equation (57) to the Schrödinger Equation (1) leads to the following system of ordinary differential equations

$$i\frac{d\psi_j(t)}{dt} = \frac{1}{2h^{\alpha}} \sum_{k=1}^{N-1} g_{|j-k|} \psi_k(t) + V_j \psi_j(t) \tag{60}$$

where $j = 1, \ldots, N-1$, $\psi_j(t) = \psi(x_j, t)$ and $V_j = V(x_j)$. The system Equation (60) in matrix notation becomes

$$\frac{d\mathbf{\Phi}(t)}{dt} = -iM\mathbf{\Phi}(t) \tag{61}$$

where $\mathbf{\Phi}(t) = (\psi_1(t), \psi_2(t), \ldots, \psi_{N-1}(t))$. The associated solution can be directly given in terms of the matrix exponential according to

$$\mathbf{\Phi}(t) = \exp(-iMt)\mathbf{\Phi}_0 \tag{62}$$

where $\mathbf{\Phi}_0 = (\psi_0(x_1), \psi_0(x_2), \ldots, \psi_0(x_{N-1}))$ is a column vector that contains the sampled initial wave packet $\psi_0(x)$. If MATLAB is used, the matrix exponential can be evaluated via the function `expm`. For illustration purposes, the corresponding matrix M in the case of $N = 5$ looks like

$$M = \frac{1}{2h^{\alpha}} \begin{pmatrix} g_0 & g_1 & g_2 & g_3 \\ g_1 & g_0 & g_1 & g_2 \\ g_2 & g_1 & g_0 & g_1 \\ g_3 & g_2 & g_1 & g_0 \end{pmatrix} + \begin{pmatrix} V_1 & 0 & 0 & 0 \\ 0 & V_2 & 0 & 0 \\ 0 & 0 & V_3 & 0 \\ 0 & 0 & 0 & V_4 \end{pmatrix} \tag{63}$$

Now, the structure of the matrix M can be readily continued for an arbitrary N. In the following, we use the above described numerical scheme in order to solve the space-fractional Schrödinger Equation (1). We take into account the Cauchy initial wave packet Equation (24) with $\varepsilon = 1$ and evaluate the resulting probability density for $\alpha = 1$ and $t = 5$. The size of the computational domain for approximation of the infinite domain is set to $L = 30$. Figure 6 displays the comparison between the analytical solution Equation (31) and the matrix approach Equation (62) for the case of a free particle.

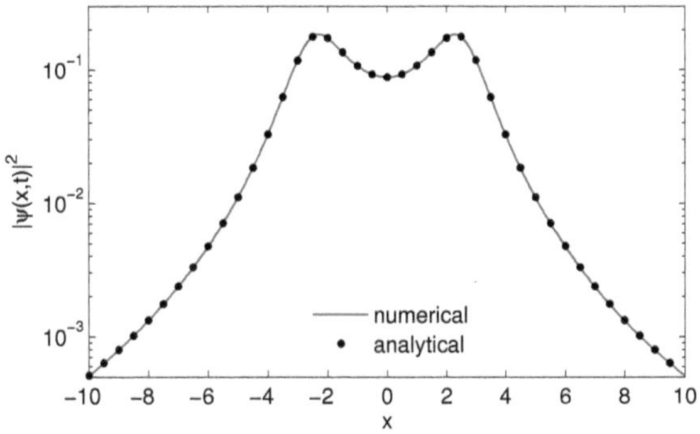

Figure 6. Comparison between the analytical solution Equation (31) and the matrix approach Equation (62) for the case of a free particle.

Next, we consider the case of the linear potential with $\beta = 0.5$. The resulting probability density belonging to this process is shown in Figure 7. In contrast to the free particle, the probability density is no longer symmetric and additionally exhibits some oscillations.

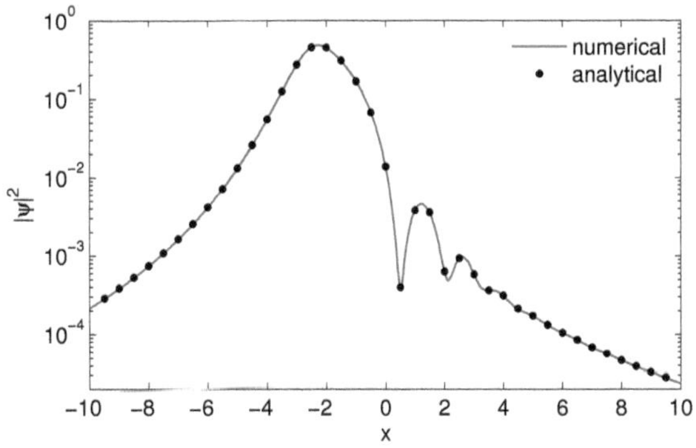

Figure 7. Comparison between the analytical solution Equation (25) and the matrix approach Equation (62) for the case of a liner potential with $\beta = 0.5$.

We repeat the last numerical experiment for the case of a stronger accelerated particle with $\beta = 4$. The resulting probability density, which is depicted in Figure 8, is given by a smooth curve without oscillations. Furthermore, in contrast to the free particle case, the probability density has not been splitted.

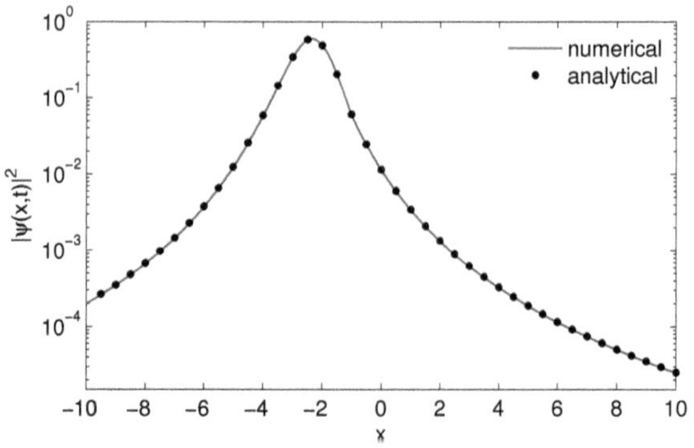

Figure 8. Comparison between the analytical solution Equation (25) and the matrix approach Equation (62) for the case of a liner potential with $\beta = 4$.

The numerical experiments outlined above confirm both the correctness of the analytical solutions as well as the good quality of the matrix approach. Besides these comparisons, we additionally have performed numerical experiments for other fractional orders α. As a result, we observed in all cases the same good agreement between the analytical solution Equation (10) and the matrix approach Equation (62). It should also be noted that the above matrix approach can be readily extended for solving the space-time fractional Schrödinger equation. More precisely, if we replace the classical time derivative occurring in Equation (1) by the Caputo derivative [24] of order $0 < \nu \leq 1$

$$D_t^\nu f(t) = \frac{1}{\Gamma(1-\nu)} \int_0^t \frac{f'(\tau)}{(t-\tau)^\nu} \, d\tau \tag{64}$$

one obtains the following system of fractional differential equations

$$D_t^\nu \mathbf{\Phi}(t) = -iM\mathbf{\Phi}(t) \tag{65}$$

The solution to this system is formally given by $\mathbf{\Phi}(t) = E_\nu(-iMt^\nu)\mathbf{\Phi}_0$, where $E_\nu(A) = \sum_{n=0}^{\infty} A^n/\Gamma(1+\nu n)$ is the Mittag-Leffler matrix function. It can be computed under consideration of the eigenvalue decomposition $A = U\Lambda U^{-1}$, where U is a matrix containing the eigenvectors of A. For $A \in \mathbb{C}^{n\times n}$ with eigenvalues $\Lambda = \operatorname{diag}(\lambda_1, \lambda_2, \ldots, \lambda_n)$, we have $E_\nu(A) = UE_\nu(\Lambda)U^{-1}$ together with the diagonal matrix

$$E_\nu(\Lambda) = \begin{pmatrix} E_\nu(\lambda_1) & 0 & \cdots & 0 \\ 0 & E_\nu(\lambda_2) & \cdots & 0 \\ \vdots & \vdots & \ddots & \vdots \\ 0 & 0 & \cdots & E_\nu(\lambda_n) \end{pmatrix} \tag{66}$$

As a final numerical experiment, we consider the space-time fractional Schrödinger equation

$$iD_t^\alpha \psi(x,t) = (-\Delta)^{\frac{\alpha}{2}} \psi(x,t)/2, \quad x \in \mathbb{R}, \quad t > 0 \tag{67}$$

for the joint fractional order $\alpha \in (0,1)$. In this case, the wave function in momentum space can be written as $\psi(\omega,t) = \psi_0(\omega)G(\omega,t)$, where $G(\omega,t) = E_\alpha(-i|\omega|^\alpha t^\alpha/2)$. The corresponding Green's function in space domain is available in elementary form as

$$G(x,t) = \frac{1}{2\pi} \int_{-\infty}^{\infty} E_\alpha(-i|\omega|^\alpha t^\alpha/2) \exp(i\omega x) \, d\omega = \frac{i}{2\pi} \frac{|x|^{\alpha-1} t^\alpha \sin(\alpha\pi/2)}{|x|^{2\alpha} + i|x|^\alpha t^\alpha \cos(\alpha\pi/2) - t^{2\alpha}/4} \tag{68}$$

We note that this result can be derived in a similar manner as the fundamental solution to the neutral-fractional wave equation [25]. The response to an arbitrary initial stage is given by the convolution

$$\psi(x,t) = \frac{it^{\alpha}\sin(\alpha\pi/2)}{2\alpha\pi}\int_0^{\infty}\frac{\psi_0(x-u^{1/\alpha})+\psi_0(x+u^{1/\alpha})}{u^2+iut^{\alpha}\cos(\alpha\pi/2)-t^{2\alpha}/4}\,du \tag{69}$$

In most cases, this integral must be evaluated numerically. However, for the Cauchy initial wave packet Equation (24) the integration can be carried out analytically and summarized as

$$\psi(x,t) = \sqrt{\frac{\varepsilon}{2\pi}}\left[\frac{(\varepsilon+ix)^{\alpha-1}}{(\varepsilon+ix)^{\alpha}+it^{\alpha}/2}+\frac{(\varepsilon-ix)^{\alpha-1}}{(\varepsilon-ix)^{\alpha}+it^{\alpha}/2}\right] \tag{70}$$

In Figure 9 we have evaluated the solution of the space-time fractional Schrödinger equation for two different fractional orders subject to the initial stage Equation (20). The analytical solution Equation (69) is denotes by filled dots, whereas the solid line is the numerical solution obtained by means of the Mittag-Leffler matrix function. In both cases the wave function is evaluated for the time value $t = 5$.

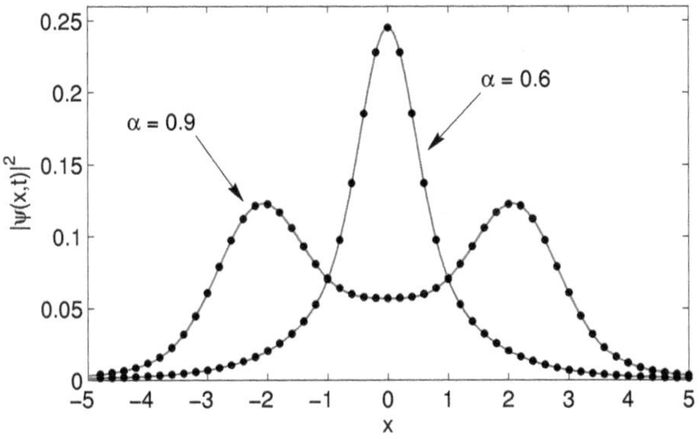

Figure 9. Comparison between the analytical solution Equation (69) and the matrix approach for the case of the Gaussian initial wave packet Equation (20) with $\sigma = 0.5$.

5. Conclusions and Discussion

In this article, we considered the space-fractional Schrödinger equation with the Riesz space-fractional derivative in the presence of the linear potential. The general solution to the 1D Schrödinger equation in momentum space has been obtained in closed form, which allows the determination of other measurable quantities such as the spatial moments. Furthermore, we investigated the time evolution of a Cauchy wave packet in detail. Besides the 1D Schrödinger equation we also addressed the case of two spatial dimensions. The obtained equations have been illustrated and verified by comparisons with an independent recently proposed numerical scheme. Consequently, the derived equations are therefore also useful for verification of other numerical approaches which are designed for solving the fractional Schrödinger equation. Besides the space-fractional time-dependent Schrödinger equation solutions to the associated time-independent counterpart are also of importance. Concerning the relationship between the space-fractional time-dependent and time-independent Schrödinger equation we refer to the article [26].

Author Contributions: Both authors have worked together in order to prepare the present manuscript.

Conflicts of Interest: The authors declare no conflict of interest.

References

1. Laskin, N. Fractional quantum mechanics. *Phys. Rev. E* **2000**, *62*, 3135.
2. Laskin, N. Fractional Schrödinger equation. *Phys. Rev. E* **2002**, *66*, 056108.

3. Metzler, R.; Klafter, J. The random walk's guide to anomalous diffusion: A fractional dynamics approach. *Phys. Rep.* **2000**, *339*, 1–77.
4. Mainardi, F.; Luchko, Y.; Pagnini, G. The fundamental solution of the space-time fractional diffusion equation. *Fract. Calc. Appl. Anal.* **2001**, *4*, 153–192.
5. Metzler, R.; Klafter, J. The restaurant at the end of the random walk: Recent developments in the description of anomalous transport by fractional dynamics. *J. Phys. A Math. Gen.* **2004**, *37*, R161.
6. Debnath, L.; Bhatta, D. *Integral Transforms and Their Applications*; CRC Press: New York, NY, USA, 2014.
7. Sandev, T.; Metzler, R.; Tomovski, Z. Correlation functions for the fractional generalized Langevin equation in the presence of internal and external noise. *J. Math. Phys.* **2014**, *55*, 023301.
8. Machida, M. The time-fractional radiative transport equation—Continuous-time random walk, diffusion approximation, and Legendre-polynomial expansion. Available online: http://arxiv.org/abs/1602.05382 (accessed on 25 April 2016).
9. Mathai, A.M.; Saxena, R.K.; Haubold, H.J. *The H-Function: Theory and Applications*; Springer Verlag: New York, NY, USA, 2010.
10. Zhang, Y.; Liu, X.; Belić, M.R.; Zhong, W.; Zhang, Y.; Xiao, M. Propagation Dynamics of a Light Beam in a Fractional Schrödinger Equation. *Phys. Rev. Lett.* **2015**, *115*, 180403.
11. Zhang, Y.; Zhong, H.; Belić, M.R.; Ahmed, N.; Zhang, Y.; Xiao, M. Diffraction-free beams in fractional Schrödinger equation. Available online: http://arxiv.org/abs/1512.08671 (accessed on 25 April 2016).
12. Zhang, Y.; Zhong, H.; Belić, M.R.; Zhu, Y.; Zhong, W.; Zhang, Y.; Christodoulides, D.N.; Xiao, M. $\mathcal{PT}$ symmetry in a fractional Schrödinger equation. *Laser Photonics Rev.* **2016**, doi:10.1002/lpor.201600037.
13. Longhi, S. Fractional Schrödinger equation in optics. *Opt. Lett.* **2015**, *40*, 1117–1120.
14. Robinett, R.W. Quantum mechanical time-development operator for the uniformly accelerated particle. *Am. J. Phys.* **1996**, *64*, 803–807.
15. Guedes, I. Solution of the Schrödinger equation for the time-dependent linear potential. *Phys. Rev. A* **2001**, *63*, 034102.
16. Plastino, A.R.; Tsallis, C. Nonlinear Schroedinger equation in the presence of uniform acceleration. *J. Math. Phys.* **2013**, *54*, 041505.
17. Sandev, T.; Petreska, I.; Lenzi, E.K. Time-dependent Schrödinger-like equation with nonlocal term. *J. Math. Phys.* **2014**, *55*, 092105.
18. Luchko, Y. Fractional Schrödinger equation for a particle moving in a potential well. *J. Math. Phys.* **2013**, *54*, 012111.
19. Al-Saqabi, B.; Boyadjiev, L.; Luchko, Y. Comments on employing the Riesz-Feller derivative in the Schrödinger equation. *Eur. Phys. J. Spec. Top.* **2013**, *222*, 1779–1794.
20. Luchko, Y. Wave-diffusion dualism of the neutral-fractional processes. *J. Comput. Phys.* **2015**, *293*, 40–52.
21. Adams, D.; Hedberg, L. *Function Spaces and Potential Theory*; Springer-Verlag: Berlin, Germany, 1999.
22. Popolizio, M. A matrix approach for partial differential equations with Riesz space fractional derivatives. *Eur. Phys. J. Spec. Top.* **2013**, *222*, 1975–1985.
23. Dubbeldam, J.L.A.; Tomovski, Z.; Sandev, T. Space-Time Fractional Schrödinger Equation With Composite Time Fractional Derivative. *Fract. Calc. Appl. Anal.* **2015**, *18*, 1179–1200.
24. Debnath, L. Recent applications of fractional calculus to science and engineering. *Int. J. Math. Math. Sci.* **2003**, *2003*, 3413–3442.
25. Luchko, Y. Fractional wave equation and damped waves. *J. Math. Phys.* **2013**, *54*, 031505.
26. Dong, J. Scattering problems in the fractional quantum mechanics governed by the 2D space-fractional Schrödinger equation. *J. Math. Phys.* **2014**, *55*, 032102.

3

Tight State-Independent Uncertainty Relations for Qubits

Alastair A. Abbott [1,*], Pierre-Louis Alzieu [1,2], Michael J. W. Hall [3] and Cyril Branciard [1,*]

[1] Institut Néel, CNRS and Université Grenoble Alpes, 38042 Grenoble Cedex 9, France
[2] Ecole Normale Supérieure de Lyon, 69342 Lyon, France; pierrelouis.alzieu@ens-lyon.fr
[3] Centre for Quantum Computation and Communication Technology (Australian Research Council), Centre for Quantum Dynamics, Griffith University, Brisbane 4111, Australia; michael.hall@griffith.edu.au
* Correspondence: alastair.abbott@neel.cnrs.fr (A.A.A.); cyril.branciard@neel.cnrs.fr (C.B.)

Academic Editors: Paul Busch, Takayuki Miyadera and Teiko Heinosaari

Abstract: The well-known Robertson–Schrödinger uncertainty relations have state-dependent lower bounds, which are trivial for certain states. We present a general approach to deriving tight state-independent uncertainty relations for qubit measurements that completely characterise the obtainable uncertainty values. This approach can give such relations for any number of observables, and we do so explicitly for arbitrary pairs and triples of qubit measurements. We show how these relations can be transformed into equivalent tight entropic uncertainty relations. More generally, they can be expressed in terms of any measure of uncertainty that can be written as a function of the expectation value of the observable for a given state.

Keywords: uncertainty relations; state-independence; quantum measurement

---

1. Introduction

One of the most fundamental features of quantum mechanics is the fact that it is impossible to prepare states that have sufficiently precise simultaneous values of incompatible observables. The most well-known form of this statement is the position-momentum uncertainty relation $\Delta x \Delta p \geq \hbar/2$, which places a lower bound on the product of standard deviations of the position and momentum observables, for a particle in any possible quantum state. This relation was first formalised by Kennard [1] during the formative years of quantum mechanics following Heisenberg's discussion of his "uncertainty principle" [2].

This "uncertainty relation" was quickly generalised by Robertson [3] to arbitrary pairs of incompatible (*i.e.*, non-commuting) observables A and B into what is now the textbook uncertainty relation. Let A and B be two observables and $[A, B] = AB - BA$ their commutator. If the standard deviations ΔA and ΔB for a system in the state ρ are defined as:

$$\Delta A = \sqrt{\langle A\rangle^2 - \langle A^2\rangle}, \qquad \Delta B = \sqrt{\langle B\rangle^2 - \langle B^2\rangle}, \tag{1}$$

where $\langle \cdot \rangle = \mathrm{Tr}[\rho \cdot]$, then Robertson's uncertainty relation can be expressed as:

$$\Delta A\, \Delta B \geq \left| \left\langle \tfrac{1}{2i}[A, B] \right\rangle \right|. \tag{2}$$

These uncertainty relations express a quantitative statement about the measurement statistics for A and B when they are measured many times, separately, on identically-prepared quantum systems.

Such relations are hence sometimes called *preparation uncertainty relations*, since they propose fundamental limits on the measurement statistics for any state preparation.

This is in contrast to Heisenberg's original discussion of his uncertainty principle, which he expressed as the inability to *simultaneously measure* incompatible observables with arbitrary accuracy. As such, quantum uncertainty relations have a long history of being misinterpreted as statements about joint measurements. It is only much more recently that progress has been made in formalising *measurement uncertainty relations* that quantify measurement disturbance in this way, although there continues to be some debate as to the appropriate measure of measurement (in)accuracy and of disturbance [4–10].

The recent interest in measurement uncertainty relations has highlighted an oft-overlooked aspect of Robertson's inequality (2): its *state dependence*. Indeed, the right-hand side of Equation (2) depends on the expectation value $\left|\left\langle \frac{1}{2i}[A,B]\right\rangle\right|$, which itself depends on the state ρ of the system and may be zero, even for non-commuting A and B. To illustrate this, consider a spin-$\frac{1}{2}$ particle and the measurement of Pauli-spin operators $\sigma_x, \sigma_y, \sigma_z$. Robertson's inequality gives us:

$$\Delta\sigma_x \, \Delta\sigma_y \geq \left|\left\langle \tfrac{1}{2i}[\sigma_x, \sigma_y]\right\rangle\right| = |\langle\sigma_z\rangle|, \tag{3}$$

where the right-hand side is zero for any state $\rho = \frac{1}{2}(\mathbf{1} + r_x\sigma_x + r_y\sigma_y)$ (with $r_x^2 + r_y^2 \leq 1$ and with $\mathbf{1}$ denoting the identity operator), even though for $\rho \neq \frac{1}{2}(\mathbf{1} \pm \sigma_x)$ and $\rho \neq \frac{1}{2}(\mathbf{1} \pm \sigma_y)$, both $\Delta\sigma_x$ and $\Delta\sigma_y$ are strictly positive.

Robertson's inequalities, like other historical inequalities such as those due to Schrödinger [11], therefore often tell us little about the accuracy with which one can prepare a state with respect to two incompatible observables A and B. There are two distinct issues with such inequalities: their triviality for certain states and the state dependence itself. The first issue can, in some cases, be avoided by considering more complicated expressions or different measures of incompatibility [12,13]. This approach can be used to give non-trivial state-dependent uncertainty relations, which have the property that they can be experimentally verified without knowing the observables A and B and, thus, are of interest for device-independent cryptography [12]. However, one may equally be interested in knowing how accurately one can prepare a state with respect to two given incompatible observables A and B; that is, in characterising the "minimum uncertainty" states of a system. In such a situation, one ideally wants an uncertainty relation that depends on the state of the system only via the (operationally defined) measures of uncertainty, which ensures that the relation is an operational statement constraining the uncertainties directly and can be evaluated without prior knowledge of the system's state. This is indeed the case, for example, with the position-momentum uncertainty relation described earlier. It thus makes sense to look for tighter, *state-independent* relations capable of addressing these issues, and it is this situation we tackle in this paper.

1.1. Entropic Uncertainty Relations

It has long been known that an alternative form of uncertainty relation can be given by considering the entropies of the observables, rather than their standard deviations [14]. It was Deutsch who first realised that such entropic uncertainty relations can be used to provide state-independent relations [15] and, thus, avoid the problems with the traditional relations discussed above.

Rather than placing lower bounds on the product of variances of two observables A and B, entropic uncertainty relations generally place bounds on the sum of the entropies of A and B. Although many different entropies can be used to formulate such inequalities, perhaps the most well known one, due to Maassen and Uffink [16], makes use of the Shannon entropy. If A has a spectral decomposition $A = \sum_{i=1}^{d} a_i P_i$ in terms of projectors P_i (with $\sum_{i=1}^{d} P_i = \mathbf{1}$), then the Shannon entropy of A for a system in a state ρ is defined as:

$$H(A) = -\sum_{i=1}^{d} \mathrm{Tr}[\rho\, P_i] \log\left(\mathrm{Tr}[\rho\, P_i]\right), \tag{4}$$

which can be seen as a measure of the uncertainty in A for the state ρ. Following the information-theoretical convention, we take the logarithms in base two, although any base can be used as long as the choice is consistent.

Maassen and Uffink's inequality can then be stated as:

$$H(A) + H(B) \geq -2\log c, \tag{5}$$

where $c = \max_{i,j} \mid \langle a_i \mid b_j \rangle \mid$ is the maximum overlap between the eigenvectors $|a_i\rangle$ and $|b_j\rangle$ of A and B. The lower bound is thus state *independent* and depends only on the unitary operator connecting the eigenbases of A and B. Although this is a significant conceptual improvement over state-dependent relations, it is not optimal since the bound cannot be saturated except for well-chosen A and B.

Many variations and improvements on this relation have been found (see [17] for a recent review), but tight bounds have proven elusive without imposing further restrictions. In the two-dimensional case (*i.e.*, for qubit measurements), several papers have improved on this bound [18–20] to determine the optimal lower bound on the sum $H(A) + H(B)$ for arbitrary qubit observables A and B. This result was further generalised to a range of higher dimensional systems in [21]. However, these bounds are not tight in the sense that, although there exist states that saturate the bound, there exist pairs of entropy values $(H(A), H(B))$ that satisfy the uncertainty relation, but are not permitted by quantum mechanics.

In order to fully characterise the obtainable uncertainties, it is thus necessary to consider functions of $H(A)$ and $H(B)$ beyond their sum; indeed, there is no *a priori* reason why one should only consider entropic uncertainty relations based on the sum $H(A) + H(B)$. Much more recent work [22] has made progress in this direction working with more general Rényi entropies and presents several conjectures and numerical results beyond two dimensions.

1.2. State-Independent Uncertainty Relations for Standard Deviations

The growth of interest in measurement uncertainty relations has prompted renewed interest in the possibility of state-independent uncertainty relations for the standard deviations of observables, rather than entropic relations [23,24]. Particular attention has been devoted to understanding the simplest case of qubit uncertainty relations, and we continue this line of research in this paper.

It will be convenient to use the Bloch sphere representation. For the two-dimensional case of qubits, an arbitrary ± 1-valued observable, a "Pauli observable", A can be written $A = \mathbf{a} \cdot \boldsymbol{\sigma}$, where $\boldsymbol{\sigma} = (\sigma_x, \sigma_y, \sigma_z)^{\mathsf{T}}$, and $\mathbf{a}$ is a unit vector. Similarly, an arbitrary state ρ can be written $\rho = \frac{1}{2}(\mathbf{1} + \mathbf{r} \cdot \boldsymbol{\sigma})$, where $|\mathbf{r}| = 1$ if ρ is a pure state and $|\mathbf{r}| < 1$ for mixed states.

Busch *et al.* [25] proved two state-independent uncertainty relations for arbitrary Pauli measurements $A = \mathbf{a} \cdot \boldsymbol{\sigma}$ and $B = \mathbf{b} \cdot \boldsymbol{\sigma}$, showing that:

$$\Delta A + \Delta B \geq |\mathbf{a} \times \mathbf{b}| \tag{6}$$

and:

$$(\Delta A)^2 + (\Delta B)^2 \geq 1 - |\mathbf{a} \cdot \mathbf{b}|. \tag{7}$$

Although these state-independent relations can be saturated by certain states, neither are tight in the sense discussed in the previous section. That is, there exist pairs of values $(\Delta A, \Delta B)$ that are allowed by the relations, but not realisable by any quantum state ρ.

In this paper, we aim precisely to provide *tight* state-independent uncertainty relations, so as to fully characterise the set of allowed values of $(\Delta A, \Delta B)$, for all possible states ρ, which we shall call the "*uncertainty region*". In [26], this is done for the case that $|\mathbf{a} \cdot \mathbf{b}| = 0$ (*i.e.*, for orthogonal spin directions), and it was shown that the bound (7) is tight in this specific case. The authors further gave a tight relation for three orthogonal spin directions as well as generalisations, both for pairs and triples of orthogonal spin observables, to higher dimensions, including the asymptotic behaviour, although these higher-dimensional results are not tight.

The restriction to orthogonal spin observables, however, is a rather strong one. The general case for arbitrary pairs of qubit measurements was completely characterised in [24] using geometric methods, in particular the fact that $|\theta_{ra} - \theta_{rb}| \leq \theta_{ab} \leq \theta_{ra} + \theta_{rb}$, where θ_{ra} is the angle between $\mathbf{r}$ and $\mathbf{a}$, *etc.* This method leads to the state-independent uncertainty relation:

$$\Delta A\,\Delta B \geq \left|\sqrt{1-(\Delta A)^2}\sqrt{1-(\Delta B)^2} - |\mathbf{a}\cdot\mathbf{b}|\right|. \tag{8}$$

Although the authors discuss some explicit instances of three-observable relations, their method does not readily lead to a generalised form of three-observable relations. We note also that Equation (8) was proven in [24] for pure states (and written explicitly for $\mathbf{a}\cdot\mathbf{b} \geq 0$ only), and another version was given for any fixed value of $|\mathbf{r}|$. With our equivalent version given in Equation (21) below, it is straightforward to see that Equation (8) also holds for mixed states.

In this paper, we present a different approach to deriving tight state-independent qubit uncertainty relations. This approach not only leads to a simpler derivation of the relation (8), but is immediately generalisable to give relations for three or more arbitrary observables. It can also be used for other uncertainty measures beyond standard deviations, such as entropic measures. This offers a unified approach to completely characterising the possible state-independent uncertainty relations in two-dimensional Hilbert space.

2. A Unified Approach to Qubit Uncertainty Relations

We present here a general method to derive tight state-independent uncertainty relations for qubits. The approach is based on the fact that, in the case of a binary measurement A, the expectation value $\langle A\rangle$ contains all of the information about the uncertainty in the measurement: for instance, both ΔA and $H(A)$ can be expressed as simple functions of $\langle A\rangle$. We will thus start by giving relations characterising the set of allowed values $(\langle A\rangle, \langle B\rangle)$, before translating these relations into ones in terms of standard deviations or entropies.

A formal characterisation of the set of possible uncertainty values was given by Kaniewski *et al.* in [12] for the more general case of binary-valued measurements in arbitrary dimensional Hilbert spaces. Although the characterisation they give, which is formulated in terms of the expectation values of anticommutators, leads to the results we present in this section (specifically, Lemmas 1–3 below, albeit in a different mathematical framework), Kaniewski *et al.* use it to derive state-*dependent* entropic uncertainty relations for this generalised scenario, which bound the sum of the entropies considered and are hence not tight in the sense we consider. In this paper, our goal is instead to use this characterisation to derive general forms of tight, state-independent uncertainty relations for qubits.

For simplicity, we will only consider Pauli measurements (with eigenvalues ±1), although, as we will discuss in Section 6, our results can straightforwardly be generalised to any projective measurements and even, with a little effort, to binary-valued positive-operator valued measures (POVMs). We shall start with the case of two observables, before generalising to any number of measurements.

2.1. For Two Observables

Let us first consider two arbitrary Pauli observables $A = \mathbf{a}\cdot\boldsymbol{\sigma}$ and $B = \mathbf{b}\cdot\boldsymbol{\sigma}$ and define the matrix:

$$M = \begin{pmatrix} \mathbf{a}^{\mathsf{T}} \\ \mathbf{b}^{\mathsf{T}} \end{pmatrix} \tag{9}$$

with $\mathbf{a}$ and $\mathbf{b}$ as rows representing the measurement directions. In the Bloch sphere representation, we have $\langle A \rangle = \mathrm{Tr}[\rho\, A] = \mathbf{a} \cdot \mathbf{r}$ and $\langle B \rangle = \mathrm{Tr}[\rho\, B] = \mathbf{b} \cdot \mathbf{r}$, so that:

$$M\mathbf{r} = \begin{pmatrix} \mathbf{a} \cdot \mathbf{r} \\ \mathbf{b} \cdot \mathbf{r} \end{pmatrix} = \begin{pmatrix} \langle A \rangle \\ \langle B \rangle \end{pmatrix} := \mathbf{u}\,. \tag{10}$$

Although M is not invertible, one can always find the Moore–Penrose pseudoinverse M^{+} such that $M^{+}M$ is an orthogonal projection onto the range of M^{T}, that is, the subspace spanned by $\{\mathbf{a}, \mathbf{b}\}$ [27]. This implies the following crucial lemma, which will be the basis of the uncertainty relations we derive in the following sections (this lemma is equivalent, for qubit measurements, to the "ellipsoid condition" given in [12]).

Lemma 1. *For any pair of Pauli observables $A = \mathbf{a} \cdot \boldsymbol{\sigma}$ and $B = \mathbf{b} \cdot \boldsymbol{\sigma}$ with M and $\mathbf{u}$ as defined in Equations* (9) *and* (10)*, and M^{+} the pseudoinverse of M, every quantum state $\rho = \frac{1}{2}(\mathbf{1} + \mathbf{r} \cdot \boldsymbol{\sigma})$ satisfies:*

$$|M^{+}\mathbf{u}| = |(M^{+}M)\mathbf{r}| \leq |\mathbf{r}|, \tag{11}$$

where equality is obtained if and only if $\mathbf{r}$ lies in $\mathrm{Span}\{\mathbf{a}, \mathbf{b}\}$.

In a more explicit form, this inequality implies the following relation for the two expectation values $\langle A \rangle$ and $\langle B \rangle$:

Lemma 2. *For any pair of Pauli observables $A = \mathbf{a} \cdot \boldsymbol{\sigma}$ and $B = \mathbf{b} \cdot \boldsymbol{\sigma}$, every quantum state $\rho = \frac{1}{2}(\mathbf{1} + \mathbf{r} \cdot \boldsymbol{\sigma})$ satisfies:*

$$\begin{aligned} |\, \langle A \rangle\, \mathbf{a} - \langle B \rangle\, \mathbf{b}\, |^2 \;&=\; \langle A \rangle^2 + \langle B \rangle^2 - 2\,(\mathbf{a} \cdot \mathbf{b})\, \langle A \rangle\, \langle B \rangle \\ &\leq\; \big(1 - (\mathbf{a} \cdot \mathbf{b})^2\big)\, |\mathbf{r}|^2 \;\leq\; 1 - (\mathbf{a} \cdot \mathbf{b})^2 = |\mathbf{a} \times \mathbf{b}|^2\,. \end{aligned} \tag{12}$$

In the case where $|\mathbf{a} \cdot \mathbf{b}| < 1$, the first and second inequalities are saturated if and only if $\mathbf{r} \in$ $\mathrm{Span}\{\mathbf{a}, \mathbf{b}\}$ *and if and only if ρ is a pure state, respectively.*

Proof. Let us choose a basis in the Bloch sphere such that $\mathbf{a}$ and $\mathbf{b}$ can be written $\mathbf{a} = \big(\sqrt{\tfrac{1+\mathbf{a}\cdot\mathbf{b}}{2}}, \sqrt{\tfrac{1-\mathbf{a}\cdot\mathbf{b}}{2}}, 0\big)^{\mathsf{T}}$ and $\mathbf{b} = \big(\sqrt{\tfrac{1+\mathbf{a}\cdot\mathbf{b}}{2}}, -\sqrt{\tfrac{1-\mathbf{a}\cdot\mathbf{b}}{2}}, 0\big)^{\mathsf{T}}$, and assume first that $|\mathbf{a} \cdot \mathbf{b}| < 1$. Then M has linearly-independent rows, and its pseudoinverse can be obtained using the relation $M^{+} = M^{\mathsf{T}}(MM^{\mathsf{T}})^{-1}$. We thus have:

$$M^{+} = \frac{1}{\sqrt{1 - (\mathbf{a} \cdot \mathbf{b})^2}} \begin{pmatrix} \sqrt{\frac{1-\mathbf{a}\cdot\mathbf{b}}{2}} & \sqrt{\frac{1-\mathbf{a}\cdot\mathbf{b}}{2}} \\ \sqrt{\frac{1+\mathbf{a}\cdot\mathbf{b}}{2}} & -\sqrt{\frac{1+\mathbf{a}\cdot\mathbf{b}}{2}} \\ 0 & 0 \end{pmatrix} \quad \text{and} \quad M^{+}M = \begin{pmatrix} 1 & 0 & 0 \\ 0 & 1 & 0 \\ 0 & 0 & 0 \end{pmatrix}, \tag{13}$$

where we see that $M^{+}M$ is a projection onto $\mathrm{Span}\{\mathbf{a}, \mathbf{b}\}$, as expected. The first inequality in Equation (12) is then obtained by squaring Equation (11), multiplying it by $\big(1 - (\mathbf{a} \cdot \mathbf{b})^2\big)$ and expanding; like Equation (11), it is saturated when $\mathbf{r} \in \mathrm{Span}\{\mathbf{a}, \mathbf{b}\}$. The second inequality follows directly from $|\mathbf{r}| \leq 1$ and is saturated when ρ is a pure state.

In the (trivial) case where $\mathbf{a} \cdot \mathbf{b} = \pm 1$ (*i.e.*, $A = \pm B$), one can easily check that the left-hand side of Equation (12) is zero, so that the relation still holds. □

Relation (12), as also noted in [12], shows that the set of allowed values for $\langle A\rangle$ and $\langle B\rangle$ forms an ellipse in the $(\langle A\rangle, \langle B\rangle)$-plane, as depicted in Figure 1 for $\mathbf{a}\cdot\mathbf{b}=0$ and $\mathbf{a}\cdot\mathbf{b}=\frac{1}{2}$ (*cf.* Figure 1 in [12]). As can be seen, this ellipse becomes a circle for $\mathbf{a}\cdot\mathbf{b}=0$ and degenerates into the line segment given by $\langle A\rangle = \pm\langle B\rangle$ when $\mathbf{a}\cdot\mathbf{b}=\pm 1$. Note also that the first inequality in Equation (12) characterises concentric ellipses for fixed maximal values of $|\mathbf{r}|$ (see Figure 1).

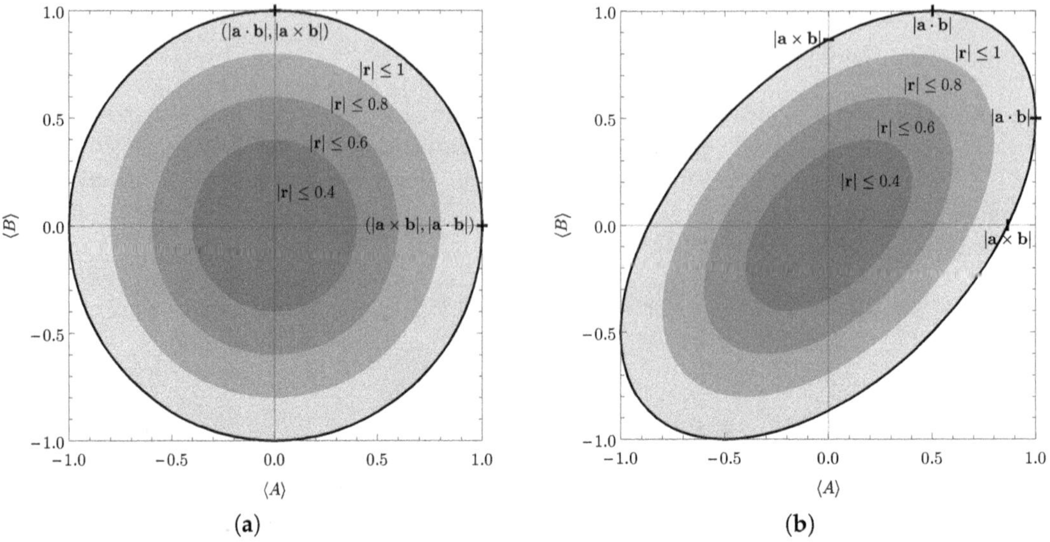

Figure 1. The ellipse delimited by the thick black line corresponds to the set of all permitted values $(\langle A\rangle, \langle B\rangle)$ for two fixed Pauli operators $A=\mathbf{a}\cdot\sigma$ and $B=\mathbf{b}\cdot\sigma$ with **(a)** $\mathbf{a}\cdot\mathbf{b}=0$ and **(b)** $\mathbf{a}\cdot\mathbf{b}=\frac{1}{2}$ for all possible states ρ, defined by the relation (12). The values $(\langle A\rangle, \langle B\rangle)$ on its boundary saturate Equation (12). The darker concentric ellipses represent the regions of permitted values for mixed states with bounded Bloch vector norms $|\mathbf{r}| \leq 0.8, 0.6$ and 0.4, characterised using the first inequality in Equation (12).

We emphasise that the relation (12) is tight. To verify (for the non-trivial case that $|\mathbf{a}\cdot\mathbf{b}|<1$) that any pair $(\langle A\rangle, \langle B\rangle)$ satisfying Equation (12) can be attained, take for instance:

$$\mathbf{r} = \frac{1}{1-(\mathbf{a}\cdot\mathbf{b})^2}\Big[\langle A\rangle\left(\mathbf{a}-(\mathbf{a}\cdot\mathbf{b})\,\mathbf{b}\right)+\langle B\rangle\left(\mathbf{b}-(\mathbf{a}\cdot\mathbf{b})\,\mathbf{a}\right)\Big]. \tag{14}$$

This gives the desired values for $\langle A\rangle$ and $\langle B\rangle$ and is indeed a valid Bloch vector (*i.e.*, its norm is at most one) if and only if Equation (12) is satisfied. It clearly lies in the **ab**-plane and, thus, saturates the first inequality in Equation (12). To saturate both inequalities in Equation (12) for a given value of $\langle A\rangle$, one can take:

$$\mathbf{r}_\pm = \langle A\rangle\,\mathbf{a} \pm \sqrt{\frac{1-\langle A\rangle^2}{1-(\mathbf{a}\cdot\mathbf{b})^2}}\left(\mathbf{b}-(\mathbf{a}\cdot\mathbf{b})\,\mathbf{a}\right), \tag{15}$$

which characterises the pure states ($|\mathbf{r}_\pm|=1$) in the **ab**-plane and gives the desired value for $\langle A\rangle$.

2.2. *Generalisation to More Observables*

Our approach, based on Lemma 1, generalises easily to more than two observables. Remarkably, when the various observables A_i span the whole Bloch sphere, it provides *exact relations* on the expectation values $\langle A_i\rangle$, rather than just inequalities.

More specifically, let us consider n observables $A_1=\mathbf{a}_1\cdot\sigma,\ldots,A_n=\mathbf{a}_n\cdot\sigma$ and define $M=(\mathbf{a}_1^T,\ldots,\mathbf{a}_n^T)^\mathsf{T}$, M^+ its pseudoinverse, and $\mathbf{u}=(\langle A_1\rangle,\ldots,\langle A_n\rangle)^\mathsf{T}$. Then, the relation (11) of Lemma 1 holds unchanged, with equality if and only if $\mathbf{r}$ lies in $\mathrm{Span}\{\mathbf{a}_1, \ldots, \mathbf{a}_n\}$. This straightforwardly implies (after squaring Equation (11)) the following relation:

Lemma 3. *For any n Pauli observables $A_1, \ldots, A_n$ with Bloch vectors $\mathbf{a}_1, \ldots, \mathbf{a}_n$ spanning the whole Bloch sphere, every quantum state $\rho = \frac{1}{2}(\mathbb{1} + \mathbf{r} \cdot \boldsymbol{\sigma})$ satisfies the relation:*

$$\sum_{1 \le i,j \le n} m_{i,j} \langle A_i \rangle \langle A_j \rangle = |\mathbf{r}|^2 \le 1, \tag{16}$$

where the coefficients $m_{i,j}$ are the elements of the symmetric matrix $(M^+)^\mathsf{T} M^+$, with M^+ the pseudoinverse of the matrix $M = (\mathbf{a}_1^\mathsf{T}, \ldots, \mathbf{a}_n^\mathsf{T})^\mathsf{T}$.

If the n Bloch vectors $\mathbf{a}_1, \ldots, \mathbf{a}_n$ do not span the whole Bloch sphere, then the equality in Equation (16) must be replaced by an inequality (with $|\mathbf{r}|^2$ upper-bounding the left-hand side), which is saturated if and only if $\mathbf{r} \in \mathrm{Span}\{\mathbf{a}_1, \ldots, \mathbf{a}_n\}$.

This relation shows that, in the general case, the set of allowed values for $(\langle A_1 \rangle, \ldots, \langle A_n \rangle)$ lies on an n-dimensional ellipsoid [12]; when at least three of the Bloch vectors are linearly independent (*i.e.*, when the n Bloch vectors span the whole Bloch sphere), all pure states give points on the surface of the ellipsoid, while mixed states give interior points.

Although this relation is satisfied for any state ρ, for $n > 3$ observables (or, more generally, if n exceeds the dimension of $\mathrm{Span}\{\mathbf{a}_1, \ldots, \mathbf{a}_n\}$), it is not tight, since there exist values $(\langle A_1 \rangle, \ldots, \langle A_n \rangle) := \mathbf{u}$ that satisfy it, but are not obtainable by any quantum state ρ. Specifically, $\mathbf{u}$ is realisable if and only if there exists a quantum state with Bloch vector $\mathbf{r}$ such that $M\mathbf{r} = \mathbf{u}$. For $n = 3$ observables with $\mathbf{a}_1, \mathbf{a}_2, \mathbf{a}_3$ linearly independent, one has $MM^+ = \mathbb{1}$, and one may simply take $\mathbf{r} = M^+\mathbf{u}$, so Equation (16) is tight in this case. However, if the Bloch vectors $\mathbf{a}_1, \ldots, \mathbf{a}_n$ are not linearly independent (as, in particular, is the case for $n > 3$), then such an $\mathbf{r}$ exists if and only if $MM^+\mathbf{u} = \mathbf{u}$.

One can understand this by noting that, once $\langle A_1 \rangle, \langle A_2 \rangle, \langle A_3 \rangle$ are determined (assuming, without loss of generality, that $\mathbf{a}_1, \mathbf{a}_2, \mathbf{a}_3$ are linearly independent), then this uniquely determines each $\langle A_i \rangle$ for $i > 3$. Thus, only three of the expectation values can be considered free variables, whereas Equation (16) has $n - 1$ free variables. The requirement that $MM^+\mathbf{u} = \mathbf{u}$ expresses this further constraint. In other words, we obtain a tight relation for $n \ge 4$ when Equation (16) is supplemented by the further constraint that the expectation values $(\langle A_1 \rangle, \ldots, \langle A_n \rangle) := \mathbf{u}$ also satisfy $MM^+\mathbf{u} = \mathbf{u}$.

Let us provide some explicit examples of relations based on Lemma 3. Consider first the case of $n = 3$ Pauli observables $A = \mathbf{a} \cdot \boldsymbol{\sigma}$, $B = \mathbf{b} \cdot \boldsymbol{\sigma}$ and $C = \mathbf{c} \cdot \boldsymbol{\sigma}$ with linearly-independent Bloch vectors: we obtain, after multiplication by $V^2 = (\,\mathbf{a} \cdot (\mathbf{b} \times \mathbf{c})\,)^2 > 0$, the relation:

$$\begin{aligned}
&|\,(\mathbf{b} \times \mathbf{c}) \langle A \rangle + (\mathbf{c} \times \mathbf{a}) \langle B \rangle + (\mathbf{a} \times \mathbf{b}) \langle C \rangle\,|^2 \\
&\quad = |\mathbf{b} \times \mathbf{c}|^2 \langle A \rangle^2 + |\mathbf{a} \times \mathbf{c}|^2 \langle B \rangle^2 + |\mathbf{a} \times \mathbf{b}|^2 \langle C \rangle^2 + 2\,(\mathbf{b} \times \mathbf{c}) \cdot (\mathbf{c} \times \mathbf{a}) \langle A \rangle \langle B \rangle \\
&\qquad + 2\,(\mathbf{b} \times \mathbf{c}) \cdot (\mathbf{a} \times \mathbf{b}) \langle A \rangle \langle C \rangle + 2\,(\mathbf{c} \times \mathbf{a}) \cdot (\mathbf{a} \times \mathbf{b}) \langle B \rangle \langle C \rangle \;=\; V^2\,|\mathbf{r}|^2 \;\le\; V^2.
\end{aligned} \tag{17}$$

(To calculate the matrix $(M^+)^\mathsf{T} M^+$ one can, for instance, parametrise $\mathbf{a}$ and $\mathbf{b}$ as in the proof of Lemma 2 and define $\mathbf{c} = \frac{1}{\sqrt{1-(\mathbf{a}\cdot\mathbf{b})^2}}\big((\mathbf{a} \cdot \mathbf{c} + \mathbf{b} \cdot \mathbf{c})\sqrt{\tfrac{1-\mathbf{a}\cdot\mathbf{b}}{2}}, (\mathbf{a} \cdot \mathbf{c} - \mathbf{b} \cdot \mathbf{c})\sqrt{\tfrac{1+\mathbf{a}\cdot\mathbf{b}}{2}}, V\big)^\mathsf{T}$.)

This relation *is* tight, and the quantum state with Bloch vector $\mathbf{r} = M^+(\langle A \rangle, \langle B \rangle, \langle C \rangle)^\mathsf{T}$ has the required expectation values for any $(\langle A \rangle, \langle B \rangle, \langle C \rangle)$ satisfying Equation (17). In the special case of three orthogonal measurements ($\mathbf{a} \cdot \mathbf{b} = \mathbf{a} \cdot \mathbf{c} = \mathbf{b} \cdot \mathbf{c} = 0$, $V^2 = 1$), we thus find the well-known relation:

$$\langle A \rangle^2 + \langle B \rangle^2 + \langle C \rangle^2 \;=\; |\mathbf{r}|^2 \;\le\; 1\,. \tag{18}$$

Consider as another example $n = 4$ observables with Bloch vectors pointing to the vertices of a regular tetrahedron. In that case, we find (one can use here, for instance, the parametrisation $\mathbf{a}_1 = (1,1,1)^\mathsf{T}/\sqrt{3}$, $\mathbf{a}_2 = (1,-1,-1)^\mathsf{T}/\sqrt{3}$, $\mathbf{a}_3 = (-1,1,-1)^\mathsf{T}/\sqrt{3}$, $\mathbf{a}_4 = (-1,-1,1)^\mathsf{T}/\sqrt{3}$):

$$3 \sum_{1\le i\le 4} \langle A_i\rangle^2 - \sum_{1\le i\ne j\le 4} \langle A_i\rangle \langle A_j\rangle = \frac{16}{3}\,|\mathbf{r}|^2 \le \frac{16}{3}. \tag{19}$$

For any quantum state ρ, the expectation values for these observables must further satisfy $\langle A_4\rangle = -\langle A_1\rangle - \langle A_2\rangle - \langle A_3\rangle$. Thus, Equation (19) can be made tight by further imposing this constraint or simply replacing $\langle A_4\rangle$ by this expression in the relation.

3. Uncertainty Relations in Terms of Standard Deviations

The relations for the expectation values presented in the previous section now allow us to proceed with the main aim of the paper and derive tight state-independent uncertainty relations in terms of standard deviations. To do so, we note that any Pauli operator A satisfies $A^2 = \mathbb{1}$, so that $\langle A^2\rangle = 1$, and therefore, the standard deviation ΔA is simply related to the expectation value $\langle A\rangle$ by:

$$(\Delta A)^2 = 1 - \langle A\rangle^2 \quad \text{and} \quad \langle A\rangle = \pm\sqrt{1-(\Delta A)^2}. \tag{20}$$

Note that because of the $\pm$ sign above, some care needs to be taken when applying the previous relations to derive uncertainty relations for standard-deviations.

3.1. Uncertainty Relation for Two Pauli Observables

Let us start again with two Pauli observables. Using $(\mathbf{a}\cdot\mathbf{b})\langle A\rangle\langle B\rangle \le |(\mathbf{a}\cdot\mathbf{b})\langle A\rangle\langle B\rangle| = |\mathbf{a}\cdot\mathbf{b}|\sqrt{1-(\Delta A)^2}\sqrt{1-(\Delta B)^2}$ and reordering the terms, the following relation follows directly from Lemma 2:

Theorem 4. *For any pair of Pauli observables $A = \mathbf{a}\cdot\boldsymbol{\sigma}$ and $B = \mathbf{b}\cdot\boldsymbol{\sigma}$, every quantum state $\rho = \frac{1}{2}(\mathbb{1}+\mathbf{r}\cdot\boldsymbol{\sigma})$ satisfies the state-independent uncertainty relation:*

$$(\Delta A)^2 + (\Delta B)^2 + 2\,|\mathbf{a}\cdot\mathbf{b}|\sqrt{1-(\Delta A)^2}\sqrt{1-(\Delta B)^2} \ge 2 - \left(1-(\mathbf{a}\cdot\mathbf{b})^2\right)|\mathbf{r}|^2 \ge 1 + (\mathbf{a}\cdot\mathbf{b})^2. \tag{21}$$

In the case where $|\mathbf{a}\cdot\mathbf{b}| < 1$, the first inequality above is saturated if and only if $\mathbf{r} \in \mathrm{Span}\{\mathbf{a},\mathbf{b}\}$ and $(\mathbf{a}\cdot\mathbf{b})\langle A\rangle\langle B\rangle \ge 0$, and the second one if and only if ρ is a pure state.

One can easily check that the relation (21) is equivalent to Equation (8) and to the relation given in [24] for a fixed value of $|\mathbf{r}|$. The uncertainty region it defines is shown in Figure 2 for the cases where $\mathbf{a}\cdot\mathbf{b} = 0$ and $|\mathbf{a}\cdot\mathbf{b}| = \frac{1}{2}$. It may be interesting to note that, in the $((\Delta A)^2, (\Delta B)^2)$-plane, this region corresponds to the convex hull of an ellipse and the point $(1,1)$.

It is worth pointing out that, using the relation:

$$\sqrt{1-(\mathbf{a}\cdot\mathbf{b})^2} = |\mathbf{a}\times\mathbf{b}| = \left|\tfrac{1}{2i}[A,B]\right| \tag{22}$$

where, in the last term, $|\cdot|$ denotes the operator norm, Equation (21) can be expressed in terms of this commutator, a measure of the incompatibility of A and B, instead of the inner product $\mathbf{a}\cdot\mathbf{b}$. Interestingly, one can thus visualise this incompatibility as the area of the parallelogram defined by $\mathbf{a}$ and $\mathbf{b}$.

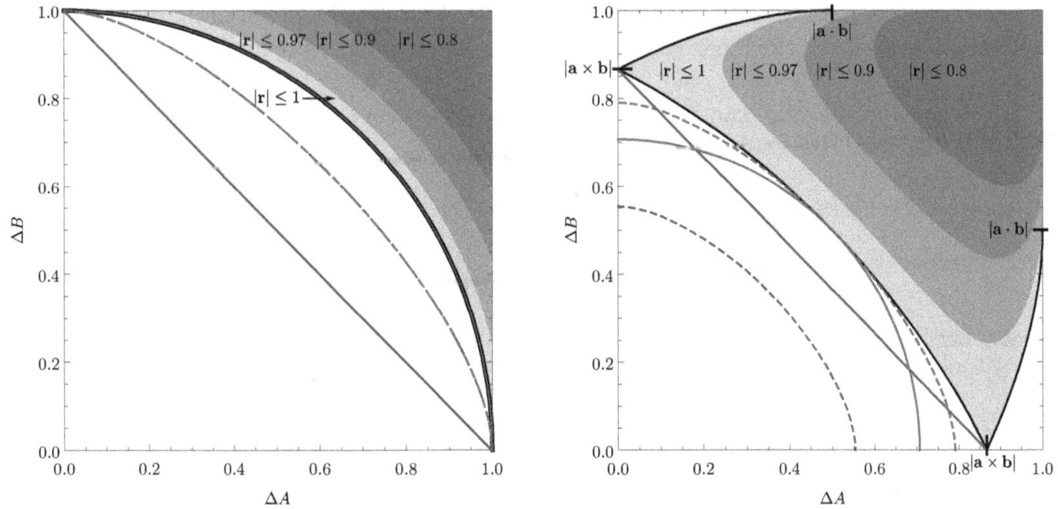

(a) (b)

Figure 2. Representation of the uncertainty regions (filled areas) characterised by Equation (21), shown here (**a**) for $\mathbf{a} \cdot \mathbf{b} = 0$ and (**b**) for $|\mathbf{a} \cdot \mathbf{b}| = \frac{1}{2}$. The values on the thick black curves saturate the uncertainty relation (21). The darker areas represent the allowed values of $(\Delta A, \Delta B)$ for mixed states with bounded Bloch vector norms $|\mathbf{r}| \leq 0.97, 0.9$ and 0.8, characterised using the first inequality in Equation (21). We compare here these uncertainty regions to the bounds given by Equations (6) and (7) (blue and red curves, which touch the uncertainty region for $(\Delta A, \Delta B) = (|\mathbf{a} \times \mathbf{b}|, 0)$ or $(0, |\mathbf{a} \times \mathbf{b}|)$ and for $(\Delta A, \Delta B) = (\sqrt{\frac{1-|\mathbf{a}\cdot\mathbf{b}|}{2}}, \sqrt{\frac{1-|\mathbf{a}\cdot\mathbf{b}|}{2}})$, respectively), as well as the two entropic uncertainty relations given by Equation (5) and in [18–21] (blue and red dashed curves, respectively), which we translate in terms of standard deviations using $H(A) = h_2(\frac{1+\sqrt{1-(\Delta A)^2}}{2})$ (with h_2 the binary entropy function; see Section 4). Our relation is clearly stronger than all of these bounds. Note that in the special case where $\mathbf{a} \cdot \mathbf{b} = 0$, the bounds given by Equations (21) and (7) coincide, as do the two entropic uncertainty relations.

Theorem 4 provides a *tight* state-independent uncertainty relation. One can indeed verify, as we did for Lemma 2, that any pair of values $(\Delta A, \Delta B)$ satisfying Equation (21) can be attained, for instance (in the non-trivial case $|\mathbf{a} \cdot \mathbf{b}| < 1$), by the state with Bloch vector:

$$\mathbf{r} = \frac{1}{1-(\mathbf{a}\cdot\mathbf{b})^2}\Big[\sqrt{1-(\Delta A)^2}\,(\mathbf{a}-(\mathbf{a}\cdot\mathbf{b})\,\mathbf{b}) + \tau\sqrt{1-(\Delta B)^2}\,(\mathbf{b}-(\mathbf{a}\cdot\mathbf{b})\,\mathbf{a})\Big] \tag{23}$$

with $\tau = \mathrm{sgn}(\mathbf{a} \cdot \mathbf{b})$, which is a valid Bloch vector (*i.e.*, its norm is at most one) if and only if Equation (21) is satisfied. Note that it saturates the first inequality in Equation (21). To saturate both inequalities in Equation (21) for a given value of ΔA, consider the pure states in the **ab**-plane with Bloch vectors:

$$\mathbf{r}_\pm = \sqrt{1-(\Delta A)^2}\,\mathbf{a} \pm \tau\frac{\Delta A}{\sqrt{1-(\mathbf{a}\cdot\mathbf{b})^2}}\,(\mathbf{b}-(\mathbf{a}\cdot\mathbf{b})\,\mathbf{a})\,. \tag{24}$$

It can be checked that $\mathbf{r}_+$ always saturates both inequalities in Equation (21), while $\mathbf{r}_-$ does so only when $\Delta A \leq |\mathbf{a}\cdot\mathbf{b}|$, $\Delta A = 1$ or $\mathbf{a}\cdot\mathbf{b} = 0$ (see Figure 2). Note that in any case, $\Delta B(\mathbf{r}_+) \leq \Delta B(\mathbf{r}_-)$ (with equality for $\Delta A = 0$, $\Delta A = 1$ or $\mathbf{a}\cdot\mathbf{b} = 0$).

The Bloch vectors $\mathbf{r}_+$ and $\mathbf{r}_-$ hence completely characterise the boundary of the uncertainty region in the space $\Delta A \times \Delta B$ where Equation (21) is saturated; the boundary is completed by the line segments $(\Delta A \geq |\mathbf{a}\cdot\mathbf{b}|, \Delta B = 1)$ and $(\Delta A = 1, \Delta B \geq |\mathbf{a}\cdot\mathbf{b}|)$; see Figure 2. Often, one is interested only in the *monotone closure* of the uncertainty region (as in [22,26]); that is, the closure under increasing either coordinate of the set of realisable pairs $(\Delta A, \Delta B)$. The lower bound of this uncertainty region is obtained by the state with Bloch vector $\mathbf{r}_+$ for $\Delta A \le \sqrt{1-(\mathbf{a}\cdot\mathbf{b})^2}$, since for this state $\Delta B = 0$ when $\Delta A = \sqrt{1-(\mathbf{a}\cdot\mathbf{b})^2} = |\mathbf{a}\times\mathbf{b}|$.

This monotone closure can also be characterised by an uncertainty relation that follows from Equation (21):

Theorem 5. *The monotone closure of Equation* (21) *is given by:*

$$(\Delta A)^2 + (\Delta B)^2 + 2\,|\mathbf{a}\cdot\mathbf{b}|\,\Delta A\,\Delta B \;\ge\; 1-(\mathbf{a}\cdot\mathbf{b})^2. \tag{25}$$

Proof. Let us first show that Equation (25) is equivalent to Equation (21) if $(\Delta A)^2 + (\Delta B)^2 \le 1 - (\mathbf{a}\cdot\mathbf{b})^2$. In this case, then $(\Delta A)^2 + (\Delta B)^2 \le 1+(\mathbf{a}\cdot\mathbf{b})^2$ also, and we can write Equation (21) as:

$$\begin{aligned} 4\,(\mathbf{a}\cdot\mathbf{b})^2 \left[1-(\Delta A)^2\right]\left[1-(\Delta B)^2\right] - \left[1+(\mathbf{a}\cdot\mathbf{b})^2-(\Delta A)^2-(\Delta B)^2\right]^2 \\ = 4\,(\mathbf{a}\cdot\mathbf{b})^2\,(\Delta A)^2\,(\Delta B)^2 - \left[1-(\mathbf{a}\cdot\mathbf{b})^2-(\Delta A)^2-(\Delta B)^2\right]^2 \;\ge\; 0. \end{aligned} \tag{26}$$

Thus, still under the assumption that $(\Delta A)^2 + (\Delta B)^2 \le 1-(\mathbf{a}\cdot\mathbf{b})^2$, we have:

$$2\,|\mathbf{a}\cdot\mathbf{b}|\,\Delta A\,\Delta B \;\ge\; 1-(\mathbf{a}\cdot\mathbf{b})^2-(\Delta A)^2-(\Delta B)^2, \tag{27}$$

which is precisely Equation (25).

Equation (25) clearly defines its own monotone closure: if $(\Delta A, \Delta B)$ satisfy it, then so do any $(\Delta A', \Delta B')$ with $\Delta A' \ge \Delta A$ and $\Delta B' \ge \Delta B$. Since, as we just showed, Equation (21) is equivalent to Equation (25) in the region where $(\Delta A)^2 + (\Delta B)^2 \le 1-(\mathbf{a}\cdot\mathbf{b})^2$, its monotone closure in that region is given by Equation (25). Furthermore, since all points with $(\Delta A)^2 + (\Delta B)^2 = 1-(\mathbf{a}\cdot\mathbf{b})^2$ satisfy Equation (25) and, hence, also (the equivalent, for these points) Equation (21), then the whole region where $(\Delta A)^2 + (\Delta B)^2 \ge 1-(\mathbf{a}\cdot\mathbf{b})^2$ is in the monotone closure of Equation (21). All together, the monotone closure of Equation (21) is thus composed of the points in the region where $(\Delta A)^2 + (\Delta B)^2 \le 1-(\mathbf{a}\cdot\mathbf{b})^2$, which satisfy Equation (25), and of all points in the region $(\Delta A)^2 + (\Delta B)^2 \ge 1-(\mathbf{a}\cdot\mathbf{b})^2$; since the latter clearly also satisfy Equation (25), this equation is sufficient to fully characterise the monotone closure of Equation (21). □

For a fixed value of $|\mathbf{r}|$, one can similarly show (e.g., by replacing $(\Delta A)^2$ by $(1-\frac{1-(\Delta A)^2}{|\mathbf{r}|^2})$ and $(\Delta B)^2$ by $(1-\frac{1-(\Delta B)^2}{|\mathbf{r}|^2})$ in the calculations) that the monotone closure is given by:

$$(\Delta A)^2 + (\Delta B)^2 + 2\,|\mathbf{a}\cdot\mathbf{b}|\,\sqrt{|\mathbf{r}|^2-(1-(\Delta A)^2)}\,\sqrt{|\mathbf{r}|^2-(1-(\Delta B)^2)} \;\ge\; 2-\left(1+(\mathbf{a}\cdot\mathbf{b})^2\right)|\mathbf{r}|^2. \tag{28}$$

Note that the relation given by Equation (25) can readily be used to derive the weaker relations given by Equations (6) and (7), which are thus in turn seen to follow from Equation (21). In particular, we can obtain Equation (6) by noting that:

$$(\Delta A + \Delta B)^2 \;\ge\; (\Delta A)^2 + (\Delta B)^2 + 2\,|\mathbf{a}\cdot\mathbf{b}|\,\Delta A\,\Delta B \;\ge\; 1-(\mathbf{a}\cdot\mathbf{b})^2 \tag{29}$$

and thus:

$$\Delta A + \Delta B \;\ge\; \sqrt{1-(\mathbf{a}\cdot\mathbf{b})^2} = |\mathbf{a}\times\mathbf{b}|. \tag{30}$$

Equation (7) can be obtained by seeing that:

$$(1+|\mathbf{a}\cdot\mathbf{b}|)\left[(\Delta A)^2+(\Delta B)^2\right] \;\ge\; (\Delta A)^2 + (\Delta B)^2 + 2\,|\mathbf{a}\cdot\mathbf{b}|\,\Delta A\,\Delta B \;\ge\; 1-(\mathbf{a}\cdot\mathbf{b})^2 \tag{31}$$

and thus:

$$(\Delta A)^2 + (\Delta B)^2 \geq 1 - |\mathbf{a} \cdot \mathbf{b}|. \tag{32}$$

Figure 2 also shows these two bounds in comparison to the tight relation Equation (21) that we derived. Note that if A and B are orthogonal spin measurements (*i.e.*, $\mathbf{a} \cdot \mathbf{b} = 0$), then Equation (7) reduces to Equation (21) and is tight, but for all other cases, neither Equation (6) nor Equation (7) are tight: these bounds are only obtained by states with $(\Delta A, \Delta B) = (|\mathbf{a} \times \mathbf{b}|, 0)$ or $(0, |\mathbf{a} \times \mathbf{b}|)$ for Equation (6) and $\Delta A = \Delta B = \sqrt{\frac{1-|\mathbf{a}\cdot\mathbf{b}|}{2}}$ for Equation (7).

We note finally that it is also possible to express the relation (25) in a further form that may be useful in understanding the lower bound of the uncertainty region. Specifically, Equation (25) is satisfied if and only if:

$$(\Delta A)^2 + (\Delta B)^2 \geq 1 \quad \text{or} \quad \Delta A \sqrt{1 - (\Delta B)^2} + \Delta B \sqrt{1 - (\Delta A)^2} \geq \sqrt{1 - (\mathbf{a} \cdot \mathbf{b})^2}, \tag{33}$$

where the lower bound of the uncertainty region lies, as we saw before, in the region $(\Delta A)^2 + (\Delta B)^2 \leq 1$ and is expressed by the second half of Equation (33). (While finishing this manuscript, we became aware that the second part of Equation (33) had been independently derived by P. Busch using a geometric argument [28].) The proof that this alternative form is equivalent to Equation (25) mirrors one given in [29], so we do not give it here, but it is important to note that the second half of this relation is violated by some allowable uncertainty pairs (e.g., $\Delta A = \Delta B = 1$), and hence, the first disjunction is essential for the relation to be valid for all pairs $(\Delta A, \Delta B)$.

Using Equation (22), we note that the right-hand side of the second inequality in Equation (33) is precisely $|\frac{1}{2i}[A, B]|$, so the monotone closure of the uncertainty region can be expressed as a relation *only* on the uncertainties bounded by a function of the commutator of A and B.

3.2. *Uncertainty Relations for n Pauli Observables*

Similarly to the previous case of two measurements, one can now use Lemma 3 to derive state-independent uncertainty relations for more observables. Namely, using $\langle A_i \rangle = \pm\sqrt{1 - (\Delta A_i)^2}$, we obtain:

Theorem 6. *For any n Pauli observables $A_1, \ldots, A_n$ with Bloch vectors $\mathbf{a}_1, \ldots, \mathbf{a}_n$ spanning the whole Bloch sphere, every quantum state $\rho = \frac{1}{2}(\mathbf{1} + \mathbf{r} \cdot \sigma)$ satisfies the relation:*

$$\exists \tau_1, \ldots, \tau_n = \pm 1, \quad \sum_{1\leq i\leq n} m_{i,i} (\Delta A_i)^2 - \sum_{1\leq i\neq j\leq n} \tau_i \tau_j m_{i,j} \sqrt{1 - (\Delta A_i)^2} \sqrt{1 - (\Delta A_j)^2}$$
$$= \Big(\sum_{1\leq i\leq n} m_{i,i}\Big) - |\mathbf{r}|^2 \geq \Big(\sum_{1\leq i\leq n} m_{i,i}\Big) - 1, \tag{34}$$

where the $m_{i,j}$ coefficients are the elements of the symmetric matrix $(M^+)^\mathsf{T} M^+$, with M^+ the pseudoinverse of the matrix $M = (\mathbf{a}_1^\mathsf{T}, \ldots, \mathbf{a}_n^\mathsf{T})^\mathsf{T}$.

The relation (34) can furthermore be written without the existential quantifier as an inequality ($\geq$) instead of an equality if one replaces all $-\tau_i \tau_j m_{i,j}$ by $+|m_{i,j}|$. This inequality then also holds if the Bloch vectors do not span the whole Bloch sphere and is saturated if and only if $\mathbf{r} \in \mathrm{Span}\{\mathbf{a}_1, \ldots, \mathbf{a}_n\}$ and $m_{i,j} \langle A_i \rangle \langle A_j \rangle \leq 0$ for all $i \neq j$.

Note that the quantifiers "$\exists \tau_i$" have been introduced in Equation (34) to make the relation state-independent (*i.e.*, so that it is a constraint that can be evaluated solely on the (measurable) ΔA_i's, where no other terms depend on the quantum state). In practice, however, if ρ is known, one can simply take the signs to be $\tau_i = \mathrm{sgn}\langle A_i \rangle$.

As for Lemma 3, Theorem 6 does not give tight relations for $n > 3$. However, it can easily be made tight by requiring, in addition, that $(\tau_1\sqrt{1-(\Delta A_1)^2},\ldots,\tau_n\sqrt{1-(\Delta A_n)^2}) := \mathbf{u}$ satisfies $MM^+\mathbf{u} = \mathbf{u}$. Crucially, note that this further condition is state independent, just like Equation (34).

Let us illustrate again this relation for some specific examples. For the case of $n = 3$ Pauli observables A, B and C with linearly-independent Bloch vectors, we obtain the tight relation (we recover, in particular, a relation given in [24] for the specific case $\mathbf{a}\cdot\mathbf{c} = \mathbf{b}\cdot\mathbf{c} = 0$, although this relation omitted the quantifier "$\exists\,\tau_{ab} = \pm 1$"):

$$\begin{aligned}
&\forall\rho,\ \exists\,\tau_A,\tau_B,\tau_C = \pm 1,\\
&\quad |\mathbf{b}\times\mathbf{c}|^2\,(\Delta A)^2 + |\mathbf{a}\times\mathbf{c}|^2\,(\Delta B)^2 + |\mathbf{a}\times\mathbf{b}|^2\,(\Delta C)^2\\
&\quad\ -2\,\tau_A\,\tau_B\,(\mathbf{b}\times\mathbf{c})\cdot(\mathbf{c}\times\mathbf{a})\sqrt{1-(\Delta A)^2}\sqrt{1-(\Delta B)^2}\\
&\quad\ -2\,\tau_A\,\tau_C\,(\mathbf{b}\times\mathbf{c})\cdot(\mathbf{a}\times\mathbf{b})\sqrt{1-(\Delta A)^2}\sqrt{1-(\Delta C)^2}\\
&\quad\ -2\,\tau_B\,\tau_C\,(\mathbf{c}\times\mathbf{a})\cdot(\mathbf{a}\times\mathbf{b})\sqrt{1-(\Delta B)^2}\sqrt{1-(\Delta C)^2}\\
&\quad\ = |\mathbf{a}\times\mathbf{b}|^2 + |\mathbf{a}\times\mathbf{c}|^2 + |\mathbf{b}\times\mathbf{c}|^2 - V^2\,|\mathbf{r}|^2 \ \geq\ 2 - 2\,(\mathbf{a}\cdot\mathbf{b})(\mathbf{a}\cdot\mathbf{c})(\mathbf{b}\cdot\mathbf{c})
\end{aligned} \tag{35}$$

where:

$$V^2 = (\,\mathbf{a}\cdot(\mathbf{b}\times\mathbf{c})\,)^2 = |\mathbf{a}\times\mathbf{b}|^2 + |\mathbf{a}\times\mathbf{c}|^2 + |\mathbf{b}\times\mathbf{c}|^2 + 2(\mathbf{a}\cdot\mathbf{b})(\mathbf{a}\cdot\mathbf{c})(\mathbf{b}\cdot\mathbf{c}) - 2 > 0. \tag{36}$$

It is interesting to note that V corresponds to the volume of the parallelepiped defined by the Bloch vectors $\mathbf{a}$, $\mathbf{b}$ and $\mathbf{c}$ and, hence, is a measure of the mutual incompatibility of the observables. When the three measurements are orthogonal ($\mathbf{a}\cdot\mathbf{b} = \mathbf{a}\cdot\mathbf{c} = \mathbf{b}\cdot\mathbf{c} = 0$, $V^2 = 1$), then we get [24,26,30]:

$$(\Delta A_1)^2 + (\Delta A_2)^2 + (\Delta A_3)^2 \ =\ 3 - |\mathbf{r}|^2 \ \geq\ 2. \tag{37}$$

In the case of the regular tetrahedron measurements described earlier, we find:

$$\begin{aligned}
&\forall\rho,\ \exists\,\tau_1,\ldots,\tau_4 = \pm 1,\\
&\quad 3\sum_{1\leq i\leq 4}(\Delta A_i)^2 + \sum_{1\leq i\neq j\leq 4}\tau_i\,\tau_j\,\sqrt{1-(\Delta A_i)^2}\sqrt{1-(\Delta A_j)^2} \ =\ 12 - \frac{16}{3}\,|\mathbf{r}|^2 \ \geq\ \frac{20}{3}.
\end{aligned} \tag{38}$$

4. Entropic Uncertainty Relations

In the case of qubits, the Shannon entropy of a Pauli observable A as defined in Equation (4) can be directly expressed in terms of the expectation value $\langle A\rangle$, namely:

$$H(A) = h_2\Big(\frac{1+\langle A\rangle}{2}\Big) = h_2\Big(\frac{1-\langle A\rangle}{2}\Big), \tag{39}$$

where h_2 is the binary entropy function defined as $h_2(p) = -p\log p - (1-p)\log(1-p)$. Denoting by h_2^{-1} the inverse function of h_2 restricted to the domain $p\in[0,\frac{1}{2}]$, one can invert this relation and obtain:

$$\langle A\rangle = \pm f(H(A)) \quad \text{with} \quad f(x) := 1 - 2\,h_2^{-1}(x) \geq 0. \tag{40}$$

This allows us now to express the relations of Lemmas 2 and 3 in terms of the Shannon entropies $H(A_i)$. Note that because of the $\pm$ sign above, the same care as before must be taken when transforming these relations into entropic ones. In fact, the exact same analysis as in the previous section (which we shall not repeat explicitly) can be carried out: one can simply replace all terms $\sqrt{1-(\Delta A)^2}$ by $f(H(A))$ and the variances $(\Delta A)^2$ by $1 - f(H(A))^2$.

For instance, one obtains the following tight relation for two Pauli observables (similarly to Theorem 4, with the same conditions for saturation):

$$f(H(A))^2 + f(H(B))^2 - 2\,|\mathbf{a}\cdot\mathbf{b}|\, f(H(A))\, f(H(B)) \leq (1 - (\mathbf{a}\cdot\mathbf{b})^2)\,|\mathbf{r}|^2 \leq 1 - (\mathbf{a}\cdot\mathbf{b})^2. \quad (41)$$

Figure 3 shows the bound given by Equation (41) in the space of Shannon entropies $H(A), H(B)$ for $\mathbf{a}\cdot\mathbf{b} = 0$ and $|\mathbf{a}\cdot\mathbf{b}| = \frac{1}{2}$, in comparison to the weaker Massen–Uffink bound Equation (5), with the maximum overlap given here by $c = \sqrt{\frac{1+|\mathbf{a}\cdot\mathbf{b}|}{2}}$, as well as the better, but still not tight, bound given in [18–21].

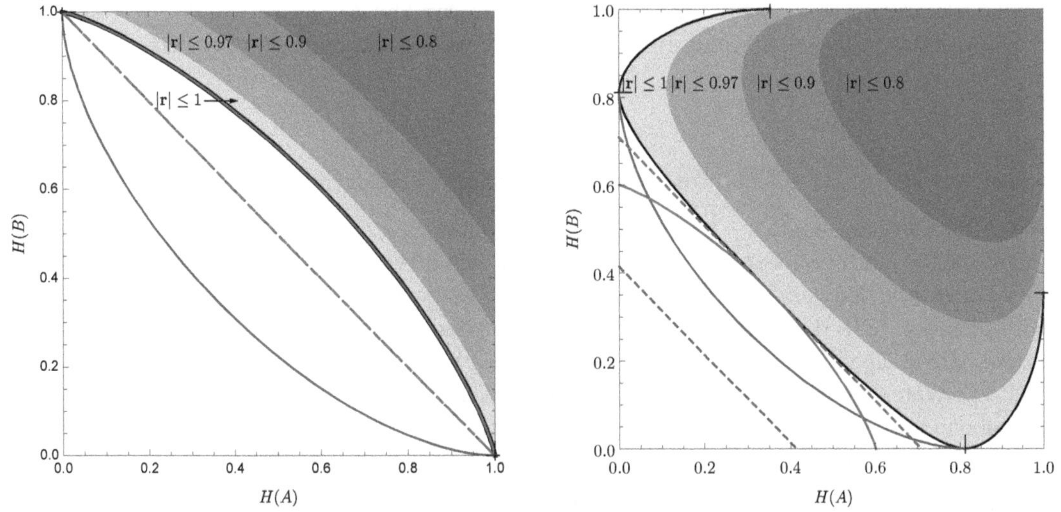

(a) (b)

Figure 3. Analogous figure to Figure 2, with the uncertainty regions and relations now shown in the Shannon entropy domain, for (**a**) $\mathbf{a}\cdot\mathbf{b} = 0$ and (**b**) $|\mathbf{a}\cdot\mathbf{b}| = \frac{1}{2}$, *i.e.*, for maximum overlaps $c = \sqrt{\frac{1+|\mathbf{a}\cdot\mathbf{b}|}{2}} = \frac{1}{\sqrt{2}}$ and $c = \frac{\sqrt{3}}{2}$, respectively. The filled area corresponds to relation (41) (which is saturated for the values of $(H(A), H(B))$ on the thick black curve), with the darker areas representing values attainable for mixed states with bounded Bloch vector norms. The dashed blue and red curves correspond to the Maassen–Uffink bound (5) and to that of [18–21], respectively. In (**b**), the former is clearly suboptimal (it coincides with the latter only for $\mathbf{a}\cdot\mathbf{b} = 0$), while the latter is the optimal bound that can be put on the sum $H(A) + H(B)$ and is thus tangential to the uncertainty region characterised by Equation (41) (it touches it at the point $H(A) = H(B) = h_2(\frac{1+|\mathbf{a}\cdot\mathbf{b}|}{2})$ for $|\mathbf{a}\cdot\mathbf{b}| \gtrsim 0.391$ and on two symmetric points when $|\mathbf{a}\cdot\mathbf{b}| \lesssim 0.391$; the critical value $|\mathbf{a}\cdot\mathbf{b}| \simeq 0.391$ corresponds to $\arccos|\mathbf{a}\cdot\mathbf{b}| \simeq 0.585$ [19], or $c \simeq 0.834$ [21]). The solid blue and red curves correspond to the bounds (6) and (7), respectively, translated in terms of entropies using $\Delta A = \sqrt{1 - f(H(A))^2}$ (see text).

5. Higher Dimensional Systems

One may wonder whether our approach can be generalised to give state-independent uncertainty relations for qutrits or higher dimensional systems. As we will see, while it remains possible to obtain (not necessarily tight) relations on the expectation values of a set of observables, the peculiar geometry of the generalised Bloch sphere for qutrits (and beyond), as well as the increased number of eigenvalues of the observables, prevents us from expressing these relations in terms of standard deviations or entropies, as we could do for qubits.

In a d-dimensional Hilbert space, an arbitrary traceless observable A can be expressed in terms of a set of $d \times d$ traceless Hermitian matrices λ_i, $1 \leq i \leq d^2 - 1$, which generate the group SU(d). These operators satisfy the commutation relations:

$$[\lambda_i, \lambda_j] = 2i\, f_{ijk}\, \lambda_k, \quad \{\lambda_i, \lambda_j\} = \frac{4}{d}\, \delta_{ij} \mathbb{1} + 2\, d_{ijk}\, \lambda_k\,, \tag{42}$$

where $[\cdot, \cdot]$ and $\{\cdot, \cdot\}$ are the commutator and anticommutator, δ_{ij} is the Kronecker delta, f_{ijk} and d_{ijk} are antisymmetric and symmetric structure constants of SU(d), respectively, and where the summation over repeated indices is implicit. For qutrits, the operators λ_1 to λ_8 are the Gell–Mann matrices, for example. As for the two-dimensional case, taking $\boldsymbol{\lambda} = (\lambda_1, \ldots, \lambda_{d^2-1})^{\mathsf{T}}$, we can write any traceless observable A as $A = \mathbf{a} \cdot \boldsymbol{\lambda}$ and, thus, represent it by its generalised Bloch vector $\mathbf{a}$.

An arbitrary state ρ can similarly be written in terms of its generalised Bloch vector $\mathbf{r}$ as:

$$\rho = \frac{1}{d}\mathbb{1} + \frac{1}{2}\mathbf{r} \cdot \boldsymbol{\lambda}\,, \tag{43}$$

where now $|\mathbf{r}| \leq \sqrt{2(1 - \frac{1}{d})}$ (with equality for pure states). However, for $d \geq 3$, it is *not* the case that *any* vector $\mathbf{r}$ with $|\mathbf{r}| \leq \sqrt{2(1 - \frac{1}{d})}$ represents a valid quantum state [31,32]: the set of valid quantum states (*i.e.*, the Bloch vector space) is a strict subset of the unit sphere in d dimensions.

The expectation value of A in the state ρ can still be expressed as $\langle A \rangle = \mathbf{a} \cdot \mathbf{r}$, so that Lemma 1, with M and $\mathbf{u}$ defined as before, remains valid for higher dimensional systems. This allows one to derive state-independent relations for the expectation values $(\langle A_1 \rangle, \ldots, \langle A_n \rangle)$, as we did for qubits in Section 2. Note that these relations may not be tight, as the vectors $\mathbf{r}$ saturating Equation (11) may not correspond to valid quantum states. It is also worth noting that the "ellipsoid condition" of Kaniewski *et al.* [12], which is a higher-dimensional version of Lemma 1 for *binary-valued* measurements, is state-dependent for $d \geq 3$, so it also fails to give the type of state-independent uncertainty relation we would like in this scenario.

Contrary to the case of qubits, however, for $d \geq 3$, one cannot directly translate these relations to express them solely in terms of standard deviations or entropies; indeed, because of the larger number of eigenvalues and of the geometry of the generalised Bloch sphere, the expectation value $\langle A \rangle$ does not contain all of the information about the uncertainty of A. For instance, the relation between $(\Delta A)^2$, $\mathbf{a}$ and $\mathbf{r}$ is not simply given by Equation (20) (where $\langle A \rangle = \mathbf{a} \cdot \mathbf{r}$), but by:

$$(\Delta A)^2 = \frac{2}{d}|\mathbf{a}|^2 + \mathbf{a}' \cdot \mathbf{r} - (\mathbf{a} \cdot \mathbf{r})^2, \tag{44}$$

where $\mathbf{a}'$ is a d-dimensional vector with components $a'_k = a_i\, a_j\, d_{ijk}$ [24]. Thus, the uncertainty of A for a state ρ no longer depends only on the angle between the Bloch vectors $\mathbf{r}$ and $\mathbf{a}$. Furthermore, in contrast to qubit operators, for three-or-more-level systems there are pairs of non-commuting observables that have a common eigenstate and, hence, can simultaneously have zero variance. In general, there is no simple analytical description of the Bloch space for qudits [31,32], so it seems implausible to give generalised forms of tight, state-independent uncertainty relations for such systems.

For certain choices of A and B, a complete analysis of the set of obtainable values for ΔA and ΔB is nevertheless tractable (at least for qutrits), and it is possible to give tight state-independent uncertainty relations. Similarly, for well-chosen A and B, a more general higher dimensional analysis is possible if one is prepared to settle for relations that are not tight. In [26], for example, such behaviour is analysed for angular momentum observables in orthogonal directions. Such an approach, however, lacks the generality of the approach possible for qubits and is necessarily, at least in part, *ad hoc*.

6. Discussion

By exploiting the relationship between the expectation values of Pauli observables and standard measures of uncertainty (such as the standard deviation), we have derived tight state-independent uncertainty relations for Pauli measurements on qubits. These uncertainty relations completely characterise the allowed values of uncertainties for such observables. Furthermore, we give the bounds on all of these relations in terms of the norm $|\mathbf{r}|$ of the Bloch vector representing the state $\rho = \frac{1}{2}(\mathbb{1} + \mathbf{r} \cdot \boldsymbol{\sigma})$, which is directly linked to the purity of the state, so that, if a bound on this is known, tighter (partially state-dependent) uncertainty relations can be obtained; for pure states $|\mathbf{r}| = 1$, and the most general form is recovered. The approach we take is general, and although we explicitly give tight uncertainty relations for arbitrary pairs and triples of Pauli observables, it can be used to give tight uncertainty relations for sets of arbitrarily many observables.

While we have focused on giving these uncertainty relations in terms of the standard deviations and variances of the observables, we showed how these can easily be rewritten in terms of Shannon entropies to give tight entropic uncertainty relations and did so explicitly for pairs of observables. These relations can furthermore be translated into uncertainty relations for any measure of uncertainty that depends only on the probability distribution, $\{\frac{1}{2}(1 + \mathbf{a} \cdot \mathbf{r}),\ \frac{1}{2}(1 - \mathbf{a} \cdot \mathbf{r})\}$, of an observable $A = \mathbf{a} \cdot \boldsymbol{\sigma}$ for a state $\rho = \frac{1}{2}(\mathbb{1} + \mathbf{r} \cdot \boldsymbol{\sigma})$, such as Rényi entropies. Indeed, one may reasonably argue that the product $\mathbf{a} \cdot \mathbf{r} = \langle A \rangle$ is the *only* parameter that an uncertainty measure for A can depend on and, thus, that our approach covers all kinds of preparation uncertainty relations for qubits.

Although we have given explicit uncertainty relations only for Pauli observables, it is simple to extend them to arbitrary qubit measurements. To do so, note that one can write any observable A in a two-dimensional Hilbert space as $A = \alpha \mathbb{1} + \mathbf{a} \cdot \boldsymbol{\sigma}$, with $\alpha \in \mathbb{R}$ and $\mathbf{a} \in \mathbb{R}^3$. Assuming $|\mathbf{a}| > 0$ (as otherwise A is simply proportional to the identity operator, and one trivially has $\Delta A = 0$ for all states ρ), the observable $\tilde{A} = \tilde{\mathbf{a}} \cdot \boldsymbol{\sigma}$ with $\tilde{\mathbf{a}} = \mathbf{a}/|\mathbf{a}|$ is a Pauli observable, and we have $\langle \tilde{A} \rangle = \frac{\langle A \rangle - \alpha}{|\mathbf{a}|}$ and $\Delta \tilde{A} = \frac{\Delta A}{|\mathbf{a}|}$. One can thus give an uncertainty relation involving A by writing the corresponding relation that we derived for $\tilde{A}$ and then replacing $\tilde{\mathbf{a}}$ by $\mathbf{a}/|\mathbf{a}|$ and $\langle \tilde{A} \rangle$ or $\Delta \tilde{A}$ by the appropriate expression given above; one can proceed similarly for the other observables in question.

Finally, we note that, although we have not done so here, it is also possible to go beyond projective measurements and give similar relations for positive-operator valued measures (POVMs) for qubits with binary outcomes. The two elements of any such POVM can be written in the form $A_{\pm} = \frac{1}{2}[\mathbb{1} \pm (\alpha \mathbb{1} + \mathbf{a} \cdot \boldsymbol{\sigma})]$, where $|\alpha| + |\mathbf{a}| \leq 1$. Attaching, for simplicity, the output values ± 1 to the two POVM outcomes (this can easily be generalised), we can then define an "effective operator" $A = A_+ - A_-$, such that the expectation value of the POVM outcomes can be simply written as $\langle A \rangle = \alpha + \mathbf{a} \cdot \mathbf{r}$. A similar construction to the one that led to Lemma 1 can thus be used, this time utilising the "effective operators" and defining $\mathbf{u} = (\langle A \rangle - \alpha, \ldots)^{\mathsf{T}}$. Lemmas 2 and 3 then still hold, after suitably substituting $\langle A \rangle$ for $\langle A \rangle - \alpha$ (and similarly for the other observables in the relation). To translate these into standard deviations, one can then use once more (for our choice of outcomes, ± 1) $\Delta A = \sqrt{1 - \langle A \rangle^2}$, where ΔA here is (in a slight abuse of notation) the POVM standard deviation, not that of the "effective operator" (for a non-projective measurement, the standard deviation should *not* be calculated as $\sqrt{\langle A^2 \rangle - \langle A \rangle^2}$ using the "effective operator" A as, in general, $A^2 \neq \mathbb{1}$, despite the POVM outcomes being ± 1), which thus leads to state-independent uncertainty relations for POVMs. Entropic uncertainty relations can similarly be obtained in this fashion.

Acknowledgements: Alastair A. Abbott and Cyril Branciard acknowledge financial support from the "Retour Post-Doctorants" program (ANR-13-PDOC-0026) of the French National Research Agency. Cyril Branciard also acknowledges the support of a Marie Curie International Incoming Fellowship (PIIF-GA-2013-623456) from the European Commission. Michael J. W. Hall was supported by the ARC Centre of Excellence CE110001027.

Author Contributions: Cyril Branciard conceived and supervised the study. All authors contributed to the technical analysis and development of the results. Alastair A. Abbott and Cyril Branciard wrote the initial version of the paper. Alastair A. Abbott, Cyril Branciard and Michael J. W. Hall performed subsequent revisions.

Conflicts of Interest: The authors declare no conflict of interest.

References

1. Kennard, E.H. Zur Quantenmechanik einfacher Bewegungstypen. *Z. Phys.* **1927**, *43*, 172–198.
2. Heisenberg, W. Über den anschaulichen Inhalt der quantentheoretischen Kinematik und Mechanik. *Z. Phys.* **1927**, *43*, 172–198.
3. Robertson, H.P. The uncertainty principle. *Phys. Rev.* **1929**, *34*, 163–164.
4. Ozawa, M. Universally valid reformulation of the Heisenberg uncertainty principle on noise and disturbance in measurement. *Phys. Rev. A* **2003**, *67*, 042105.
5. Hall, M.J.W. Prior information: How to circumvent the standard joint-measurement uncertainty relation. *Phys. Rev. A* **2004**, *69*, 052113.
6. Branciard, C. Error-tradeoff and error-disturbance relations for incompatible quantum measurements. *Proc. Natl. Acad. Sci. USA* **2013**, *110*, 6742–6747.
7. Busch, P.; Lahti, P.; Werner, R.F. Proof of Heisenberg's error-disturbance relation. *Phys. Rev. Lett.* **2013**, *111*, 160405.
8. Ozawa, M. Disproving Heisenberg's error-disturbance relation. **2013**, arXiv:1308.3540.
9. Dressel, J.; Nori, F. Certainty in Heisenberg's uncertainty principle: Revisiting definitions for estimation errors and disturbance. *Phys. Rev. A* **2014**, *89*, 022106.
10. Busch, P.; Lahti, P.; Werner, R.F. Colloquium: Quantum root-mean-square error and measurement uncertainty relations. *Rev. Mod. Phys.* **2014**, *86*, 1261.
11. Schrödinger, E. Zum Heisenbergschen Unschärfeprinzip. *Sitzungsber. Preuss. Akad. Wiss. Phys. Math. Kl.* **1930**, *14*, 296–303.
12. Kaniewski, J.; Tomamichel, M.; Wehner, S. Entropic uncertainty from effective anticommutators. *Phys. Rev. A* **2014**, *90*, 012332.
13. Maccone, L.; Pati, A.K. Stronger uncertainty relations for all incompatible observables. *Phys. Rev. Lett.* **2014**, *113*, 260401.
14. Hirschman, I.I., Jr. A note on entropy. *Am. J. Math.* **1957**, *79*, 152–156.
15. Deutsch, D. Uncertainty in quantum mechanics. *Phys. Rev. Lett.* **1983**, *50*, 631–633.
16. Maassen, H.; Uffink, J.B.M. Generalized entropic uncertainty relations. *Phys. Rev. Lett.* **1988**, *60*, 1103.
17. Coles, P.J.; Berta, M.; Tomamichel, M.; Wehner, S. Entropic uncertainty relations and their applications. **2015**, arXiv:1511.04857.
18. Garrett, A.J.M.; Gull, S.F. Numerical study of the information uncertainty principle. *Phys. Lett. A* **1990**, *151*, 453–458.
19. Sánchez-Ruiz, J. Optimal entropic uncertainty relation in two-dimensional Hilbert space. *Phys. Lett. A* **1998**, *244*, 189–195.
20. Ghirardi, G.; Marinatto, L.; Romano, R. An optimal entropic uncertainty relation in a two-dimensional Hilbert space. *Phys. Lett. A* **2003**, *317*, 32–36.
21. De Vicente, J.I.; Sánchez-Ruiz, J. Improved bounds on entropic uncertainty relations. *Phys. Rev. A* **2008**, *77*, 042110.
22. Abdelkhalek, K.; Schwonnek, R.; Maassen, H.; Furrer, F.; Duhme, J.; Raynal, P.; Englert, B.G.; Werner, R.F. Optimality of entropic uncertainty relations. **2015**, arXiv:1509.00398.
23. Huang, Y. Variance-based uncertainty relations. *Phys. Rev. A* **2012**, *86*, 024101.
24. Li, J.L.; Qiao, C.F. Reformulating the quantum uncertainty relation. *Sci. Rep.* **2015**, *5*, 12708.
25. Busch, P.; Lahti, P.; Werner, R.F. Heisenberg uncertainty for qubit measurements. *Phys. Rev. A* **2014**, *89*, 012129.
26. Dammeier, L.; Schwonnek, R.; Werner, R.F. Uncertainty relations for angular momentum. *New J. Phys.* **2015**, *17*, 093046.
27. Horn, R.A.; Johnson, C.R. *Matrix Analysis*; Cambridge University Press: Cambridge, UK, 1985.
28. Busch, P. Qubit uncertainty tutorial; 12th International Workshop on Quantum Physics and Logic, Oxford, UK, 17 July 2015. Available online: https://www.cs.ox.ac.uk/qpl2015/slides/busch-tutorial.pdf (accessed on 22 February 2016).

29. Branciard, C. Deriving tight error-trade-off relations for approximate joint measurements of incompatible quantum observables. *Phys. Rev. A* **2014**, *89*, 022124.
30. Hofmann, H.F.; Takeuchi, S. Violation of local uncertainty relations as a signature of entanglement. *Phys. Rev. A* **2003**, *68*, 032103.
31. Bertlmann, R.A.; Krammer, P. Bloch vector for qudits. *J. Phys. A Math. Theor.* **2008**, *41*, 235303.
32. Kimura, G. The Bloch vector for *N*-level systems. *Phys. Lett. A* **2003**, *314*, 339–349.

4

Dynamics and the Cohomology of Measured Laminations

Carlos Meniño Cotón

Instituto de Matemática, Universidade Federal do Rio de Janeiro, Rio de Janeiro 21941-909, Brazil; carlos.meninho@gmail.com

Academic Editor: Yuli B. Rudyak

Abstract: In this paper, the interconnection between the cohomology of measured group actions and the cohomology of measured laminations is explored, the latter being a generalization of the former for the case of discrete group actions and cocycles evaluated on abelian groups. This relation gives a rich interplay between these concepts. Several results can be adapted to this setting—for instance, Zimmer's reduction of the coefficient group of bounded cocycles or Fustenberg's cohomological obstruction for extending the ergodicity of a $\mathbb{Z}$-action to a skew product relative to an S^1 evaluated cocycle. Another way to think about foliated cocycles is also shown, and a particular application is the characterization of the existence of certain classes of invariant measures for smooth foliations in terms of the L^∞-cohomology class of the infinitesimal holonomy.

Keywords: foliations; cohomology; group action; foliated cocycles; invariant measures

---

1. Introduction

One of the most important tools to study the dynamical properties of an ergodic group action is given by the cohomology of its measurable 1-cocycles. For instance, Furstenberg's example (see [1]) of a minimal dynamical system on the torus which is minimal but non-Lebesgue strictly ergodic can be interpreted as an obstruction for some continuous 1-cocycles to be cohomologically trivial in the measurable cohomology. Zimmer's results relative to semisimple Lie group ergodic actions and their lattices are stated in terms of the cohomology of their actions [2]. Again, 1-cocycles carry their own importance in terms of rigidity and amenability [3,4]. Zimmer pointed out the geometric interpretation of the cohomology of group actions as the cohomology of a measurable lamination in the 1980s , which led to a Mostow rigidity theorem for some kind of measured foliations [5]. Cocycles associated with pseudogroup actions or holonomy grupoids are the direct generalization of cocycles associated to group actions where the infinitesimal holonomy cocycle is the best example (see e.g., [6,7]).

The measurable cohomology of foliations can be traced back to the works of Connes [8,9] and Heitsch and Lazarov [10], and it was studied in full generality by Bermudez and Hector [11–13]. In [14], the author introduced the singular version of this cohomology which was a missing piece in order to apply the full power of algebraic topology to this setting. The author was initially motivated by the relation between the cup product of this cohomology and the tangential Lusternik–Schnirelmann category of measured foliations [15]. It is required to mention that, in this work, we deal with generalized measurable laminations in the sense that leaves could not be manifolds but only path connected and locally compact Polish spaces, for instance connected and locally compact graphs or CW-complexes (see e.g. [16] for formal definitions). We shall use the word *lamination* for such generic foliations and the word *foliation* if every leaf is a manifold.

The purpose of this paper is to extend another bridge between these concepts. It is important to note that only abelian groups are considered here as coefficient groups, whilst non-abelian coefficients can be considered for group actions or foliated cocycles (e.g., Γ-structures).

The paper is organized in five sections. The first section serves as an introduction to the definitions of singular and simplicial cohomology of measured laminations and the cohomology of group actions. It is shown that, for a given group action of a discrete group, there exists a measurable lamination where its first cohomology group is isomorphic to the cohomology of that action. In fact, it is possible to obtain a measurable foliation with this property.

The second section is devoted to prove an analogue of a Zimmer's result (see [2]) for ergodic group actions: if an R-evaluated 1-cocycle $\omega : T \times G \to R$ satisfies that $\omega(\{t\} \times G)$ is precompact for every $t \in T$, then ω is cohomologous to a 1-cocycle evaluated in a compact subgroup of R. This result is extended to higher dimensional cohomology groups.

In the third section, some classical results due to Furstenberg (see [1]) are translated to our setting of measurable laminations. Furstenberg shows obstructions for a $\mathbb{Z}$-action given by a skew product over a strictly ergodic process to be also strictly ergodic, the obstruction is measured by the triviality of some cohomological classes. This result is extended to the case of measurable cohomology of laminations. In this section, it is also introduced the first group of continuous cohomology—it is no surprise that the obstruction for a skew product to be minimal is given by the triviality of some cohomology classes in the continuous category.

The fourth section deals with the infinitesimal holonomy cocycle associated with a smooth foliation. This cocycle induces a closed $\mathbb{R}$-valued 1-cochain in an associated measurable lamination. Observe that the Lebesgue measure is always quasi-invariant for any (transversely) smooth foliation. For Lebesgue ergodic smooth foliations, the triviality of this cocycle in L^∞-cohomology is equivalent to the existence of invariant measures in the Lebesgue class with L^∞ Radon-Nikodyn derivative, and it is also shown that this is equivalent to a boundedness condition on the infinitesimal holonomy.

The final section gives further comments and other possible generalizations of well-known results of cohomological dynamics, as the Livsic theorem, to the setting of leafwise cohomology.

2. Cohomology of Measured Laminations

The purpose of this section is to introduce the singular and simplicial cohomology of measurable laminations. It will be necessary to introduce a wider category of objects that are able to manage the dual character of measurable laminations (the measurable structure and the leaf topology). A *measurable topological space*, or MT-*space*, is a set X endowed with a σ-algebra and a topology. Usually, measure theoretic concepts will refer to the σ-algebra of X, and topological concepts will refer to its topology; in general, the σ-algebra is different from the Borel σ-algebra induced by the topology. An MT-*map* between MT-spaces is a measurable and continuous map. An MT-*isomorphism* is a map between MT-spaces, which is a measurable isomorphism and a homeomorphism.

Obvious examples are topological spaces with the Borel σ-algebra and measurable spaces with the discrete topology. Let X and Y be MT-spaces. Suppose that there exists a measurable embedding $i : X \to Y$ that maps measurable sets to measurable sets. Then, X is called an MT-*subspace* of Y. The product $X \times Y$ is also an MT-space with the product topology and σ-algebra.

Let "$\sim$" be an equivalent relation on an MT-space X. The quotient $X/\sim$ is an MT-space with the quotient topology and the σ-algebra generated by the projections of measurable saturated sets of X.

A *Polish space* is a completely metrizable and separable topological space. A *standard Borel space* is a measurable space isomorphic to a Borel subset of a Polish space. Let S be a standard Borel space and let P be a Polish space and let us consider the Borel σ-algebra on P. $P \times S$ will be endowed with the structure of MT-space defined by the product σ-algebra and the product of the discrete topology on S and the topology of P.

A *measurable chart* on an MT-space X is an MT-isomorphism $\varphi : U \to P \times S$, where U is open and measurable in X, S is a standard Borel space, and P is a locally compact, connected and locally path connected Polish space; let us remark that P and T depend on the chart. The sets $\varphi^{-1}(P \times \{*\})$ are called *plaques* of φ, and the sets $\varphi^{-1}(\{*\} \times S)$ are called *associated transversals* of φ. A *measurable atlas* on X is a countable family of measurable charts whose domains cover X. A *measurable lamination* is an MT-space that admits a measurable atlas. Observe that we always consider countable atlases; therefore, the ambient space is also a standard space. The connected components of X are called *leaves*. An example of measurable lamination is a usual foliation with the ambient Borel σ-algebra and the leaf topology. The notation $\mathcal{F}$ will be used to denote the collection of leaves and $\|\mathcal{F}\|$ will denote the underlying MT-space. For every $x \in \|\mathcal{F}\|$, the leaf of $\mathcal{F}$ that contains x will be denoted by L_x.

According to this definition, the leaves are second countable connected manifolds, but they may not be Hausdorff. Therefore, in what follows, it will be assumed that $\|\mathcal{F}\|$ is Hausdorff, locally path connected and locally compact. Under these conditions, leaves are locally compact and path connected Polish spaces.

In the case where the plaques of each chart are homeomorphic to a Euclidean ball, we shall use the term *measurable foliation*. In the case where the leaves are graphs the term *measurable graph* will be used.

A measurable subset $T \subset X$ is called a *transversal* if its intersection with each leaf is countable [10]. Let $\mathcal{T}(X)$ be the family of transversals of X. This set is closed under countable unions and intersections, but it is not σ-algebra. A transversal meeting all leaves is called *complete*.

A *measurable holonomy transformation* is a measurable isomorphism $\gamma : T \to T'$, for $T, T' \in \mathcal{T}(X)$, which maps each point to a point in the same leaf. A *transverse invariant measure* on X is a σ-additive map, $\mu : \mathcal{T}(X) \to [0, \infty]$, invariant by measurable holonomy transformations. The classical definition of transverse invariant measure in the context of foliated spaces is a measure on topological transversals invariant by holonomy transformations (see e.g., [6]). These two notions of transverse invariant measures agree for foliated spaces [8]. When the sets of null measure are invariant by holonomy transformations, the measure μ will be called *quasi-invariant*. A *measured foliation* is a pair $(\|\mathcal{F}\|, \mu)$ where μ is a finite measure over some complete transversal and quasi-invariant for $\mathcal{F}$. Fix a complete transversal T of $\mathcal{F}$ so that (T, μ) is a finite measure space. Given a set $A \subset M$, let $\mathrm{sat}(A)$ denote the set given by the union of all the leaves meeting A, this set will be called the *saturation* of A.

Definition 1. The measured foliation $(\mathcal{F}, \mu)$ will be called *ergodic* if for every measurable subset $A \subset T$, the set $\mathrm{sat}(A) \cap T$ (which is always measurable) has null or full measure. This is equivalent to say that every leafwise constant measurable map is almost everywhere constant.

Remark 1. In the world of measured laminations, holonomy transformations can be still interpreted as the slice of a transversal following a leafwise path [14]. For this, it is always assumed that our atlas is *regular, i.e.*, plaques are precompact spaces in the corresponding leaf, every intersecting pair of measurable charts of a given atlas is contained in another measurable chart (not necessarily in the given atlas), and any chart meets a finite number of other charts. However, the concept of germ of a measurable holonomy transformation makes no sense since transverse topology is needed, *i.e.*, from a measurable point of view, the holonomy group of a leaf is trivial. In a measured lamination, the ergodic components play the role of minimal sets for foliations; therefore, ergodicity will be a natural hypothesis with which to work.

Example 1 (Measurable suspensions). Let P be a locally compact, connected, locally path connected and semi-locally 1-connected Polish space, and let S be a standard space. Let $\mathrm{Meas}(S)$ denote the group of measurable transformations of S. Let

$$h : \pi_1(P, x_0) \to \mathrm{Meas}(S)$$

be a homomorphism. Let $\widetilde{P}$ be the universal covering of P and consider the action of $\pi_1(P, x_0)$ on the MT-space $\widetilde{P} \times S$ given by

$$g \cdot (x, t) = (xg^{-1}, h(g)(t)) .$$

The corresponding quotient MT-space, denoted as $\widetilde{P} \times_h S$, will be called the MT-*suspension* of h. $\widetilde{P} \times_h S$ is a measurable lamination, $\{*\} \times S$ is a complete transversal, and its leaves are covering spaces of P. See Figure 1 for a clarifying picture.

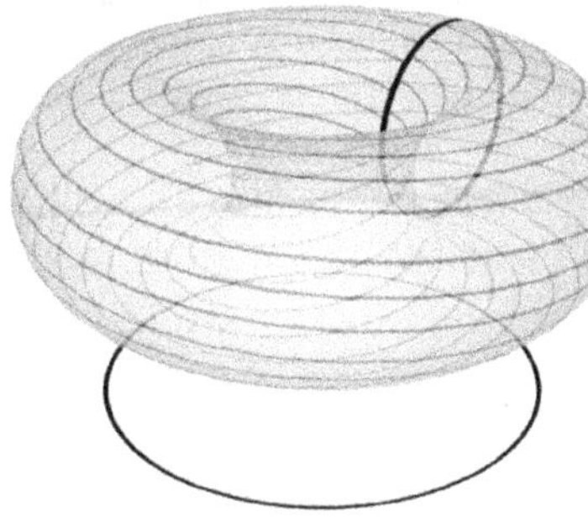

Figure 1. The simplest suspension of a circle transformation over S^1. Bold circles represent the fiber and base spaces.

The measurable cohomology of MT-spaces is the natural version of a cohomology theory, which takes into account the mixing of their topology and measurable structure. It is also supposed that the considered coefficient groups, which will be denoted by R, are standard abelian groups, *i.e.*, an admissible coefficient group is an abelian group and a standard space where the product and the inverse maps are measurable. In the world of group actions, cocycles evaluated in non-abelian groups (e.g., $Gl(n,\mathbb{R})$ or $\mathrm{Diff}(S^1)$) are of great importance, however the introduction of a non-commutative cohomology theory is beyond the purpose of this work.

Definition 2 (Measurable simplex). A *measurable simplex* is the MT-space induced by the product $\triangle^N \times S$ where S is a standard space and $\triangle^N$ is the canonical N-simplex. A *measurable singular simplex* on $\|\mathcal{F}\|$ is an MT-map $\sigma : \triangle^N \times S \to \|\mathcal{F}\|$.

Let ω be a usual *singular n-cochain* over a coefficient group R, as usual, an inversion of orientation is translated to an inversion in R. It is said that ω is *measurable* if $\omega_\sigma : S \to R$, $s \mapsto \omega(\sigma_{|\triangle \times \{s\}})$, is measurable for all measurable singular n-simplices σ. The set of measurable cochains is a subcomplex of the complex of usual cochains since the coboundary operator δ preserves the measurability. This measurable subcomplex is denoted by $C^*_{\mathrm{MT}}(\mathcal{F}, R)$, and the coboundary operator restricted to this complex is also denoted by δ.

The *singular measurable cohomology* is defined as usual by

$$H^n_{\mathrm{MT}}(X,R) = \mathrm{Ker}\,\delta_n / \mathrm{Im}\,\delta_{n-1}\,.$$

Any MT-map $f : \|\mathcal{F}\| \to \|\mathcal{G}\|$ defines a cochain map $f^* : C^*_{\mathrm{MT}}(\mathcal{G}, R) \to C^*_{\mathrm{MT}}(\mathcal{F}, R)$ by $f^*(\omega)(\sigma) = \omega(f \circ \sigma)$. Since it commutes with the coboundary operator, it induces a homomorphism between measurable cohomology groups, $f^* : H^*_{\mathrm{MT}}(Y,R) \to H^*_{\mathrm{MT}}(X,R)$.

The importance of the measurable singular cohomology is the fact that it has substantial theoretical advantages, which allows for adapting easily classical results from algebraic topology as excision, functoriality, homotopy invariance, Mayer–Vietoris or cup product in relative cohomology—another bonus is that it can be applied to every MT-space. This topic is treated deeply in [14]. However, for explicit calculus, simplicial cohomology is easier to be analyzed. They are fully described in [11–13] and, in [14], it is shown that simplicial and singular measurable versions are isomorphic, in particular, it does not depend on the (measurable) simplicial description of leaves.

Definition 3 (Measurable triangulation [11]). A *measurable triangulation* of a measurable lamination $\mathcal{F}$ is the assocation of a structure of simplicial complex to each leaf of $\mathcal{F}$, it will be denoted by $\mathcal{T} = \bigsqcup_{L\in\mathcal{F}} \mathcal{T}_L$. Measurability must be understood in the following way. For each n, the set of barycenters of n-simplices, which will be denoted by $\mathcal{B}^n$, is a (measurable) transversal and the function $\sigma^n : \triangle^n \times \mathcal{B}^n \to \|\mathcal{F}\|$, mapping a barycenter $b \in \mathcal{B}^n$ to the embedding $\sigma^n_b : \triangle^n \to L_b$ given by the triangulation $\mathcal{T}_{L_b}$, is also measurable, where $\triangle^n$ is the canonical n-simplex. The term *measurable simplicial lamination* will be also used for a measurable lamination endowed with a measurable triangulation.

Let b a barycenter of some n-simplex in $\mathcal{T}$, then $\triangle_b$ will denote the simplex containing b, *i.e.*, the map $\sigma^n_b : \triangle^n \to L_b$, $\partial\triangle^n$ will denote the set of faces of the canonical n-simplex and $\partial\triangle_b = \{\sigma^n_{b|F})\}_{F\in\partial\triangle^n}$.

Remark 2. A simplicial complex does not need to be a manifold. This is the case for measurable graphs.

A measurable simplicial lamination will be denoted by the pair $(\|\mathcal{F}\|, \mathcal{T})$. An *n-cochain* over a standard abelian group R is a measurable map

$$\omega : \{\pm\} \times \mathcal{B}^n \to R$$

such that $\omega(-,b) = -\omega(+,b)$. The orientation $(+,b)$ refers to the orientation induced by the standard orientation of the canonical n-simplex in the domain of σ^n_b and $(-,b)$ to the inverse orientation. In order to relax the notation, we use only the notation b for signed barycenters and $-b$ should denote the inverse orientation on the corresponding simplex. In this sense, we can define the set ∂b formed by the face barycenters of $\triangle_b$ with a positive sign if the orientation agrees with the orientation in $\{+\} \times \mathcal{B}^{k-1}$ or negative otherwise. Analogously, we shall use the signed notation for oriented simplices, where $-\triangle_b$ will denote that simplex with the opposite orientation. Clearly, a measurable n-cochain is determined by the values on $\{+\} \times \mathcal{B}^n$, in this sense n-cochains can be interpreted as measurable maps $\omega : \mathcal{B}^n \to R$. This relaxed notation will be used only for the cases $n = 0$ and $n = 1$. Let $C^n(\mathcal{T}, R)$ denote the set of simplicial n-cochains; this set is endowed with a group structure induced by R. We define the coboundary operator $\delta : C^n(\mathcal{T}, R) \to C^{n+1}(\mathcal{T}, R)$ as usual by $\delta\omega : \{\pm\} \times \mathcal{B}^{n+1} \to R$,

$$\delta\omega(b) = \sum_{b'\in\partial b} \omega(b') .$$

Clearly, $\delta^2 = 0$ and we can define the cohomology groups as usual:

$$H^n(\mathcal{F}, \mathcal{T}, R) = \operatorname{Ker}\delta_n / \operatorname{Im}\delta_{n-1} .$$

Remark 3. In the case of a measured lamination $(\|\mathcal{F}\|, \mu)$, the above definitions can be applied in an almost everywhere sense (a.e. in what follows). This means that exact cocycles satisfy $\omega = \delta\theta$ a.e., and this leads to the concept of a.e. cohomology or measured cohomology. The cohomology groups in this sense will be denoted by $H^n(\mathcal{F}, \mu, R)$.

In a similar sense, when $R = \mathbb{R}$ or $\mathbb{C}$, it can be introduced the L^p cohomology, which we denote by $H^n_{L^p}(\mathcal{F}, \mu, R)$, $1 \le p \le \infty$. For a locally compact R, the L^∞ cohomology is obtained by considering cochains which are taking its values a.e. in a compact subset of R.

It is also obvious that the definition of simplicial cohomology can be extended to the case of polygonal subdivisions on the leaves, where simplices are replaced by compact linear regions of the Euclidean space.

3. Cohomology of Measurable Group Actions and Measurable Laminations

The cohomology of measured group actions was introduced to provide invariants useful to detect the wild dynamical behavior shown by non-amenable group actions [4]. For the amenable case, a measured group action is orbit equivalent to a $\mathbb{Z}$-action [17] and measurable dynamics are not so interesting.

Let G be a locally compact separable group acting (on the right) by measurable isomorphisms on a measured standard Borel space (S,μ), where μ is a quasi-invariant measure. Let R be a standard group (not necessarily conmutative). A 1-cocycle for the action is a map

$$\omega : S \times G \to R$$

such that $\omega(s,gh) = \omega(s,g)\omega(sg,h)$ for all $s \in S$ and $g,h \in G$ and it is said that ω and θ are cohomologous if there exists a measurable function $f : S \times R$ such that $f(s)\omega(s,g)f(sg)^{-1} = \theta(s,g)$ a.e. $s \in S$ and all $g \in G$, when R is abelian this is written as $\omega(s,g) - \theta(s,g) = f(sg) - f(t)$.

The right action will be denoted by $a : S \times G \to S$, however, the notation will be ignored if it is clear in the context ($sg := a(s,g)$). Let $H^1(S \times_a G, \mu, R)$ denote the set of 1-cocycles up to cohomology.

Recall that only abelian coefficient groups R and finitely generated G are considered in this work, the remaining cases (G a continuous group and R non-abelian) are of great importance ([2,18]) but they will not be treated in this work.

Assuming G finitely generated and R abelian, let $S_G = \{g_1, \ldots, g_n\}$ be a minimal system of generators of G and let Z_n be the wedge union of n circles. Of course, $\pi_1(Z_n) = *_1^n\mathbb{Z}$ with generators $e_1, \ldots, e_n$ given by the loops of each circle component. Let us consider the homomorphism

$$h : \pi_1(Z_n, *) \to \mathrm{Meas}(S)\,,\ e_i \mapsto h(e_i)(s) = a(s,g_i)\,,$$

which allows for defining the MT-suspension $X_a = \widetilde{Z}_n \times_h S$. Clearly, X_a is a measurable simplicial lamination, and, moreover, a measurable graph where the 0-simplices correspond with $\{*\} \times S \equiv S$, and the 1-simplices are provided by the lifts of the generators e_i to each leaf, their barycenters are in a natural bijection with n copies of T. A positive orientation on the edges is provided by $[t \to tg]$.

Now, we want to attach 2-cells to this measurable graph in the following way. Whenever we have a non trivial irreducible relation $g_{i_1}^{\pm 1} \ldots g_{i_k}^{\pm 1} = 1$, a chain of 2-cells bounded by the respective edges in X_a is attached. This is done by attaching a 2-cell to the loop given by the relation in a minimal set of relations. Let $\widehat{X}_a$ be the resulting measurable cellular lamination. To obtain a simplicial structure on leaves, we can add edges joining the barycenter of each added 2-cell to each vertex in its boundary. It is clear that these triangulations perform a structure of measurable triangulation on $\widehat{X}_a$ which will be denoted by $\mathcal{T}_a$.

Remark 4. It is not important, for the definition of $\widehat{X}_a$, the fact that the sets of generators and relations are minimal (whenever they perform a presentation of the given group). However, the space $\widehat{X}_a$ depends strongly on the choice of generators and relations (even minimal). This can be seen directly in the points $s \in S$ where the isotropy group is trivial, the leaf of $\widehat{X}_a$ meeting s is homeomorphic to the Cayley graph of G associated to S_G where every simple loop is bounded by at least one chain of 2-cells.

When the isotropy group is not trivial, interesting topology can appear. For instance, assume that some generator e satisfies the relation $e^2 = 1$ and $h(e)(s) = s$ for some s. Thus, in the construction of the leaf passing through s, we have to attach a 2-cell to the loop induced by e but in such a way that its boundary is a double covering of that loop. This produces a projective plane embedded in that leaf. It is clear that different relations can produce other embedded surfaces without any limitations.

The main point of this construction is the following observation: although the topology of the leaves can change wildly, the first cohomology group of leaves is invariant.

Proposition 4. *Let G be a finitely generated group acting by measurable isomorphisms on a standard measure space (S,μ) and let R be a standard abelian coefficient group. Then $H^1(S \times_a G, \mu, R)$ and $H^1(\widehat{X}_a, \mu, R)$ are isomorphic.*

Proof. Let us consider the above CW-structure $\mathcal{T}_a$ on the leaves of $\widehat{X}_a$ and identify the set of 1-barycenters $\mathcal{B}^1$ of $\mathcal{T}_a$ with $S \times \{g_1, \ldots, g_n\}$. Therefore, a 1-cocycle ω induces a simplicial 1-cochain $i^*(\omega)$ of $\widehat{X}_a$ via the restriction map. The cocycle condition implies that $\omega(s,1) = 0$ for all $s \in S$, hence $i^*(\omega)$ is closed since the attached 2-cells occur in a minimal set of relations for the group. Remark that, since 2-cells were attached on a minimal set of relations, every relation (not necessarily irreducible) induces a 1-simplicial chain bounded by a chain of 2-cells.

Clearly, 0-cocycles are naturally identified since $\mathcal{B}^0 \equiv S$. If ω is exact then $\omega(t,g) = f(tg) - f(t)$ a.e. and $f(tg) - f(t) = \delta f$ by definition. Therefore, i^* defines an injective map at the level of 1-cohomology groups.

Given a closed 1-cochain $\theta : \mathcal{B}^1 \equiv S \times \{g_1, \ldots, g_n\} \to R$, it defines an R-value on any chain of 1-simplices just by summing over the values on each simplex of the chain. We have to show that it determines a 1-cocycle $\omega : S \times G \to R$ so that $i^*(\omega) = \theta$. Let us define

$$\omega(s, g_{i_1}^{\pm 1} \ldots, g_{i_k}^{\pm 1}) := \theta(s, g_{i_1}^{\pm 1}) + \theta(sg_{i_1}^{\pm 1}, g_{i_2}^{\pm 1}) + \cdots + \theta(sg_{i_1}^{\pm 1} \ldots g_{i_{k-1}}^{\pm 1}, g_{i_k}^{\pm 1}) .$$

Observe that θ extends naturally to a map on the simplicial chains of $\mathcal{T}_a$, in fact, a complete definition of cochain is given as a homomorphism from the abelian group of simplicial chains to R, and the map θ is a way to give values for the generators of the group of simplicial chains. The above definition of ω is consistent with the value of θ in the simplicial chain given by a oriented path of edges with initial point in $t \in T$ and determined by the sequences of generators $g_{i_1}^{\pm 1} \ldots g_{i_k}^{\pm}$. However, the above equation does not guarantee the cocycle condition for ω. For an arbitrary measurable cochain, two different words representing the same element in G can give different values in R. This is the point where we use the fact that ω is closed. Let $w = w_1 \ldots w_k$, $v = v_1 \ldots v_l$ be words in S_G representing the same element $g \in G$, then wv^{-1} is a trivial relation for G, and therefore (considering $w_0 = v_0 = 1$)

$$\sum_{i=0}^{k-1} \theta(sw_0 \ldots w_i, w_{i+1}) - \sum_{i=0}^{l-1} \theta(sv_0 \ldots v_i, v_{i+1}) = 0 ,$$

since the edges provided by the word wv^{-1} bound a chain of 2-cells. The above equality means that $\omega(s,w) = \omega(s,v)$ as desired. □

Example 2. In the case where G is a free group (with the usual presentation), there are no relations and, therefore, $\widehat{X}_a = X_a$.

Example 3. The suspension of two circle rotations gives a well known foliation over T^3 with base T^2. In this case, this foliation agrees with the space $\widehat{X}_a$. This is not the case for a suspension of three circle rotations. In this case, three minimal relations appear (the commutators of each pair of generators of $\mathbb{Z}^3$) and $\widetilde{X}_a$ can be given as a 2-skeleton on the suspension foliation in T^4 with base T^3. Since the MT-spaces considered only differ by 3-cells, the first cohomology groups of $\widehat{X}_a$ and the suspension foliation on T^4 are isomorphic.

The above example suggests that under certain hypothesis we can replace $\widetilde{X}_a$ by a measurable foliation, this happens when G is the fundamental group of a (closed) manifold such that the fundamental region in the universal covering space is an n-cell. This is the case for surfaces of genus ≥ 1. However, in general, the homomorphism $h_a : \pi_1(B) \to \mathrm{Meas}(S)$ is not a monomorphism, this means that non-trivial words in $\pi_1(B)$ (or in G) can act trivially, which leads to a non-trivial (in cohomology) loop on every leaf, even when h_a is a monomorphism it could happen that a non-trivial word induces an a.e. trivial map leading to a similar issue. This can be represented by a measurable 1-cochain (evaluating by a constant $\neq 0$ in these loops and 0 otherwise). Thus, non-trivial classes represent non-trivial dynamics whenever the action of $\pi_1(B)$ is a.e. faithfull or essentially free, this depends on our meaning of "non-trivial dynamics". In the a.e. faithful case, non-trivial classes with trivial dynamics can arise from a set of fixed points of positive measure, but interesting things can be happening outside of this set. In the essentially free case, a non-trivial cohomology class is always detecting dynamics, but this is a more restrictive condition. Observe that, for a.e. faithful (resp. essentially free), it is equivalent to say that fixed points in S have non-full (resp. null) measure for any element of G.

When G is finitely presented, then it is well known that there exists a smooth closed 4-manifold with fundamental group isomorphic to G. Although the fundamental region for this 4-manifold could not be diffeomorphic to a disk, this region must be simply connected and the previous argument is easily adapted. This can be summarized as follows:

Proposition 5. *Let G be a group with finite presentation acting by measurable isomorphisms in (S,μ). Let R be a standard abelian group, then there exists a measured foliation $(\|\mathcal{F}\|,\mu)$ of dimension 4 such that $H^1(S\times G,\mu,R)$ is isomorphic to $H^1(\mathcal{F},\mu,R)$. This foliation is given by the suspension process of the G-action over a closed 4-manifold with a fundamental group isomorphic to G.*

4. Reduction of the Coefficient Group

Assume for this section that R is a topological abelian group which is a locally compact Polish space.

Definition 6. A cocycle $\omega : S\times G\to R$ is said to be *bounded* if the set $A(s)=a(\{s\}\times G)$ is precompact for all $s\in S$.

We want to adapt the following result of Zimmer.

Proposition 7 (Zimmer [2]). *Let $a : S\times G\to S$ be a measured ergodic action on a standard measure space (S,μ). If ω is a bounded cocycle then ω is cohomologous to a cocycle evaluated on a compact subgroup of R.*

The next definition is introduced in order to adapt the notion of bounded cochain.

Definition 8. Let $\mathcal{T}^1(s)$ be the set of (finite) simplicial paths formed by an ordered finite family of oriented edges of the 1-skeleton of $\mathcal{T}$ with initial point in $s\in\mathcal{B}^0$, so that the end point of an edge is the initial point of the next one (with the eventual exception of the last edge). A (measurable) cochain $\omega:\mathcal{B}^1\to R$ is called *bounded* if $\omega(\mathcal{T}(s))$ is a precompact set for all $s\in S$. It is said that ω is *uniformly bounded* if there exists $F\subset\mathcal{B}^0$ a measurable set of full measure so that $\bigcup_{s\in F}\mathcal{T}^1(s)$ is a precompact set.

Remark 5. In the above section, where the cohomology of group actions was translated to the language of the cohomology of measurable laminations, the role of the edges of the 1-skeleton is the analogue of the generators of the group. Thus, the 1-dimensional simplicial paths can be thought as an analogue of the elements of the group.

Proposition 9. *Let ω be a measurable 1-cochain of an ergodic simplicial lamination $(\|\mathcal{F}\|,\mathcal{T},\mu)$. If ω is a bounded cochain, then ω is cohomologous to a closed cochain evaluated on a compact subgroup of R.*

Proof. The proof is a direct application of Zimmer's original argument [2]. Since R is Polish, the space C of compact sets in R is a complete and separable metric space with the Hausdorff metric. Let $b\in\{\pm\}\times\mathcal{B}^1$, and let $i(b),e(b)\in\mathcal{B}^0$ denote the initial and end points of the edge containing b with the induced orientation. Since $\omega:\{\pm\}\times\mathcal{B}^1\to R$ is a measurable function, the map $\mathcal{K}:\{\pm\}\times\mathcal{B}^1\to C$, $b\mapsto\overline{\omega(\mathcal{T}^1(i(b)))}$ is measurable. Clearly, $\omega(b)+\overline{\omega(\mathcal{T}^1(e(b)))}=\overline{\omega(\mathcal{T}^1(i(b)))}$. If $K\in C$ and there exists a sequence $x_n\in R$ with $x_n+K\to K'$ in C, then, by compactness, there exists a subsequence converging to $x\in R$ such that $x+K=K'$. Therefore, the orbits of C under the natural R action are closed. As a consequence, the quotient space $\mathcal{C}/R$ is Polish. Since $\omega(b) + \overline{\omega(\mathcal{T}^1(e(b)))} = \overline{\omega(\mathcal{T}^1(i(b)))}$, the sets $\overline{\omega(\mathcal{T}^1(e(b)))}, \overline{\omega(\mathcal{T}^1(i(b)))}$ represent the same equivalence class in $\mathcal{C}/R$.

We want to show that there exists an equivalence class $[B] \in \mathcal{C}/R$ with $\omega(\mathcal{T}^1(s)) \in [B]$ for almost every $s \in \mathcal{B}^0$. Remark that $\mathcal{B}^0$ is a complete transversal and $(\mathcal{B}^0, \mu)$ is ergodic relative to the action of the measurable holonomy pseudogroup, the measurable map

$$\overline{\mathcal{K}} : \mathcal{B}^0 \to \mathcal{C}/R, \; s \mapsto \left[\overline{\omega(\mathcal{T}^1(s))}\right]$$

is constant on each orbit (which is the intersection of each leaf with $\mathcal{B}^0$) and thus it is a.e. constant by ergodicity. Let $e_B : R \to [B] \subset \mathcal{C}, r \mapsto r + B$, and let $\mathbf{s} : [B] \to R$ be a measurable section, *i.e.*, for all $D \in [B]$, $\mathbf{s}(D) + B = D$ (and $B = -\mathbf{s}(D) + D$). The function $\varphi(s) = -\mathbf{s}(\overline{\omega(\mathcal{T}^1(s))})$ satisfies, for almost every $b \in \mathcal{B}^1$:

$$\begin{aligned} \varphi(i(b)) + (\omega(b) + (-\varphi(e(b)) + B)) - \varphi(i(b)) + (\omega(b) + \overline{\omega(\mathcal{T}^1(e(b)))}) = \\ = \varphi(i(b)) + \overline{\omega(\mathcal{T}^1(i(b)))} = B\,. \end{aligned}$$

Therefore, ω is cohomologous, via φ, to a cochain θ which takes values a.e. in the stabilizer of B, $\mathrm{Stab}(B)$, which is compact since B is compact. By redefining θ in a negligible set by 0 the proof is complete. □

Corollary 10. *Let ω be a measurable* 1*-cochain of an ergodic simplicial lamination* $(\|\mathcal{F}\|, \mathcal{T}, \mu)$. *If ω is a uniformly bounded cochain then ω is L^∞-cohomologous to a closed cochain evaluated on a compact subgroup of R.*

Proof. Observe that a uniformly bounded cochain is necessarily an element of $L^\infty(\mu)$. All the arguments in the above proof works verbatim, and it is only required to check that the 0-cochain $\varphi(s) = -\mathbf{s}(\overline{\omega(\mathcal{T}^1(s))})$ is essentialy bounded. Since $\omega(b) + \overline{\omega(\mathcal{T}^1(e(b)))} = \overline{\omega(\mathcal{T}^1(i(b)))}$, it follows that the image of φ is contained in $\overline{\bigcup_{s \in F} \mathcal{T}^1(s)}$ a.e. $s \in \mathcal{B}^0$ which is a compact set by hypothesis. □

Remark 6. Observe that the cochains are not required to be closed which is the natural translation of cocycles. This comes from the fact that any measurable cochain satisfies for free a cocycle condition since they can be seen as a homomorphism on the group of 1-simplicial chains, and this is enough for the above argument to work. From another point of view, arbitrary cochains are closed cochains relative to the measurable lamination defined by the 1-skeleton of $\mathcal{T}$.

The above proposition has an analogue for higher dimensional cochains, and this will need the introduction of new technology.

Definition 11. A *k-simplicial path* starting at $b = b_0 \in \{\pm\} \times \mathcal{B}^{k-1}$ is a finite and ordered family of pairs $(\triangle_0^k, b_0), \ldots, (\triangle_m^k, b_m)$ where $\triangle_i^k$ are oriented k-simplices and b_i are (signed) barycenters of $(k-1)$-simplices for all i and for some $m \in \mathbb{N}$, such that

- $b_i \in \partial\triangle_i^k$ and the sign of b_i is determined by the induced orientation.
- $-b_{i+1} \in \partial\triangle_i^k$ and $b_i, -b_{i+1}$ represent different faces of $\triangle_i^k$ for all $i \geq 0$.

A k-simplicial path is called even (resp. odd) whenever $m+1$ is even (resp. odd). If $\triangle_0^k = \triangle_m^k$ then the k-simplicial path will be called *k-simplicial loop*.

Definition 12. For $b \in \mathcal{B}^{k-1}$, let $\mathcal{T}^k(b)$ the set of k-simplicial paths starting at b. A measurable 1-cochain is bounded if $\omega(\mathcal{T}^k(b))$ is precompact for all $b \in \mathcal{B}^{k-1}$.

Definition 13. A measured simplicial lamination is said to be *k-coherent* if, for a.e. leaf, any pair of different non-oriented $(k-1)$-simplices in the same leaf is connected by a k simplicial path and a.e. leaf contains simplices of dimension k. Observe that this is the case for measured simplicial foliations. For $k = 1$, coherency is automatic since leaves are path connected by definition (whenever a.e. leaf is not a singleton).

Definition 14. A measurable simplicial lamination $(\mathcal{F}, \mathcal{T})$ is said to be *k-evenly triangulated* if there exists a measurable set $T_k \subset \{+\} \times \mathcal{B}^{k-1}$ so that for every $b \in \{+\} \times \mathcal{B}^k$, ∂b contains exactly one element of $T_k \sqcup -T_k$.

Remark 7. A barycentric subdivision of any triangulation is always even in the above sense. This is provided by the fact that the barycentric subdivision of any old simplex can be seen as a subdivision by attached pairs of new k-simplices, the faces (with a choice of orientation) between these attached pairs perform the set T_k. This is the reason to use the word "evenly" for this kind of triangulation.

Proposition 15. *Let ω be a measurable k-cochain, $k \geq 2$, of an ergodic, k-coherent and k-evenly triangulated measured lamination $(\|\mathcal{F}\|, \mathcal{T}, \mu)$. If ω is a bounded cochain then ω is uniformly bounded and cohomologous to a k-cochain evaluated on a compact subgroup of R.*

Proof. As before, let $\mathcal{C}$ denote the Polish space given by the compact sets of R with the Hausdorff metric. Let $b \in \mathcal{B}^k$ and set $\partial b = \{i_0(b), \ldots, i_k(b)\}$ an enumeration of face barycenters with the induced orientations. It is easy to check the following equation

$$-\omega(b) + \overline{\omega(\mathcal{T}^k(i_j(b))} = \bigcup_{l \neq j} \overline{\omega(\mathcal{T}^k(-i_l(b)))}\,.$$

By a symmetric argument we get the following chain of equations:

$$\begin{aligned}\overline{\omega(\mathcal{T}^k(i_j(b))} &= \omega(b) - \omega(b) + \overline{\omega(\mathcal{T}^k(i_j(b)))} = \\ &= \omega(b) + \bigcup_{l \neq j} \overline{\omega(\mathcal{T}^k(-i_l(b)))} = \\ &= \bigcup_{l \neq j} (\omega(b) + \overline{\omega(\mathcal{T}^k(-i_l(b)))}) = \bigcup_{l} \overline{\omega(\mathcal{T}^k(i_l(b)))}\,.\end{aligned}$$

Thus $\overline{\omega(\mathcal{T}^k(i_j(b))} = \bigcup_l \overline{\omega(\mathcal{T}^k(i_l(b)))}$ for all $j \in \{0, \ldots, k\}$. This is only possible if $\overline{\omega(\mathcal{T}^k(i_j(b)))} = \overline{\omega(\mathcal{T}^k(i_l(b)))}$ for all j, l. An intuitive way to see this relation is by observing that a simplicial path $\mathcal{P}$ starting at $i_j(b)$ induces another simplicial path $\widehat{\mathcal{P}} = ((\triangle_b, i_l(b)), ((-\triangle_b, -i_m(b)), \mathcal{P})$ starting at $i_l(b)$ for pairwise different j, l, m, observe that this is only true for $k \geq 2$ since an intermediate third face, $i_m(b)$, is needed to perform $\widehat{\mathcal{P}}$. Now, consider the involution $i : \mathcal{C} \to \mathcal{C}$, $K \mapsto -K$. As before $\mathcal{C}/i$ is Polish. Since $\overline{\omega(\mathcal{T}^k(-s))} = -\overline{\omega(\mathcal{T}^k(s))}$, the map

$$\overline{\mathcal{K}} : \{\pm\} \times \mathcal{B}^{k-1} \to \mathcal{C}/i,\ s \mapsto \left[\overline{\omega(\mathcal{T}^k(s))}\right]$$

is a.e. constant if the measurable simplicial lamination is k-coherent. By ergodicity, there exists B, a compact subset of R, such that $\left[\overline{\omega(\mathcal{T}^k(s))}\right] = [B]$ for a.e. $s \in \{\pm\}\mathcal{B}^{k-1}$. Therefore, ω is uniformly bounded.

Now, a simple calculation shows that for a.e. $b \in \{\pm\} \times \mathcal{B}^k$, $\omega(b) + B = -B$ or $\omega(b) - B = B$ depending on whether $B = \overline{\omega(\mathcal{T}^k(t))}$ or $B = \overline{\omega(\mathcal{T}^k(-t))}$. If b and c are barycenters of contiguous k-simplices with compatible orientations (they form a k-simplicial path), then $\omega(b) + \omega(c) + B = B$ and thus $\omega(b) + \omega(c) \in \mathrm{Stab}(B)$ which is a compact subgroup of R. If $(\triangle_b, \triangle_c, \triangle_d)$ is a k-simplicial path of length 3, then a similar argument shows $\omega(b) - \omega(d) \in \mathrm{Stab}(B)$ for a.e. b, d in these conditions. More generally, if $\triangle_b, \triangle_c$ are connected by a k-simplicial path then either $\omega(b) + \omega(c)$ or $\omega(b) - \omega(c)$ belong to $\mathrm{Stab}(B)$ according to the length of the path being even or odd. Moreover, by ergodicity, either $\omega(c) + \omega(b)$ or $\omega(c) - \omega(b)$ belong to $\mathrm{Stab}(B)$ for a.e. $b, c \in \{\pm\} \times \mathcal{B}^k$ (not necessarily in the same leaf).

Assume that there exists a k simplicial loop of even length based on a k-simplex b (b counts twice for the length of that loop), therefore, $2\omega(b) \in \mathrm{Stab}(B)$ and, as a consequence, $\omega(b) \in \mathrm{Stab}(B \cup -B)$, which is also a compact subgroup. Observe that any other k-simplex in that leaf admits a simplicial loop of even length, suppose that $\mathcal{L}$ is an even simplicial loop starting on $\triangle_b$ and ending at $\triangle_c$, then $(\mathcal{P}, \mathcal{L}, \mathcal{P}^{-1})$ is another even simplicial loop starting at c where $\mathcal{P}$ is a k-simplicial path joining $\triangle_c$ and $\triangle_b$ provided by coherency. Therefore, ω itself takes values in the compact subgroup $\mathrm{Stab}(B \cup -B)$. Observe that the measurable function $ev : \{\pm\} \times \mathcal{B}^k \to \{0,1\}$, defined as $ev(b) = 1$ iff b has a simplicial loop of even length, is leafwise constant. Therefore, if the set of these simplices is non-negligible, then ω itself takes its values a.e. in a compact subgroup of R by ergodicity.

Assume now that a.e. leaf does not admit simplicial loops of even length and $\mathcal{F}$ is k-evenly triangulated by $\mathcal{T}$. Therefore, there exists a measurable set $T_k \subset \{+\} \times \mathcal{B}^{k-1}$ so that each simplex of dimension k has exactly one face barycenter (up to orientation) in T_k. Since there are not even loops, the measurable map $\xi : \{+\} \times \mathcal{B}^k \to \{-1, 1\}$ mapping $c \in \mathcal{B}^k$ to 1 if $\omega(b) + \omega(c) \in \mathrm{Stab}(B)$ and -1 if $\omega(c) - \omega(b) \in \mathrm{Stab}(B)$, is well defined whenever $\omega(b) \notin \mathrm{Stab}(B)$ (in which case ω takes its values in $\mathrm{Stab}(B)$). This gives a partition of $\mathcal{B}^k$ into two subsets: $\xi^{-1}(\{\pm 1\})$. The main point is that each element $b \in \xi^{-1}(\{1\})$ has exactly one face barycenter in $S_k \sqcup -S_k$. Let $\widehat{T}_k \subset \{\pm\} \times \mathcal{B}^{k-1}$ be the measurable set obtained from T_k in order to match the induced orientations of $\xi^{-1}(\{1\})$, it agrees with T_k up to orientation. Let us define the measurable cochain $\varphi : \{\pm\} \times \mathcal{B}^{k-1} \to R$ as $\varphi(s) = \pm\omega(b)$ iff $s \in \pm\widehat{T}_k$ and 0 otherwise. It follows that $\omega - \delta\varphi$ is evaluated in $\mathrm{Stab}(B)$ completing the proof. □

Remark 8. Let us remark that the reduction of ω is given by a $(k-1)$-cochain in $L^\infty(\mu)$. Therefore, the Proposition 15 holds for L^∞-cohomology.

Question 1. It is unclear whether the above result holds for non k-evenly triangulated laminations. Observe that measured cohomology does not depend on the chosen triangulation [14]. Let us consider the k-skeleton, $\mathcal{T}(k)$ of $\mathcal{T}$, thus every k-cochain is closed. Barycentric subdivision provides a k-even triangulation of $\mathcal{T}(k)$, and we can consider the isomorphism between the measured simplicial cohomology of $\mathcal{T}(k)$ and its barycentric subdivision. However, it is uncertain whether this isomorphism should respect bounded k-cochains (up to cohomology).

Observe that the above proof holds when a.e. leaf admits even k-simplicial loops (even in the non k-evenly triangulated case). In the case where a.e. k-simplicial loop is odd, Lusin–Novikov theorem on the existence of measurable sections (see Theorem 18.10 in [19]) might be enough to produce a measurable set with the properties of T_k.

Example 4. Let $\mathcal{F}_{\alpha,\beta}$ consider the suspension of two rationally independent circle rotations $R_\alpha, R_\beta : \mathbb{R}/\mathbb{Z} \to \mathbb{R}/\mathbb{Z}, t \mapsto t + \alpha, t \mapsto t + \beta$. The Lebesgue measure is an invariant and ergodic measure for $\mathcal{F}$. Leaves are planes, and we can consider the measurable triangulation given by the 2-simplices with vertices $S_1(t) = [t, t+\beta, t+\alpha+\beta]$ and $S_2(t) = [t, t+\alpha, t+\alpha+\beta]$, $t \in \mathbb{R}/\mathbb{Z}$. Let ω be the $\mathbb{Z}$ evaluated measurable 2-cochain ω so that $\omega(S_1(t)) = -1$ and $\omega(S_2(t)) = 1$ relative to a compatible orientation of the foliation, see Figure 2.

Clearly, ω is bounded (every simplicial path gives value -1, 0 or 1). Therefore it is cohomologous to a 0 evaluated cochain (0 is the unique compact subgroup of $\mathbb{Z}$), *i.e.*, ω is exact. Observe that this triangulation is 2-even and T_k can be given by the diagonal edges of the triangulation. An invariant element in $\mathcal{C}/i$ is the class of $B = \{0, 1\}$ or $B = \{-1, 0\}$. This is a toy example where the steps of the proof of the Proposition 15 can be checked directly.

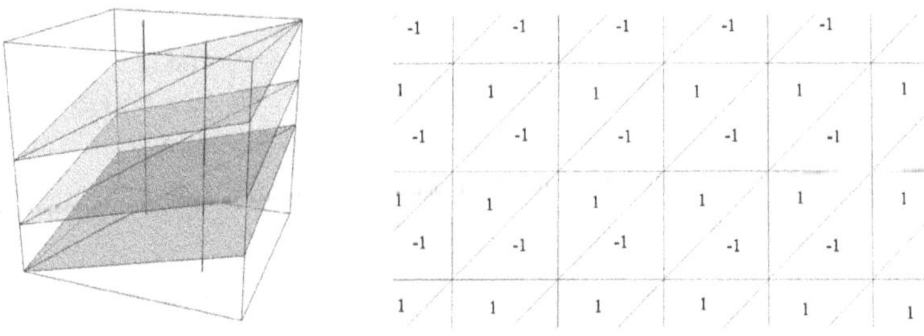

Figure 2. Minimal linear foliation on T^3 with a leafwise triangulation. Vertical segments represent, after quotient, the set of 1-barycenters $\mathcal{B}^1$. The bounded cochain given in Example 4 is represented in a generic leaf.

5. Furstenberg's Criteria for Minimality and Ergodicity of Skew Products

In this section, we generalize some classical results due to Furstenberg for continuous $\mathbb{Z}$-actions. In [1], a cohomological criterium for the strict ergodicity of a skew product over a strictly ergodic transformation is obtained. Observe that in this section we are working on topological laminations. A formal definition can be found in [6], essentially they are defined by changing in our introductory definitions of measurable laminations the word "standard" by "locally compact and Polish" and measurable by "continuous". For topological foliations, the notation $\|\mathcal{F}\|$ refers to the ambient space. Furstenberg gives also a criterium for the minimality in the setting of continuous cocycles, this allows to produce continuous transformations of the torus which are minimal but not Lebesgue ergodic.

Our aim is to define the necessary objects and show that Furstenberg's proofs can be adapted wihout major issues. This is in fact a way to show that Furstenberg's work holds for discrete actions and opens the door to higher dimensional analogues.

Furstenberg assumes that the measure μ is invariant instead of quasi-invariant. This is not restrictive for continuous flows: since leaves are homeomorphic to S^1 or $\mathbb{R}$, they satisfy the Fölner condition, the averaging process [20] on any orbit produces an invariant measure supported in the closure of that leaf. However, in general, the existence of invariant measures is a high restriction. In this section, all the considered measures will be quasi-invariant and Radon. As usual, S^1 will denote the circle as the set of unimodular complex numbers and m will denote the normalized Lebesgue measure on the circle. Although S^1 is abelian, we shall use the multiplicative notation for the operations in S^1.

Now, the Furstenberg's criterium is stated in a different way from the original. In this version, nothing is said about strict ergodicity, but this is a price to pay in order to allow the inclusion of quasi-invariant measures.

Proposition 16 (Furstenberg)**.** *Let $T_0 : \Omega_0 \to \Omega_0$ be a Borel transformation (which induces a $\mathbb{Z}$-action) on a compact Polish space Ω_0 and let μ_0 be a quasi-invariant measure which is ergodic for this action. Let $g : \Omega_0 \to S^1$ be a Borel function and let $T : \Omega_0 \times S^1 \to \Omega_0 \times S^1$, $(w_0, z) \mapsto (T_0(w_0), g(w_0)z)$. Then $\mu = \mu_0 \times m$ is a ergodic measure for T if and only if g^k is not trivial in measured cohomology for every integer $k \neq 0$, i.e., $[g^k] \neq 0$ in $H^1(X_{T_0}, \mu, S^1)$ where X_{T_0} is the suspension of T_0 over the circle.*

Proof. We refer to the original proof in [1] for more details. It relies on the following two points:

1. μ is always a quasi-invariant measure for T. This is provided by using Fubbini in the product measure μ, the quasi-invariance of μ_0 and the fact that m is fiberwise invariant.
2. Let $f \in L^2(\Omega_0 \times S^1, \mu)$ so that $Tf = f$, since μ is a product measure we can express f as a fiberwise Fourier series

$$f \equiv \sum_{k=-\infty}^{\infty} c_k(w) z^k$$

for $c_k \in L^2(\Omega_0, \mu_0)$. By using the invariance under T, one can easily check that $c_k(T_0(w)) = g(w)^{-k} c_k(w)$ for every $k \in \mathbb{Z}$. The ergodicity of T_0 shows that some c_k is non essentially zero for $k \neq 0$. Since $|g| = 1$ the saturation of the non-zero set of c_k is also non-zero, and thus has full measure. Therefore, $h(w) = c_k(w)^{-1}$ trivializes g in measured cohomology. Conversely, if $g^k = \delta h$ then the function $h(w)^{-1} z^k$ is T invariant and is not a.e. constant, so T is not ergodic.

□

Before proceeding to adapt the above proposition, it is necessary to explain the meaning of a leafwise continuous triangulation of a topological lamination. A measurable triangulation of a topological lamination is said to be *continuous* if the union of the barycenters (of all simplices) is a closed space in the ambient topology and for every convergent sequence $b_n \to b$ of barycenters, the corresponding simplices converge in the Hausdorff metric. Observe that simplices associated to b_n and b could have different dimensions. In the case of foliations on a smooth manifold where the leaves are smooth manifolds, a leafwise triangulation can be obtained by choosing a triangulation of the ambient manifold in general position with the foliation (see Figure 3). This provides a polygonal subdivision on leaves which can be refined to simplicial after a barycentric subdivision. Most of the interesting examples of topological laminations arise from minimal sets of usual foliations, so the existence of a leafwise triangulation is not so strong as it seems at first glance.

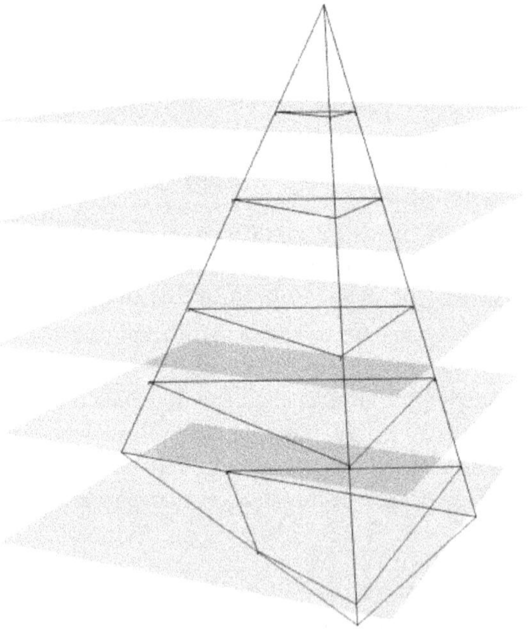

Figure 3. Transverse intersection of a planar foliation with a 3-simplex in general position.

Now, it is time to translate the necessary objects and apply Furstenberg's original proof. Of course, Ω_0 will be the set of 0-barycenters $\mathcal{B}^0$ and T_0 must be now given by oriented edges. Whenever we consider the space $\|\mathcal{F}\| \times S^1$, we can obtain a new topological foliation by translating each leaf of $\mathcal{F}$ to each fiber $\|\mathcal{F}\| \times \{z\}, z \in S^1$. This foliation can be seen as a "trivial skew product". Its ambient space can fail to be a manifold in general (since $\|\mathcal{F}\|$ is not a manifold in general), but we can at least perturb in the S^1 "direction" and ask what ergodic and topological properties of $\mathcal{F}$ can be translated to the perturbed foliation.

When we deal with suspensions, the triangulation obtained by lifting a triangulation of the base manifold provides a leafwise continuous triangulation where each $\mathcal{B}^k$, $k \geq 0$, is homeomorphic to a disjoint union of closed sets homeomorphic to the fiber space. However, in the general case, $\mathcal{B}^1$ can fail to be closed (check the Reeb component). In this case, it has accumulation points in $\mathcal{B}^0$.

Definition 17. Let us consider $\xi(\mathcal{B}^1) = \overline{\mathcal{B}^1} \setminus \mathcal{B}^1$, which is always a closed set. A continuous 1-cochain is a continuous map $g : \overline{\mathcal{B}^1} \to S^1$ such that $g_{|\xi(\mathcal{B}^1)} = 1$, *i.e.*, g vanishes in its relative boundary.

Although $\mathcal{B}^0$ is a complete transversal in the measurable sense it has, in general, singularities where bifurcations occur, this set of singularities agrees with $\xi(\mathcal{B}^1)$. Recall that now the dynamics are generated by the holonomy maps associated to a 1-simplicial path of edges, we can assume that the domain and range are contained in $\mathcal{B}^0$ whenever the initial point of the first edge and the end point of the last edge of the path are non-singular. If any of these points is singular, we can also assume that the domain and range are in $\mathcal{B}^0$ but not as open subsets, this is done by considering the *outer* and *inner* transversal relative to a singular point and a 1-simplex defined below.

Definition 18. Let U be a foliated chart neighboring a singular point $p \in \mathcal{B}^0$ and let $b \in \{\pm\} \times \mathcal{B}^1$ so that $i(b) = p$. Since simplices must be converging in the Hausdorff metric, it follows that for any plaque of U close to p, there exists a sequence of 1-simplices $\triangle_{b_P}$ which are converging in the Hausdorff metric to $\triangle_b$. We define the *outer* transversal through p relative to b, denoted by $\mathrm{out}(p,b)$, as the union of the initial points of all b_P and p. This is a locally closed set which meets each plaque close to P in a single point, so it is a transversal in the topological sense where the holonomy is well defined. Analogously, the *inner* transversal, $\mathrm{inn}(p,b)$, is defined for a signed 1-barycenter b such that $e(b) = p$. See Figure 4.

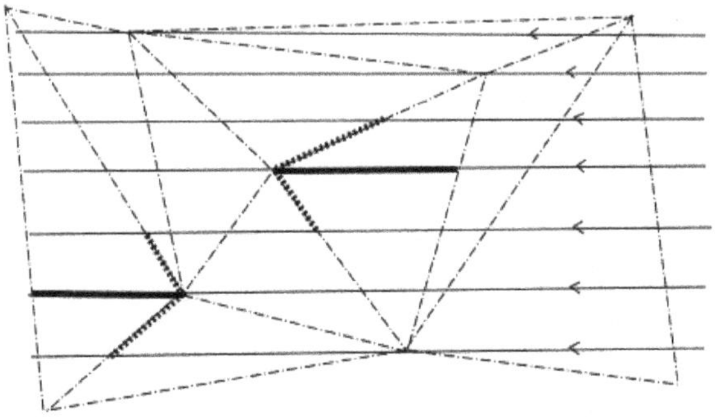

Figure 4. Outer and inner transversals associated to the extreme singular points of the bold 1-simplices. The leafwise triangulation is induced by a transverse triangulation.

Definition 19 (Leafwise skew product). Let g be a continuous S^1 evaluated closed 1-cochain on the topological 1-dimensional foliation $\mathcal{F}$. For every $t \in \mathcal{B}^0$, let $b \in \{\pm\} \times \mathcal{B}^1$ such that $i(b) = t$. The *leafwise skew product*, $\mathcal{F} \ltimes_g S^1$ is defined as the topological simplicial foliation in $\|\mathcal{F}\| \times S^1$ where the 0 simplices are identified with $\mathcal{B}^0 \times S^1$ and each (t,z) is joined with $(e(b), g(b)z)$ by an edge with this orientation. The fact that g is closed means that any 1-simplicial loop bounded by a 2-cell induces a family of 1-simplicial loops in the skew product and thus 2-cells (and higher dimensional cells) can be attached accordingly, this implies that $\|\mathcal{F} \ltimes_g S^1\| = \|\mathcal{F}\| \times S^1$. The fact that g vanishes in its relative boundary is necessary to make the holonomy of the leafwise skew product continuous.

Proposition 20. *Let $(\mathcal{F}, \mu_0)$ be an ergodic oriented topological measured simplicial lamination. Let g : $\{\pm\} \times \mathcal{B}^1 \to S^1$ be a closed 1-cochain. Then $\mu = \mu_0 \times m$ is an ergodic measure for $\mathcal{F} \ltimes_g S^1$ if and only if g^k is not trivial in measured cohomology for every integer $k \neq 0$, i.e., $[g^k] \neq 0$ in $H^1(\mathcal{F}, S^1, \mu)$.*

Proof. We follow the method of proof of Proposition 16. First, we show that $\mu_0 \times m$ is quasi-invariant. Let us consider a holonomy transformation $\gamma_g : D \times S^1 \to Q \times S^1$ of $\mathcal{F} \ltimes_g S^1$, where D, Q are local transversals which can be considered as subsets of $\mathcal{B}^0$ (see Definition 18) which are the domain and range of a holonomy transformation γ of $\mathcal{F}$ associated to a 1-simplex, so $\gamma_g(t,z) = (\gamma(t), g([t,\gamma(t)])z)$. It is clear that such maps γ_g generate the holonomy pseudogroup of $\mathcal{F} \ltimes_g S^1$. Since μ is a Radon product measure, we can estimate the measure of a set Z of null measure by product functions. Let us consider a sequence of continuous and positive functions $f_n : Q \times S^1 \to \mathbb{R}$ such that $\bigcap_n \mathrm{Supp}(f_n) = Z$ and

$$f_n(w,z) = \sum_{k=1}^{\infty} a_{n,k}(w) b_{n,k}(z)\,,$$

$$\int_{Q\times S^1} f_n(w,z)\, d\mu = \sum_{k=1}^{\infty} \int_{Q\times S^1} a_{n,k}(w) b_{n,k}(z) d\mu \xrightarrow[n\to\infty]{} 0\,.$$

We only have to show that $\int_{D\times S^1} f_n \circ \gamma_g d\mu$ converges to zero. Since $f_n(\gamma_g(w,z)) = \sum_k a_{n,k}(\gamma(w)) b_{n,k}(g(w)z)$, we have

$$\begin{aligned}
\int_{Q\times S^1} f_n \circ \gamma_g\, d\mu &= \sum_{k=1}^{\infty} \int_{Q\times S^1} a_{n,k}(\gamma(w)) b_{n,k}(g(w)z)\, d\mu = \\
&= \int_Q \left(\int_{S^1} b_{n,k}(g(w)z) dm \right) a_{n,k}(\gamma)\, d\mu_0 = \\
&= \sum_{k=1}^{\infty} \int_Q \left(\int_{S^1} b_{n,k}(z) dm \right) a_{n,k}(\gamma(w))\, d\mu_0 = \\
&= \sum_{k=1}^{\infty} \int_Q a_{n,k}(\gamma(w)) d\mu_0 \int_{S^1} b_{n,k}(z)\, dm = \\
&= \int_{D\times S^1} f_n\, d\gamma^* \mu_0 \times m\,.
\end{aligned}$$

Since μ_0 is quasi-invariant, μ is quasi-invariant for the transformation $\gamma \times \mathrm{id} : D \times S^1 \to Q \times S^1$, therefore the latest expression converges to zero as desired.

In order to check whether $\mathcal{F} \ltimes_g S^1$ is ergodic, let $f \in L^2(\mathcal{F} \ltimes_g S^1, \mu)$ (with complex values) be a leafwise constant (but a.e. non-constant) function and let $f \equiv \sum_{k\in\mathbb{Z}} c_k(x) z^k$ be the fiberwise Fourier series. At the level of the $\mathcal{B}^0$ and for a given holonomy map γ of $\mathcal{F}$ induced by an oriented 1-simplex, we obtain $c_k(\gamma(w)) = g([w,\gamma(w)])^{-k} c_k(w)$ where $[w,\gamma(w)]$ denotes the 1-simplex joining w and $\gamma(w)$. Of course c_0 is a.e. constant by ergodicity of $\mathcal{F}$ and therefore some c_k is not essentially zero for some $k \neq 0$. Now, by ergodicity of $\mathcal{F}$ and the fact that $|g| = 1$, this c_k is a.e. non-zero, and we can check that $\varphi(w) = c_k(w)^{-1}$ a.e. satisfies $g^k = \delta\varphi$ almost everywhere. Conversely, if some g^k is trivial in cohomology then $g^k = \delta\varphi$ a.e. and $f = \varphi(w)^{-1} z^k$ a.e. gives a leafwise constant function in $\mathcal{B}^0$ which is non a.e. constant. □

Example 5. Let $\mathcal{F}_{\alpha,\beta}$ be a linear foliation on T^3. Dealing with this foliation as a suspension with the polygonal structure given in Figure 2 by removing the diagonals, we obtain $\mathcal{B}^1 = S^1 \sqcup S^1$ with no singularities. Let g be a 1-cochain with constant values γ, θ on each copy of S^1, it is clear that it is a closed 1-cochain. The foliation $\mathcal{F}_{\alpha,\beta} \ltimes_g S^1$ is nothing else than a 2-dimensional linear foliation on T^4 which can be seen as the suspension over T^2 of the following conmuting transformations on T^2: $A(z_1,z_2) = (\alpha z_1, \gamma z_2)$ and $B(z_1,z_2) = (\beta z_1, \theta z_2)$. It is well known that this foliation is not ergodic (or minimal) if and only if there exists integers $k, k' \neq 0$ so that $\alpha^k = \gamma^{k'}$ and $\theta^k = \theta^{k'}$. We can check this directly, under this condition the function $f(z) = z^k$ satisfies $f(\alpha z) f(z)^{-1} = \alpha^k = \gamma^{k'}$ and $f(\beta z) f(z)^{-1} = \theta^{k'}$, so $g^{k'} = \delta f$. The converse is a little harder, but in this particular example it is easy to see that $g^k = \delta f$ implies $f(\beta\alpha^{-1} z) = (\gamma\theta^{-1})^k \cdot f(z)$ for all $z \in S^1$, since $\beta\alpha^{-1}$ is irrational by hypothesis it follows, by a Fourier series argument, that f must be of the form $f(z) = c_0 + c_m z^m$ for

$c_0, c_m \in S^1$ constants and some $m \neq 0$. Finally, the analysis of this example is finished by Furstenberg's criterium.

The next step is to work in the continuous category and show that a similar proposition holds for the continuous cohomology.

Definition 21. Let $\mathcal{F}$ be a topological simplicial lamination as above and let $\mathcal{T}$ be a leafwise continuous triangulation. The first R-evaluated continuous cohomology group for $\mathcal{F}$ is defined as the quotient group of the continuous closed 1-cochains $g : \mathcal{B}^1 \to R$ via the equivalence relation induced by the coboundary of continuous maps $f : \mathcal{B}^0 \to R$. It will be denoted by $H^1_{C^0}(\mathcal{F}, \mathcal{T}, R)$. Observe that δf vanishes in the relative boundary of $\mathcal{B}^1$.

Proposition 22. *Let $\mathcal{F}$ be a minimal oriented 1-dimensional topological foliation with a regular leafwise triangulation $\mathcal{T}$. Let $g : \mathcal{B}^1 \to S^1$ be a continuous closed 1-cochain. Then $\mathcal{F} \ltimes_g S^1$ is minimal if and only if g^k is not trivial in continuous cohomology for every integer $k \neq 0$, i.e., $[g^k] \neq 0$ in $H^1_{C^0}(\mathcal{F}, \mathcal{T}, S^1)$.*

Proof. Recall that minimality can be checked by any of the following three equivalent definitions:

1. There are no closed proper saturated sets,
2. Every leaf is dense,
3. Any continuous leafwise constant function $f : \mathcal{F} \to H$, where H is any Frechet space, is constant.

We have a natural projection $\pi : \|\mathcal{F} \ltimes_g S^1\| \to \|\mathcal{F}\|$, by minimality of $\mathcal{F}$, any non-empty closed saturated set $K \subset \|\mathcal{F} \ltimes_g S^1\|$ satisfies $\pi(K) = \|\mathcal{F}\|$. Let $f \in C(\mathcal{F} \ltimes_g S^1, \mathbb{C})$ be a leafwise constant (but non-constant) continuous function. As before, we can consider the fiberwise Fourier series $f = \sum_{k \in \mathbb{Z}} c_k(w) z^k$ where the Fourier coefficients belong to $C(\|\mathcal{F}\|, \mathbb{C})$. Let γ be a holonomy transformation induced by an oriented 1-simplex, we have $c_k(\gamma(w)) = g^{-k}([w, \gamma(w)]) c_k(w)$ since the function is supposed to be leafwise constant. By minimality of $\mathcal{F}$, c_0 is a constant λ. If $c_k = 0$ for all $k \neq 0$ then f is constant which contradicts the choice of f. Assume that some c_k is not constant zero for $k \neq 0$, then, since $|g| = 1$, the zero level set of c_k is closed and saturated for $\mathcal{F}$ and therefore it must be the emptyset. This means that $\varphi(w) = c_k(w)^{-1}$ is well defined and continuous and satisfies $g^k = \delta\varphi$. Conversely for $g^k = \delta\varphi$, the function $\varphi(w)^{-1} z^k$ is a continuous leafwise constant but non-constant. □

In fact, for the continuous cohomology we can adapt Zimmer's result on the coefficient group. As a corollary, we obtain a well known characterization of exact $\mathbb{R}$ or $\mathbb{C}$ evaluated cocycles of $\mathbb{Z}$-actions given by W. H. Gottschalk and G. A. Hedlund [21].

Proposition 23. *Let ω be a continuous 1-cochain of a minimal simplicial lamination $(\mathcal{F}, \mathcal{T}, \mu)$ evaluated in an abelian group R which is isomorphic to a discrete group, $\mathbb{R}^n$ or $\mathbb{C}^n$. If ω is a bounded cochain then ω is cohomologous to a continuous cochain evaluated on a compact subgroup of R, in the particular case of $\mathbb{R}^n$ or $\mathbb{C}^n$ the bounded cochain is, in fact, null cohomologous.*

Proof. As in Proposition 9, $\mathcal{C}$ denotes the Polish space of compact sets in R and $\mathcal{T}^1(t)$ the set of oriented 1-simplicial paths with initial point $t \in \mathcal{B}^0$. Since $\omega : \mathcal{B}^1 \to R$ is continuous and vanishes in its relative boundary the map $\mathcal{K} : \{\pm\} \times \mathcal{B}^1 \to \mathcal{C}$, $b \mapsto \overline{\mathcal{T}^1(i(b))}$ is also continuous. Clearly, $\omega(b) + \overline{\omega(\mathcal{T}^1(e(b)))} = \overline{\omega(\mathcal{T}^1(i(b)))}$. The quotient space $\mathcal{C}/R$ is a Polish since the equivalence relation is closed. Since $\omega(b) + \overline{\omega(\mathcal{T}^1(e(b)))} = \overline{\omega(\mathcal{T}^1(i(b)))}$, the sets $\overline{\omega(\mathcal{T}^1(e(b)))}, \overline{\omega(\mathcal{T}^1(i(b)))}$ represent the same equivalence class in $\mathcal{C}/R$.

The map $\overline{\mathcal{K}} : \mathcal{B}^0 \to \mathcal{C}/R$, $s \mapsto \mathcal{T}^1(s)$ is continuous and constant on each orbit (which is the intersection of each leaf with $\mathcal{B}^0$) and thus it is constant by minimality of $\mathcal{F}$, call $[B]$ this constant class and assume without loss of generality that $0 \in B$. Let $e_B : R \to [B] \subset \mathcal{C}$, $r \mapsto r + B$, the class $r + B$ is naturally identified with the coset $r + \mathrm{Stab}(B)$. Now, we study each possibility in order to obtain a map $\mathbf{s} : [B] \to R$ which is a continuous section, *i.e.*, for all $D \in [B]$, $\mathbf{s}(D) + B = D$.

- If R is a discrete abelian group, then any section is also a continuous section because the quotient space is also discrete.
- If R is $\mathbb{R}^n$ or $\mathbb{C}^n$, then $\text{Stab}(B) = \{0\}$ because these spaces does not admit non-trivial compact subgroups, thus e_B is bijective with continuous inverse. This is the desired section.

Let φ be the 0-cochain $\varphi(t) = -s(\overline{\omega(\mathcal{T}^1(t))})$. As in Proposition 9 $\omega + \delta\varphi$ takes its values in $\text{Stab}(B)$. □

Corollary 24 (Gottschalk-Hedlund [21]). *Let $h : X \to X$ be a minimal homeomorphism on a compact Hausdorff space. A continuous function $f : X \to \mathbb{C}$ is null-cohomologous if and only if the functions $\psi_k = \sum_{i=0}^{k} f \circ h^i : X \to \mathbb{C}$, $k \in \mathbb{Z}$, are uniformly bounded for all $x \in X$.*

Proof. We are in the hypothesis of Proposition 23 by considering the suspension of h and the induced simplicial structure. In this corollary, $R = \mathbb{C}$. The fact that ψ_k are uniformly bounded implies that f is a bounded cochain and so f is null cohomologous. Conversely, if f is null cohomologous, there exists $g : X \to \mathbb{C}$ continuous so that $f = g \circ h - g$. Therefore, a telescopic argument shows that $\psi_k = g \circ h^{k+1} - g$ for $k \geq 0$ and $\psi_k = g \circ h - g \circ h^k$ for $k < 0$. Thus $\|\psi_k\| \leq 2\|g\|$ for all $k \in \mathbb{Z}$ concluding the proof. □

Remark 9. Furstenberg uses this dual result to obtain a minimal but not strictly Lebesgue ergodic transformation on the torus $T^k, k \geq 2$, just by obtaining a non-trivial class in the continuous cohomology which is trivial in the measurable cohomology. This work provides the technology to perform such examples in higher dimensional foliations. Another observation is that Furstenberg's technique works in the world of discrete actions, and we are showing that in fact it works in the world of (simplicial) foliations. Another interesting option is to change the coefficient group by a compact Lie group and the Lebesgue measure by the Haar measure (although this is only interesting if non-abelian coefficient groups were allowed).

Observe also that our continuous simplicial cohomology of foliations is nothing else than the leafwise cohomology for usual foliations (via a leafwise de Rham theorem).

6. Foliated Cocycles and Invariant Measures

It is important to remark that all the dynamical information of a measurable lamination is encoded in the holonomy pseudogroup, and the holonomy pseudogroup is generated by the transverse coordinate changes of a regular foliated atlas. Given a locally finite foliated atlas $\mathcal{U}$, we can associate a measurable graph $\mathcal{G}_{\mathcal{U}}$ where each plaque in a foliated chart is identified with its barycenter and we perform an edge between barycenters associated to adjacent plaques.

Let $U_\alpha, U_\beta, U_\sigma \in \mathcal{U}$ so that $U_\alpha \cap U_\beta \cap U_\sigma \neq \emptyset$. Let $P_i \in U_i$, $i \in \{\alpha, \beta, \sigma\}$, be plaques so that $P_\alpha \cap P_\beta \cap P_\sigma \neq \emptyset$. They induce a 1-simplicial loop in the leaf which contains these plaques, these kind of loops will be called *cocycle loops*. We attach a 2-simplex on each cocycle loop, and we can perform these operation in a measurable way performing a new measurable lamination $\widehat{\mathcal{G}}_{\mathcal{U}}$. This lamination induces the same holonomy pseudogroup in $\mathcal{B}^0$. In particular, they have the same invariant measures.

Definition 25. (see e.g., [7]) An R-evaluated *foliated cocycle* relative to the atlas $\mathcal{U} = \{U_\alpha\}_{\alpha \in A}$ of the measurable lamination $\mathcal{F}$ is a family of maps $\nu = \{\gamma_{\alpha\beta} : U_\alpha \cap U_\beta \to R\}$ so that $\gamma_{\alpha\beta} \cdot \gamma_{\beta\sigma} = \gamma_{\alpha\sigma}$ on any point of $U_\alpha \cap U_\beta \cap U_\sigma$. Of course, we consider these maps only on non trivial intersections. When R is the germ of diffeomorphisms of $\mathbb{R}^q$ at the origin, then the cocycle is called a Haefliger cocycle.

Of course, Haefliger cocycles are evaluated on a non-abelian group, and so they are out of our actual discussion. An interesting example to be considered is the infinitesimal holonomy cocycle associated to a usual foliation in a smooth manifold. For a point $x \in U_\alpha \cap U_\beta$ we define $\nu = \{\gamma_{\alpha\beta}(x) = \log|\det Jh_{\alpha\beta}(x)|\}$, where $Jh_{\alpha\beta}$ denotes the Jacobian matrix relative to the transverse change of coordinates. This evaluation does not depend on the points in the same plaque, and, therefore, we can define an $\mathbb{R}$-evaluated 1-cochain in $\mathcal{G}_{\mathcal{U}}$ by $\nu([b_\alpha, b_\beta]) = \gamma_{\alpha\beta}(x_{\alpha\beta})$ for any $x_{\alpha\beta} \in P_\alpha \cap P_\beta$. This is a closed cochain in $\widehat{\mathcal{G}}_{\mathcal{U}}$ and, conversely, any closed 1-cochain in $\widehat{\mathcal{G}}_{\mathcal{U}}$ defines a foliated cocycle relative to $\mathcal{U}$ which is constant on the plaques of the atlas.

The measurable cohomology class of ν is an obstruction to the existence of invariant measures in the measure class (*i.e.*, with the same sets of null measure) of the Lebesgue measure of a smooth foliation. This is explictly done in [7,22].

Proposition 26. *Let $(M, \mathcal{F}, m)$ be a smooth measured foliation on a smooth manifold M so that m is the Lebesgue measure in $\mathcal{B}^0$. Let $\mathcal{G}_{\mathcal{U}}$ be the measurable graph associated to a regular foliated atlas $\mathcal{U}$ of $\mathcal{F}$. Then $\mathcal{F}$ has an invariant measure in the same class of m if and only if the cohomology class $[\nu] \in H^1(\mathcal{G}_{\mathcal{U}}, m, R)$ is trivial.*

Proof. Let $b \in \mathcal{B}^1$, we associate a holonomy transformaion $\gamma_b : \mathrm{out}(i(b), b) \to \mathrm{inn}(e(b), b)$. Of course, $\mathrm{out}(i(b), b)$ (resp. $\mathrm{inn}(e(b), b)$) is identified with the projection of $U_\alpha \cap U_\beta$ in an associated transversal of U_α (resp. U_β). γ_b can be seen also as a map $\gamma_b : \mathrm{inn}(i(b), b) \to \{\pm\} \times \mathcal{B}^1$, $t \mapsto [t, \gamma_b(t)]$ where $[t, \gamma_b(t)]$ denotes the barycenter of the edge between t and $\gamma_b(t)$ with the orientation $t \mapsto \gamma_b(t)$. We shall use the same notation for both maps.

Assume μ a measure in the Lebesgue class. Therefore, the Radon–Nikodyn formula provides a measurable map $h : \mathcal{B}^0 \to \mathbb{R}$ so that $e^h dm = d\mu$. The change of variable formula provides that $\gamma_b^* d\mu = e^{\nu \circ \gamma_b + h \circ \gamma_b - h} d\mu$ (Lemma 7.1.21 in [7]). If μ is invariant, then $\nu + \delta h = 0$ a.e. and so $[\nu] = 0$. Conversely, let h so that $\nu + \delta h = 0$ a.e. then the above equation shows that the measure given by $e^h dm$ is invariant and, of course, it is in the Lebesgue class. □

The infinitesimal holonomy is one of the components of the Godbillon–Vey class of a foliation and Hurder shows in [22] that the existence of an invariant measure in the Lebesgue class forces the vanishing of the Godbillon–Vey class. The Godbillon–Vey class encodes interesting data of the transverse dynamics, for instance, a deep result of Duminy shows that, for codimension 1 foliations, a non-trivial Godbillon–Vey class implies the existence of a *resilient leaf* (a non-proper leaf which has a transverse self-accumulation point given by a holonomy contraction), see [7] for a proof. Then, even for smooth foliations, the measurable cohomology is interesting on its own. At this point, we can use our work in Section 4 and obtain the following corollary.

Corollary 27. *Let $(M, \mathcal{F}, m)$ be a smooth Lebesgue ergodic foliation on a smooth manifold M. If ν is a bounded 1-cochain of $\mathcal{G}_{\mathcal{U}}$ (in the sense of Definition 8) then $\mathcal{F}$ has an invariant measure in the Lebesgue class.*

Remark 10. When the Lebesgue measure is invariant, then it is clear that ν is trivial and in particular, bounded.

Recall that, in continuous cohomology, $\mathbb{R}$-valued bounded 1-cochains are exactly the cohomologically trivial 1-cochains (see Corollary 24). This suggests that Corollary 10 is close to be an equivalence. If M is a closed manifold, then ν defines an $L^\infty(m)$ 1-cochain. Clearly, if $\nu = \delta h$ for some $h \in L^\infty(\mathcal{B}^0, \mathbb{R}; m)$ then ν is uniformly bounded since $\sup\{|x| \mid x \in \omega(\mathcal{T}^1(t)|\} \leq 2|h|_\infty$ for all $t \in \mathcal{B}^0$. The converse is also true, and it was proved in Corollary 10. These data can be encoded in the following result.

Corollary 28. *Let $(M, \mathcal{F}, m)$ be a smooth Lebesgue ergodic foliation on a smooth closed manifold M. Then ν is a uniformly bounded 1-cochain of $\mathcal{G}_{\mathcal{U}}$ (in the sense of Definition 8) if and only if there exists an invariant measure μ in the Lebesgue class so that $d\mu = e^h dm$ for some $h \in L^\infty(\mathcal{B}^0, \mathbb{R}; m)$.*

7. Conclusions

What is lost in our approach is the possibility of having non-abelian coefficient groups. This is of extreme importance in many interesting problems where the coefficient groups can be $Gl(n, \mathbb{R})$, $\mathrm{Diff}_+(S^1)$, a compact Lie group or the unitary group of a Hilbert space. For measurable graphs, the definition of the first cohomology set (in general fails to be a group) can be adapted with no further problems. However, higher dimensional cohomology seems to be much more difficult to properly define.

At this point, the reader would see that a lot of dynamical and topological results can be a target for being adapted in this framework. In this section, we state two of them.

Livsic's theorem states that the Hölder cohomology classes of a transitive Anosov diffeomorphism are determined by their values on the periodic points. More precisely:

Theorem 29 (Livsic). *[18] Let $h : M \to M$ be a transitive Anosov diffeomorphism. Let $f : M \to \mathbb{R}$ be an α-Hölder function so that $h^n(p) = p$ implies $\sum_{i=0}^{n} f \circ h^i(p) = 0$, then there exists an α-Hölder function $g : M \to \mathbb{R}$ so that $f = g \circ h - g$, i.e., f is null cohomologus in the α-Hölder cohomology.*

It is well known that the Livsic theorem holds for transitive Anosov flows, and this suggests an adaptation to foliations: periodic points are interpreted as compact leaves, and the Anosov condition is translated to the notion of Anosov pseudogroups, which are analogous to Anosov flows.

On the other hand, the Mostow rigidity theorem says that the volume of hyperbolic manifolds of dimensions greater than 2 is determined by their fundamental group. Zimmer [5] adapted the Mostow rigidity for measurable laminations coming from ergodic measured locally free actions of certain Lie groups. Thus, the natural question is how to extend Zimmer's result to a wider class of measurable laminations.

Acknowledgments: I have to express my gratitude to Steven Hurder who pointed out to me, back in 2012, that my treatment of cohomology of measurable laminations was linked in a natural way to Zimmer's work and that it would be worthy of interest to explain this relation in more detail. This work was supported by CAPES (Brazil), postdoc program PNPV 2015.

Conflicts of Interest: The author declares no conflict of interest.

References

1. Furstenberg, H. Strict ergodicity and transformation of the torus. *Am. Math. J.* **1961**, *83*, 573–601.
2. Zimmer, R.J. On the cohomology of ergodic group actions. *Israel J. Math.* **1980**, *35*, 289–300.
3. Gaboriau, D. Sur la (co-)homologie L^2 des actions préservant une mesure. *C. R. Acad. Sci. Paris Sér. I Math.* **2000**, *330*, 365–370.
4. Zimmer, R.J. On the cohomology of ergodic actions of semisimple lie groups and discrete groups. *Am. J. Math.* **1981**, *103*, 937–950.
5. Zimmer, R.J. On the Mostow rigidity theorem and measurable foliations by hyperbolic space. *Israel J. Math* **1982**, *43*, 281–290.
6. Candel, A.; Conlon, L. *Foliations I*; Graduate Studies in Mathematics; American Mathematical Society: Providence, RI, USA, 1999; Volume 23.
7. Candel, A.; Conlon, L. *Foliations II*; Graduate Studies in Mathematics; American Mathematical Society: Providence, RI, USA, 1999; Volume 60.
8. Connes, A. A survey of foliations and operator algebras. *Proc. Symp. Pure Math.* **1982**, *38*, 520–628.
9. Connes, A.; Fack, T. Morse inequalities for foliations. *C*-Algebras and Elliptic Theory*; Trends in Mathematics; Birkhäuser Verlag: Basel, Switzerland, 2006; pp. 61–72.
10. Heitsch, J.L.; Lazarov, C. Homotopy invariance of foliation Betti numbers. *Invent. Math.* **1991**, *104*, 321–347.
11. Bermúdez, M. La caratéristique d'Euler des feuilletages mesurés. *J. Funct. Anal.* **2006**, *237*, 150–175.
12. Bermúdez, M.; Hector, G. Laminations hyperfinies et revêtements. *Ergod. Theory Dyn. Syst.* **2006**, *26*, 305–339.

13. Bermúdez, M. Laminations Boréliennes. Ph.D. Thesis, Université Claude Bernard Lyon 1, Villeurbanne, France, 2002.
14. Meniño, C. Cohomology of measurable laminations. *Topol. Appl.* **2013**, *160*, 692–702.
15. Meniño, C. Measurable versions of the LS category on laminations. *Manuscr. Math.* **2014**, *144*, 135–163.
16. Hatcher, A. *Algebraic Topology*; Cambridge University Press: Cambridge, UK, 2002.
17. Connes, A.; Feldman, J.; Weiss, B. An amenable equivalence relation is generated by a single transformation. *Ergod. Theory Dyn. Syst.* **1981**, *1*, 431–450.
18. Livsic, A. Certain properties of the homology of Y-systems. *Mat. Zamet.* **1971**, *10*, 555–564.
19. Kechris, A.S. *Classical Descriptive Set Theory*; Graduate Texts in Mathematics; American Mathematical Society, Springer-Verlag: New York, NY, USA, 1994; Volume 156.
20. Plante, J.F. Foliations with measure preserving holonomy. *Ann. of Math.* **1975**, *102*, 327–361.
21. Gottschalk, W.H.; Hedlund, G.A. *Topological Dynamics*; American Mathematical Society, Colloquium Publications: Providence, RI, USA, 1955; Volume 36.
22. Hurder, H. The Godbillon measure of amenable foliations. *J. Differ. Geom.* **1986**, *23*, 347–365.

5

New Approach for Fractional Order Derivatives: Fundamentals and Analytic Properties

Ali Karcı

Department of Computer Engineering, Faculty of Engineering, İnönü University, 44280 Malatya, Turkey; adresverme@gmail.com or ali.karci@inonu.edu.tr

Academic Editor: Hari M. Srivastava

Abstract: The rate of change of any function *versus* its independent variables was defined as a derivative. The fundamentals of the derivative concept were constructed by Newton and l'Hôpital. The followers of Newton and l'Hôpital defined fractional order derivative concepts. We express the derivative defined by Newton and l'Hôpital as an ordinary derivative, and there are also fractional order derivatives. So, the derivative concept was handled in this paper, and a new definition for derivative based on indefinite limit and l'Hôpital's rule was expressed. This new approach illustrated that a derivative operator may be non-linear. Based on this idea, the asymptotic behaviors of functions were analyzed and it was observed that the rates of changes of any function attain maximum value at inflection points in the positive direction and minimum value (negative) at inflection points in the negative direction. This case brought out the fact that the derivative operator does not have to be linear; it may be non-linear. Another important result of this paper is the relationships between complex numbers and derivative concepts, since both concepts have directions and magnitudes.

Keywords: derivatives; fractional calculus; fractional order derivatives

---

1. Introduction

The asymptotic behaviors of functions can be analyzed by velocities or rates of change in functions, while very small changes occur in the independent variables. The concept of rate of change in any function *versus* change in the independent variables was defined as a derivative, and this concept attracted many scientists and mathematicians such as Newton, l'Hôpital, Leibniz, Abel, Euler, Riemann, *etc.*

Isaac Newton defined the fundamentals of classical mechanics and this study contains rates of changes of functions [1]. He collected his works in *Philosophiæ Naturalis Principia Mathematica*, a book that includes geometrical proofs, gravitational force law, and attraction of bodies [1]. He was the first scientist who concerned himself with the concept of derivatives/fluxions. On the other hand, Newton tried to determine the change in length of distance in terms of the velocity of bodies, and the change in velocity of bodies in terms of acceleration. L'Hôpital was a follower of Newton in that he defined the concept of a derivative and generalized this concept. There are other mathematicians who dealt with the concepts of derivatives/fluxions.

The first important and detailed work in differential calculus and differential geometry was done by l'Hôpital [2,3]. L'Hôpital generalized the ideas of Newton through variations on calculus. Leibniz is another scientist who expounded on differential calculus and infinitesimal calculus [4]; he mastered the mathematics of his day and developed his own calculus over the short span of a few years [4]. Newton, l'Hôpital, and Leibniz are not the only mathematicians who dealt with variations of calculus. Some mathematicians tried to explain the ratio between the two displacements of at least two variables [5], since it is important for analyzing the asymptotic behaviors of functions.

The asymptotic behaviors of functions can be regarded as the rates of displacements of functions *versus* rates of displacement of independent variables—in other words, the rates of movement of functions. The term fluxion indicates motion and the idea of the fluxional calculus developed from the concept that a geometrical magnitude was the result of continuous motion of a point, line, or plane [6]. This motion, speaking of plane curves, could be considered in reference to coordinate axes as the result of two motions, one in the direction of the X-axis and the other in the direction of the Y-axis [6]. The velocity of the X-component and the Y-component were called "fluxions" by Newton [1,6]. The velocity of a point is represented by an equation involving the fluxions x and y [6].

The problems of variation of calculus are attractive to mathematicians and there are several familiar mathematicians who focused their attention on these problems such as Newton, l'Hôpital, Leibniz, Euler, Abel, Caputo, Riemann, Grünwald, Miller, Ross, *et al.* [7–11].

The problems of variation of calculus and infinitesimal calculus are not solved completely, and there are still open problems in fractional variation of calculus. There are a lot of studies about the fractional variations of calculus [11–22]. Euler, Caputo, Riemann, Abel, *et al.* dealt with fractional variations of calculus and fractional order calculus and systems. Karci defined the fractional order derivative concept in a different way by using indefinite limits and the l'Hôpital rule [23–25].

Some popular fractional order derivative methods, such as Euler, Caputo, and Riemann-Liouville, can be summarized as follows.

The Euler method is $\frac{d^n x^m}{dx^n} = \frac{\Gamma(m+1)}{\Gamma(m-n+1)} x^{m-n}$ and its deficiencies can be illustrated for constant and identity functions. Assume that $f(x) = cx^0$ where c is a constant and $c \in R$. Assume that $n = 1$ and $\alpha = \frac{1}{4}$ and:

$$\frac{d^n x^m}{dx^n} = \frac{\Gamma(m+1)}{\Gamma(m-n+1)} x^{m-n} = \frac{d^{\frac{1}{4}} x^0}{dx^{\frac{1}{4}}} = \frac{\Gamma\left(\frac{5}{4}\right)}{\Gamma\left(\frac{3}{4}\right)} cx^{-\frac{1}{4}}$$

The derivative of any constant function is always zero; however, the result of fractional order derivative with respect to the Euler method is different from zero. Assume that $f(x) = x$, $n = 1$ and $\alpha = \frac{2}{3}$.

$$\frac{d^{\frac{2}{3}} x^1}{dx^{\frac{2}{3}}} = \frac{\Gamma(1+1)}{\Gamma\left(1 - \frac{2}{3} + 1\right)} x^{1-\frac{2}{3}} = \frac{\Gamma(2)}{\Gamma\left(\frac{4}{3}\right)} x^{\frac{1}{3}} \neq 1$$

The Riemann-Liouville method is for function $f(t)$ ${}_aD_t^\alpha f(t) = \frac{1}{\Gamma(n-\alpha)} \left(\frac{d}{dt}\right)^n \int_a^t \frac{f(v)dv}{(t-v)^{\alpha-n+1}}$. The fractional order derivative also can be applied to constant and identity functions with respect to the Riemann-Liouville method.

Assuming that $f(x) = cx^0$, $c \in R$, $n = 1$ and $\alpha = \frac{2}{3}$,

$${}_aD_t^\alpha f(t) = \frac{1}{\Gamma(n-\alpha)} \left(\frac{d}{dt}\right)^n \int_a^t \frac{f(v)dv}{(t-v)^{\alpha-n+1}} = {}_aD_t^{\frac{2}{3}} f(t) = \frac{1}{\Gamma\left(\frac{1}{3}\right)} \frac{d}{dt} \int_a^t \frac{cdv}{(t-v)^{\frac{2}{3}}} = \frac{c}{\Gamma\left(\frac{1}{3}\right)} \left(-\frac{1}{(t-a)^{\frac{2}{3}}}\right) \neq 0$$

The obtained result is inconsistent, since the result is a function of x. However, the initial function is a constant function and its derivative is zero, since there is no change in the dependent variable. The same case is valid for identity functions.

$$\begin{aligned} {}_aD_t^\alpha f(t) &= \frac{1}{\Gamma(n-\alpha)} \left(\frac{d}{dt}\right)^n \int_a^t \frac{f(v)dv}{(t-v)^{\alpha-n+1}} = {}_aD_t^{\frac{2}{3}} f(t) = \frac{1}{\Gamma\left(\frac{1}{3}\right)} \frac{d}{dt} \int_a^t \frac{xdx}{(t-x)^{\frac{2}{3}-1+1}} = \frac{1}{\Gamma\left(\frac{1}{3}\right)} \frac{d}{dt} \int_a^t \frac{xdx}{(t-x)^{\frac{2}{3}}} \\ &= \frac{1}{\Gamma\left(\frac{1}{3}\right)} \left(3a(t-a)^{\frac{2}{3}} + \frac{9}{4}(t-a)^{\frac{4}{3}}\right) \neq 1 \end{aligned}$$

The Caputo method is ${}_a^CD_t^\alpha f(t) = \frac{1}{\Gamma(\alpha-n)} \int_a^t \frac{f^{(n)}(v)dv}{(t-v)^{\alpha+1-n}}$. The Caputo method does not have inconsistency for constant functions; however, it has inconsistency for identity functions. Assuming that $f(x) = x$, $n = 1$ and $\alpha = \frac{2}{3}$,

$$ {}_{a}^{C}D_{t}^{\alpha}f(t) = \frac{1}{\Gamma(\alpha-n)}\int_{a}^{t}\frac{f^{(n)}(v)dv}{(t-v)^{\alpha+1-n}} = \frac{1}{\Gamma\left(-\frac{1}{3}\right)}\int_{a}^{t}\frac{dv}{(t-v)^{\frac{2}{3}+1-1}} = \frac{1}{\Gamma\left(-\frac{1}{3}\right)}\left(3(t-a)^{\frac{1}{3}}\right) \neq 1 $$

Due to these deficiencies, there is a need for a new approach to fractional order derivatives; this paper contains such a definition and some important properties of this approach.

This paper is organized as follows. The motivation of this paper will be presented in Section 2. Section 3 illustrates the applications of rational/irrational orders of derivatives. Section 4 is the definition and details of the new approach and puts forth the analytical results of this new approach for the derivative concept. Finally, the paper is concluded in Section 5.

2. Motivation

The rate of change of functions is an important concept to examine in mathematics. For this purpose, the concept of derivatives was identified, because the rate of change of the function gives detailed information about a system modeled by that function, and the nature of the problems or systems. For this purpose, an athlete's speed may be examined on a ski-jump ramp (Figure 1). The trajectory of movement of an athlete on the ski-jump ramp can be considered as a curve. The rate of change of the athlete's speed increases until the inflection point (Figure 1); after that point the rate of change of the athlete's speed will decrease. So, the rate of change has its maximum at the inflection point. The rate of change of the athlete's speed will be zero at point E (the local extremum point). The rate of change can be determined as shown in Figure 1, where it can be seen that the rate of change has magnitude and direction. The rate of change as seen in Figure 1 is directed to positive, and the situation of rates is seen in Figure 1. This concept and the relationships with complex numbers will be discussed in detail in subsequent sections of this paper.

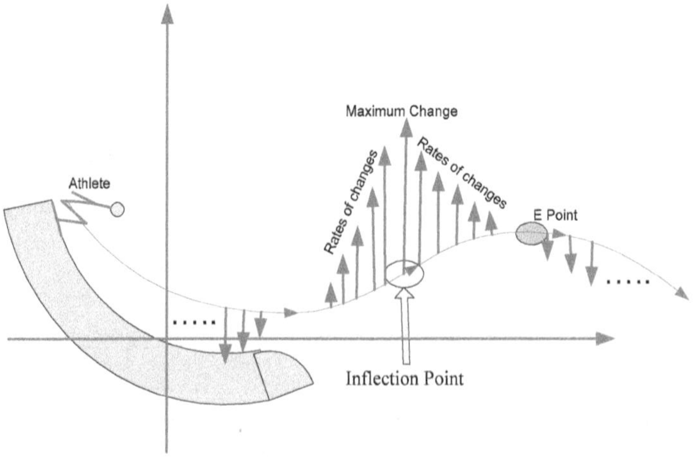

Figure 1. The movement of an athlete on the ski-jump ramp and the rates of change.

In order to make this situation more clear and understandable, trigonometric and polynomial functions can be used. To this end, sine and cosine and two polynomial functions can be examined as examples. The rate of change can be regarded as the velocity of function change.

In order to determine the behaviors of rate of change for any function, there will be very small increments/decrements in the independent variable (these increments/decrements are equal) and the response of the dependent variable to these small increments/decrements must be examined.

Assume that $f(x) = \sin(x)$ and $x \in [1, 2\pi]$. This closed interval can be divided into four closed intervals as follows:

$$I_1 = [\mathrm{Inf}_1, E_1] = \left[0, \frac{\pi}{2}\right] \; I_2 = [E_1, \mathrm{Inf}_2] = \left[\frac{\pi}{2}, \pi\right] \; I_3 = [\mathrm{Inf}_2, E_2] = \left[\pi, \frac{3\pi}{2}\right] \text{ and}$$

$$I_4 = [E_2, \mathrm{Inf}_3] = \left[\frac{3\pi}{2}, 2\pi\right]$$

The first interval $I_1 = [\mathrm{Inf}_1, E_1]$ can be examined for equal-length increments/decrements in the independent variable x such as $\Delta x_1 = \Delta x_2 = \ldots = \Delta x_{n-1} = \Delta x_n = \Delta x << 1$. The first point is inflection point $[\mathrm{Inf}_1, E_1]$; assume that $y_{\mathrm{inf1}} = \sin(x_{\mathrm{inf1}})$ is valid.

(a) $x\mathrm{inf1} \in [\mathrm{inf1}, E1]$ and $1 \leqslant i \leqslant n$

$$x_{1\mathrm{inf1}} = x_{\mathrm{inf1}} + \Delta x$$
$$y_{1\mathrm{inf1}} = y_{\mathrm{inf1}} + \Delta y_{1\mathrm{inf1}} = \sin(x_{\mathrm{inf1}} + \Delta x)$$
$$x_{2\mathrm{inf1}} = x_{1\mathrm{inf1}} + \Delta x = x_{\mathrm{inf1}} + 2\Delta x$$
$$y_{2\mathrm{inf1}} = y_{1\mathrm{inf1}} + \Delta y_{2\mathrm{inf1}} = y_{\mathrm{inf1}} + \Delta y_{1\mathrm{inf1}} + \Delta y_{2\mathrm{inf1}} = \sin(x_{1\mathrm{inf1}} + \Delta x) = \sin(x_{\mathrm{inf1}} + 2\Delta x)$$
$$x_{3\mathrm{inf1}} = x_{2\mathrm{inf1}} + \Delta x = x_{1\mathrm{inf1}} + 2\Delta x = x_{\mathrm{inf1}} + 3\Delta x$$
$$y_{3\mathrm{inf1}} = y_{2\mathrm{inf1}} + \Delta y_{3\mathrm{inf1}} = y_{1\mathrm{inf1}} + \Delta y_{2\mathrm{inf1}} + \Delta y_{3\mathrm{inf1}} = \sin(x_{2\mathrm{inf1}} + \Delta x) = \sin(x_{1\mathrm{inf1}} + 2\Delta x) = \sin(x_{\mathrm{inf1}} + 3\Delta x)$$
$$\ldots \; \ldots$$
$$x_{n\mathrm{inf1}} = x_{(n-1)\mathrm{inf1}} + \Delta x = x_{\mathrm{inf1}} + \sum_{i=1}^{n} \Delta x = x_{\mathrm{inf1}} + n\Delta x$$
$$y_{n\mathrm{inf1}} = \sin\left(x_{\mathrm{inf1}} + \sum_{i=1}^{n} \Delta x\right) = \sin(x_{\mathrm{inf1}} + n\Delta x).$$

At this point, the changes in the dependent variable $y = f(x)$ are $\Delta y_{1\mathrm{inf1}}, \Delta y_{2\mathrm{inf1}}, \ldots, \Delta y_{(n-1)\mathrm{inf1}}, \Delta y_{n\mathrm{inf1}}$ and inequalities for these changes are as follows: $\Delta y_{1\mathrm{inf1}} \geqslant \Delta y_{2\mathrm{inf1}} \geqslant \ldots . \geqslant \Delta y_{(n-1)\mathrm{inf1}} \geqslant \Delta y_{n\mathrm{inf1}}$ and $|\Delta y_{1\mathrm{inf1}}| \geqslant |\Delta y_{2\mathrm{inf1}}| \geqslant \ldots . \geqslant |\Delta y_{(n-1)\mathrm{inf1}}| \geqslant |\Delta y_{n\mathrm{inf1}}|$. So, the velocities of change can be identified as follows:

$$\frac{\Delta y_{1\mathrm{inf1}} - \Delta y_{2\mathrm{inf1}}}{\Delta x} \geqslant \frac{\Delta y_{2\mathrm{inf1}} - \Delta y_{3\mathrm{inf1}}}{\Delta x} \geqslant \ldots . \geqslant \frac{\Delta y_{(n-2)\mathrm{inf1}} - \Delta y_{(n-1)\mathrm{inf1}}}{\Delta x} \geqslant \frac{\Delta y_{(n-1)\mathrm{inf1}} - \Delta y_{n\mathrm{inf1}}}{\Delta x} \geqslant 0$$

(b) The same argument can be made for the second closed interval $\Delta x \in I2 = [E1, \mathrm{Inf2}]$, and $y = f(x) = \sin(x_{E1})$.

$$x_{1E1} = x_{E1} + \Delta x$$
$$y_{1E1} = y_{E1} + \Delta y_{1E1} = \sin(x_{E1} + \Delta x)$$
$$x_{2E1} = x_{1E1} + \Delta x = x_{E1} + 2\Delta x$$
$$y_{2E1} = y_{1E1} + \Delta y_{2E1} = y_{E1} + \Delta y_{1E1} + \Delta y_{2E1} = \sin(x_{1E1} + \Delta x) = \sin(x_{E1} + 2\Delta x)$$
$$x_{3E1} = x_{2E1} + \Delta x_3 = x_{1E1} + 2\Delta x = x_{E1} + 3\Delta x$$
$$y_{3E1} = y_{2E1} + \Delta y_{3E1} = y_{1E1} + \Delta y_{2E1} + \Delta y_{3E1} = \sin(x_{2E1} + \Delta x) = \sin(x_{1E1} + 2\Delta x) = \sin(x_{E1} + 3\Delta x)$$
$$\ldots \; \ldots$$
$$x_{nE1} = x_{(n-1)E1} + \Delta x = x_{E1} + \sum_{i=1}^{n} \Delta x = x_{E1} + n\Delta x$$
$$y_{nE1} = \sin\left(x_{E1} + \sum_{i=1}^{n} \Delta x\right) = \sin(x_{E1} + n\Delta x).$$

At this point, the changes in the dependent variable $y = f(x)$ are $\Delta y_{1E1}, \Delta y_{2E1}, \ldots, \Delta y_{(n-1)E1}, \Delta y_{nE1}$ and inequalities for these changes are as follows: $\Delta y_{1E1} \geqslant \Delta y_{2E1} \geqslant \ldots . \geqslant \Delta y_{(n-1)E1} \geqslant \Delta y_{nE1}$ and $|\Delta y_{1E1}| \leqslant |\Delta y_{2E1}| \leqslant \ldots \leqslant |\Delta y_{(n-1)E1}| \leqslant |\Delta y_{nE1}|$. So, the velocities of change can be identified as follows:

$$\frac{\Delta y_{1E1} - \Delta y_{2E1}}{\Delta x} \geqslant \frac{\Delta y_{2E1} - \Delta y_{3E1}}{\Delta x} \geqslant \ldots . \geqslant \frac{\Delta y_{(n-2)E1} - \Delta y_{(n-1)E1}}{\Delta x} \geqslant \frac{\Delta y_{(n-1)E1} - \Delta y_{nE1}}{\Delta x} \geqslant 0$$

(c) The same argument can be done for the second closed interval $\Delta x \in I3 = [\text{Inf2}, E2]$, and $y = f(x) = \sin(x_{\text{inf2}})$.

$x_{1\text{inf2}} = x_{\text{inf2}} + \Delta x$

$y_{1\text{inf2}} = y_{\text{inf2}} + \Delta y_{1\text{inf2}} = \sin(x_{\text{inf2}} + \Delta x)$

$x_{2\text{inf2}} = x_{1\text{inf2}} + \Delta x_2 = x_{\text{inf2}} + 2\Delta x$

$y_{2\text{inf2}} = y_{1\text{inf2}} + \Delta y_{2\text{inf2}} = y_{\text{inf2}} + \Delta y_{1\text{inf2}} + \Delta y_{2\text{inf2}} = \sin(x_{1\text{inf2}} + \Delta x) = \sin(x_{\text{inf2}} + 2\Delta x)$

$x_{3\text{inf2}} = x_{2\text{inf2}} + \Delta x = x_{1\text{inf2}} + 2\Delta x = x_{\text{inf2}} + 3\Delta x$

$y_{3\text{inf2}} = y_{2\text{inf2}} + \Delta y_{3\text{inf2}} = y_{1\text{inf2}} + \Delta y_{2\text{inf2}} + \Delta y_{3\text{inf2}} = \sin(x_{2\text{inf2}} + \Delta x) = \sin(x_{1\text{inf2}} + 2\Delta x) = \sin(x_{inf2} + 3\Delta x)$

… …

$$x_{n\text{inf2}} = x_{(n\text{-}1)\text{inf2}} + \Delta x = x_{\text{inf2}} + \sum_{i=1}^{n} \Delta x = x_{\text{inf2}} + n\Delta x$$

$$y_{n\text{inf2}} = \sin\left(x_{\text{inf2}} + \sum_{i=1}^{n} \Delta x\right) = \sin\left(x_{\text{inf2}} + n\Delta x\right).$$

At this point, the changes in the dependent variable $y = f(x)$ are $\Delta y_{1\text{inf2}}, \Delta y_{2\text{inf2}}, \ldots, \Delta y_{(n\text{-}1)\text{inf2}}, \Delta y_{n\text{inf2}}$ and inequalities for these changes are as follows: $\Delta y_{1\text{inf2}} \leqslant \Delta y_{2\text{inf2}} \leqslant \ldots \leqslant \Delta y_{(n\text{-}1)\text{inf2}} \leqslant \Delta y_{n\text{inf2}}$ and $|\Delta y_{1\text{inf2}}| \geqslant |\Delta y_{2\text{inf2}}| \geqslant \ldots \geqslant |\Delta y_{(n\text{-}1)\text{inf2}}| \geqslant |\Delta y_{n\text{inf2}}|$. So, the velocities of change can be identified as follows:

$$\frac{\Delta y_{1\text{inf2}} - \Delta y_{2\text{inf2}}}{\Delta x} \leqslant \frac{\Delta y_{2\text{inf2}} - \Delta y_{3\text{inf2}}}{\Delta x} \leqslant .... \leqslant \frac{\Delta y_{(n-2)\text{inf2}} - \Delta y_{(n-1)\text{inf2}}}{\Delta x} \leqslant \frac{\Delta y_{(n-1)\text{inf2}} - \Delta y_{n\text{inf2}}}{\Delta x} \leqslant 0$$

(d) The same argument can be made for the second closed interval $\Delta x \in I4 = [E2, \text{Inf2}]$, and $y = f(x) = \sin(x_{E2})$.

$x_{1E2} = x_{E2} + \Delta x$

$y_{1E2} = y_{E2} + \Delta y_{1E2} = \sin(x_{E2} + \Delta x)$

$x_{2E2} = x_{1E2} + \Delta x = x_{E2} + 2\Delta x$

$y_{2E2} = y_{1E2} + \Delta y_{2E2} = y_{E2} + \Delta y_{1E2} + \Delta y_{2E2} = \sin(x_{1E2} + \Delta x) = \sin(x_{E2} + 2\Delta x)$

$x_{3E2} = x_{2E2} + \Delta x = x_{1E2} + 2\Delta x = x_{E2} + 3\Delta x$

$y_{3E2} = y_{2E2} + \Delta y_{3E2} = y_{1E2} + \Delta y_{2E2} + \Delta y_{3E2} = \sin(x_{2E2} + \Delta x) = \sin(x_{1E2} + 2\Delta x) = \sin(x_{E2} + 3\Delta x)$

… …

$$x_{nE2} = x_{(n\text{-}1)E2} + \Delta x = x_{E2} + \sum_{i=1}^{n} \Delta x = x_{E2} + n\Delta x$$

$$y_{nE2} = \sin\left(x_{E2} + \sum_{i=1}^{n} \Delta x\right) = \sin\left(x_{E2} + n\Delta x\right).$$

At this point, the changes in the dependent variable $y = f(x)$ are $\Delta y_{1E2}, \Delta y_{2E2}, \ldots, \Delta y_{(n\text{-}1)E2}, \Delta y_{nE2}$ and inequalities for these changes are as follows: $\Delta y_{1E2} \geqslant \Delta y_{2E2} \geqslant \ldots \geqslant \Delta y_{(n\text{-}1)E2} \geqslant \Delta y_{nE2}$ and $|\Delta y_{1E2}| \geqslant |\Delta y_{2E2}| \geqslant \ldots \geqslant |\Delta y_{(n\text{-}1)E2}| \geqslant |\Delta y_{nE2}|$. So, the velocities of change can be identified as follows:

$$\frac{\Delta y_{1E2} - \Delta y_{2E2}}{\Delta x} \geqslant \frac{\Delta y_{2E2} - \Delta y_{3E2}}{\Delta x} \geqslant .... \geqslant \frac{\Delta y_{(n-2)E2} - \Delta y_{(n-1)E2}}{\Delta x} \geqslant \frac{\Delta y_{(n-1)E2} - \Delta y_{nE2}}{\Delta x} \geqslant 0$$

Similar arguments can be made for one period of a cosine function; this period was divided into four closed intervals as follows:

$$I_1 = [E_1, \text{Inf}_1] = \left[0, \frac{\pi}{2}\right] \; I_2 = [\text{Inf}_1, E_2] = \left[\frac{\pi}{2}, \pi\right] \; I_3 = [E_2, \text{Inf}_2] = \left[\pi, \frac{3\pi}{2}\right] \text{ and}$$

$$I_4 = [\text{Inf}_2, E_3] = \left[\frac{3\pi}{2}, 2\pi\right]$$

The very small increments/decrements in independent variable x can be $\Delta x_1 = \Delta x_2 = \ldots = \Delta x_{n\text{-}1} = \Delta x_n = \Delta x$. First of all, the changes in the dependent variable $y = f(x) = \cos(x)$ can be examined for the closed interval I_1. At this point, the changes in the dependent variable $y = f(x)$ are $\Delta y_{1E1}, \Delta y_{2E1}$,

$\ldots$, $\Delta y_{(n-1)E1}$, Δy_{nE1} and inequalities for these changes are as follows: $\Delta y_{1E1} \geqslant \Delta y_{2E1} \geqslant \ldots \geqslant \Delta y_{(n-1)E1} \geqslant \Delta y_{nE1}$ and $|\Delta y_{1E1}| \leqslant |\Delta y_{2E1}| \leqslant \ldots \leqslant |\Delta y_{(n-1)E1}| \leqslant |\Delta y_{nE1}|$. So, the velocities of change can be identified as follows:

$$\frac{\Delta y_{1E1} - \Delta y_{2E1}}{\Delta x} \leqslant \frac{\Delta y_{2E1} - \Delta y_{3E1}}{\Delta x} \leqslant .... \leqslant \frac{\Delta y_{(n-2)E1} - \Delta y_{(n-1)E1}}{\Delta x} \leqslant \frac{\Delta y_{(n-1)E1} - \Delta y_{nE1}}{\Delta x} \geqslant 0$$

The very small increments/decrements in independent variable x can be $\Delta x_1 = \Delta x_2 = \ldots. = \Delta x_{n-1} = \Delta x_n = \Delta x$. The changes in the dependent variable $y = f(x) = \cos(x)$ can be examined for the closed interval I_2. At this point, the changes in the dependent variable $y = f(x)$ are Δy_{1inf1}, Δy_{2inf1}, $\ldots$, $\Delta y_{(n-1)inf1}$, Δy_{ninf1} and inequalities for these changes are as follows: $\Delta y_{1inf1} \leqslant \Delta y_{2inf1} \leqslant \ldots \leqslant \Delta y_{(n-1)inf1} \leqslant \Delta y_{ninf1}$ and $|\Delta y_{1inf1}| \geqslant |\Delta y_{2inf1}| \geqslant \ldots \geqslant |\Delta y_{(n-1)inf1}| \geqslant |\Delta y_{ninf1}|$. So, the velocities of change can be identified as follows:

$$\frac{\Delta y_{1inf1} - \Delta y_{2inf1}}{\Delta x} \leqslant \frac{\Delta y_{2inf1} - \Delta y_{3inf1}}{\Delta x} \leqslant .... \leqslant \frac{\Delta y_{(n-2)inf1} - \Delta y_{(n-1)inf1}}{\Delta x} \leqslant \frac{\Delta y_{(n-1)inf1} - \Delta y_{ninf1}}{\Delta x} \leqslant 0$$

The changes in the dependent variable $y = f(x) = \cos(x)$ can be examined for the closed interval I_3. At this point, the changes in the dependent variable $y = f(x)$ are Δy_{1E2}, Δy_{2E2}, $\ldots$., $\Delta y_{(n-1)E2}$, $\leqslant \Delta y_{nE2}$ and inequalities for these changes are as follows: $\Delta y_{1E2} \leqslant \Delta y_{2E2} \leqslant \ldots \leqslant \Delta y_{(n-1)E2} \leqslant \Delta y_{nE2}$ and $|\Delta y_{1E2}| \leqslant |\Delta y_{2E2}| \leqslant \ldots \leqslant |\Delta y_{(n-1)E2}| \leqslant |\Delta y_{nE2}|$. So, the velocities of change can be identified as follows:

$$\frac{\Delta y_{1E2} - \Delta y_{2E2}}{\Delta x} \leqslant \frac{\Delta y_{2E2} - \Delta y_{3E2}}{\Delta x} \leqslant .... \leqslant \frac{\Delta y_{(n-2)E2} - \Delta y_{(n-1)E2}}{\Delta x} \leqslant \frac{\Delta y_{(n-1)E2} - \Delta y_{nE2}}{\Delta x} \leqslant 0$$

The changes in the dependent variable $y = f(x) = \cos(x)$ can be examined for the closed interval I_4. At this point, the changes in the dependent variable $y = f(x)$ are Δy_{1inf2}, Δy_{2inf2}, $\ldots$, $\Delta y_{(n-1)inf2}$, Δy_{ninf2} and inequalities for these changes are as follows: $\Delta y_{1inf2} \geqslant \Delta y_{2inf2} \geqslant \ldots \geqslant \Delta y_{(n-1)inf2} \geqslant \Delta y_{ninf2}$ and $|\Delta y_{1inf2}| \geqslant |\Delta y_{2inf2}| \geqslant \ldots \geqslant |\Delta y_{(n-1)inf2}| \geqslant |\Delta y_{ninf2}|$. So, the velocities of change can be identified as follows:

$$\frac{\Delta y_{1inf2} - \Delta y_{2inf2}}{\Delta x} \geqslant \frac{\Delta y_{2inf2} - \Delta y_{3inf2}}{\Delta x} \geqslant .... \geqslant \frac{\Delta y_{(n-2)inf2} - \Delta y_{(n-1)inf2}}{\Delta x} \geqslant \frac{\Delta y_{(n-1)inf2} - \Delta y_{ninf2}}{\Delta x} \geqslant 0$$

3. Applications of Rational/Irrational Orders

The theoretical information given in Section 2 can be verified by applications of trigonometric and polynomial functions. To this end, sine, cosine, and increasing and decreasing polynomial functions are selected. These functions are $\sin(x)$ for $x \in [0, 2\pi]$, $\cos(x)$ for $x \in [0, 2\pi]$, $x^3 - 7x^2$ for $x \in [-1000, 1000]$ and $-x^3 - 7x^2$ for $x \in [-1000, 1000]$.

Figures 2 and 3 show the rates of changes for sine functions and the obtained results support the claims of Figure 1. The red circles in both figures depict the inflection points. Figures 4 and 5 depict the same cases for the cosine function. In the case of Figures 3 and 5 while power is equal to 1, the rate of change is the same as an ordinary derivative. The domains of sine and cosine functions were determined as a period of functions. The inflection points for sine function are {(0, 0), (0, π), (0, 2π)} and large changes occur at these points for sine function (Figures 2 and 3). The inflection points for cosine function are {(π/2, 0), (3π/2, 0} and large changes occur at these points for cosine function (Figures 4 and 5).

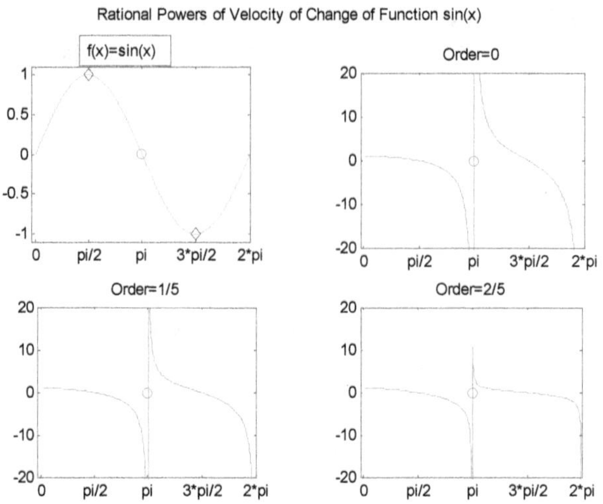

Figure 2. The rational orders of rates of change for the sine function (Orders are 0, 1/5, 2/5).

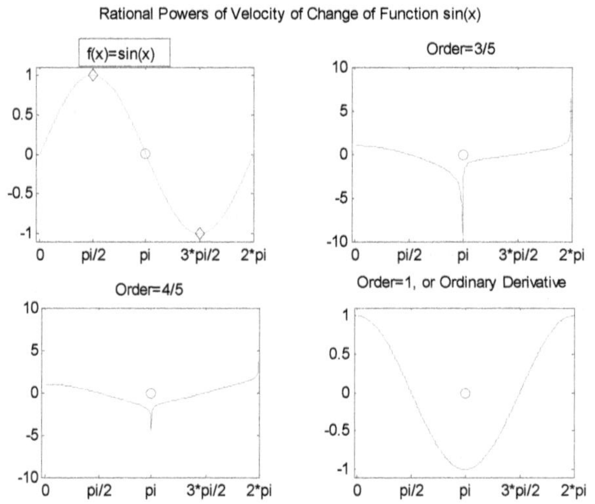

Figure 3. The rational orders of rates of change for the sine function (Orders are 3/5, 4/5, 1).

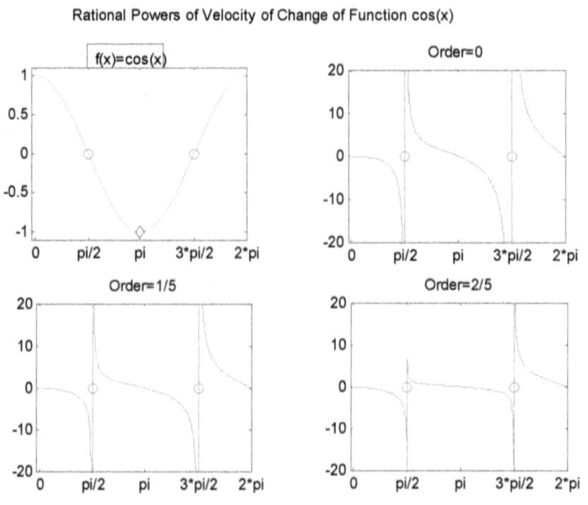

Figure 4. The rational orders of rates of change for the cosine function (Orders are 0, 1/5, 2/5).

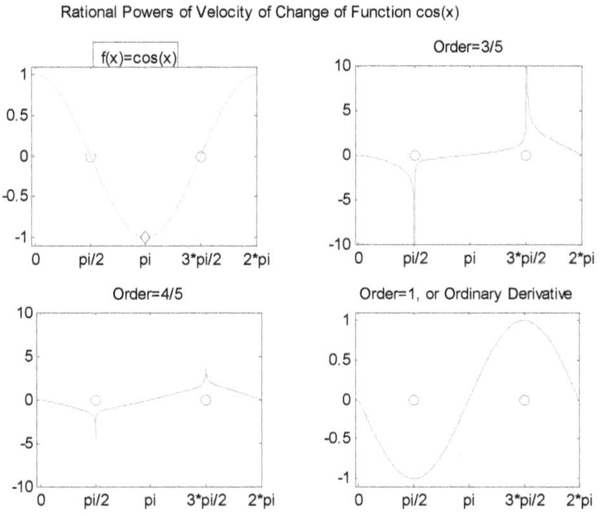

Figure 5. The rational orders of rates of change for the cosine function (Orders are 3/5, 4/5, 1).

The same idea can be argued for polynomial functions, specifically functions $f(x) = x^3 - 7x^2$ for $x \in [-1000, 1000]$ and $f(x) = -x^3 - 7x^2$ for $x \in [-1000, 1000]$. The domains for these functions were selected to cover the inflection points and extremum points. The inflection point for $f(x) = x^3 - 7x^2$ is (7/3, 1372/27); it is illustrated by a red circle. The inflection point for $f(x) = -x^3 - 7x^2$ is (−7/3, −686/27); it is also illustrated by a red circle. The black diamond points are extremum points for both functions. The approach in Section 2 was verified by applications for polynomials as seen in Figures 6–9.

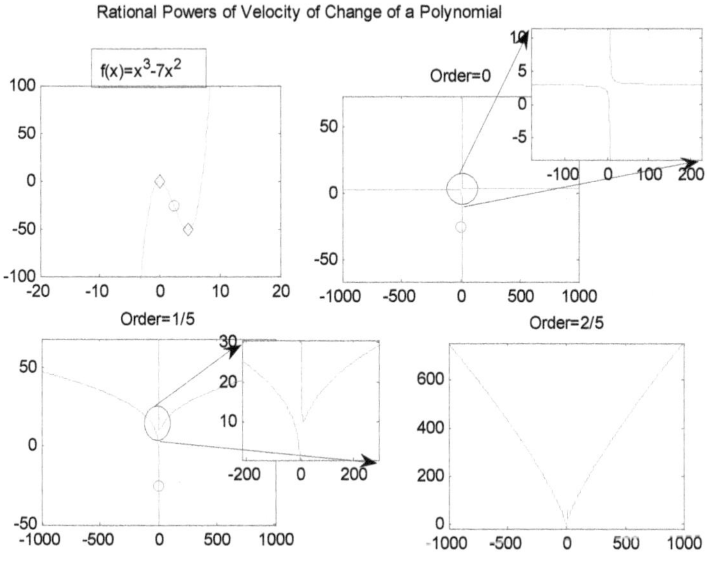

Figure 6. The rates of change for $f(x) = x^3 - 7x^2$ orders {0, 1/5, 2/5}.

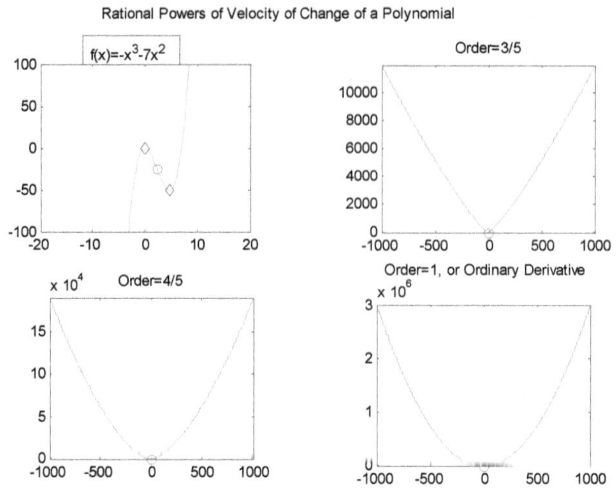

Figure 7. The rates of change for $f(x) = x^3 - 7x^2$ orders $\{3/5, 4/5, 1\}$.

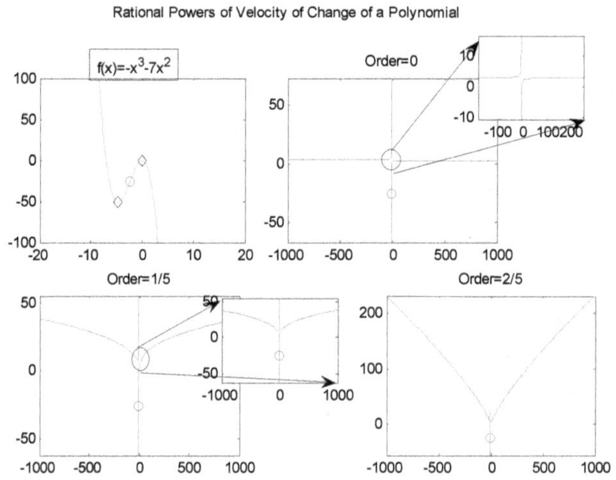

Figure 8. The rates of change for $f(x) = -x^3 - 7x^2$ orders $\{0, 1/5, 2/5\}$.

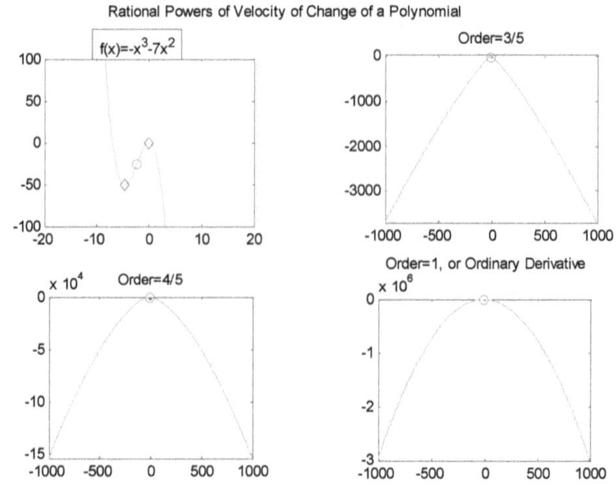

Figure 9. The rates of change for $f(x) = x^3 - 7x^2$ orders $\{3/5, 4/5, 1\}$.

A similar case can be considered for negative order, and to this end, sine function was selected for illustration. Figure 10 illustrates that in the case of negative orders, the same comments can be made for fractional order derivatives. The negative order reverses the direction of the increase/decrease, so the inequalities in Section 2 can be rephrased.

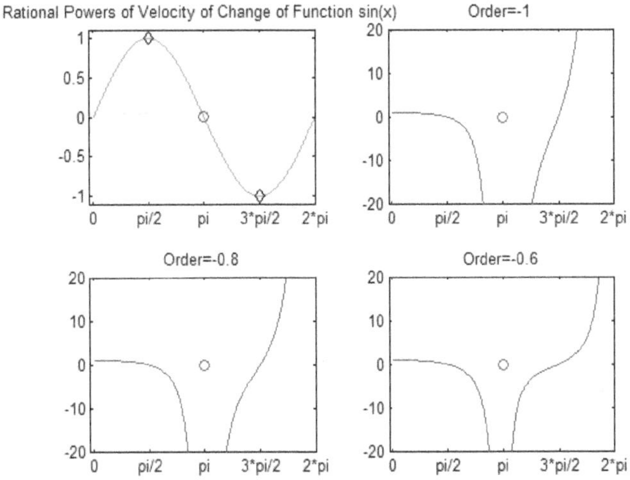

Figure 10. The rates of change for $f(x) = \sin(x)$ for orders $\{-1, -0.8, -0.6\}$.

4. Analytical Approach and Results

The meaning of derivative is the rate of change or velocity of change in the dependent variable *versus* the changes in the independent variables [23–25]. Thus, the derivative of $f(x) = cx^0$ is:

$$\lim_{h \to 0} \frac{f(x+h) - f(x)}{(x+h) - x} = \lim_{h \to 0} \frac{c - c}{h} = 0$$

In the case of an identity function, it is:

$$\lim_{h \to 0} \frac{f(x+h) - f(x)}{(x+h) - x} = \lim_{h \to 0} \frac{(x+h) - x}{(x+h) - x} = 1$$

So, the definition for rational/irrational order derivative can be considered as follows.

Definition 1. *f(x): R → R is a function, α ∈ R and the rational/irrational order derivative can be considered as follows:*

$$f^{(\alpha)}(x) = \lim_{h \to 0} \frac{f^{\alpha}(x+h) - f^{\alpha}(x)}{(x+h)^{\alpha} - x^{\alpha}} \tag{1}$$

Before handling applications, the new definitions for rational/irrational order derivatives must be rephrased. Definition 1 is a classical definition of derivative, and it has an indefinite limit such that $\frac{0}{0}$, while $h = 0$. In this case, Definition 1 can be rephrased as seen in Definition 2.

Definition 2. *Assume that f(x): R → R is a function, α ∈ R and L(.) is a l'Hôpital process. The rational/irrational order derivative of f(x) is:*

$$f^{(\alpha)}(x) = \lim_{h \to 0} L\left(\frac{f^{\alpha}(x+h) - f^{\alpha}(x)}{(x+h)^{\alpha} - x^{\alpha}}\right) = \lim_{h \to 0} \frac{\frac{d(f^{\alpha}(x+h) - f^{\alpha}(x))}{dh}}{\frac{d((x+h)^{\alpha} - x^{\alpha})}{dh}} = \frac{f'(x) f^{\alpha-1}(x)}{x^{\alpha-1}} \tag{2}$$

The new definition of rational/irrational order derivative (Definition 2) can be applied to some specific functions such as constant, identity, sine, and cosine functions. The velocity of change for a constant function is zero whatever the independent variable changes. The velocity of change for an identity function is 1 whatever the order of derivative and change of independent variable. The derivatives of sine, cosine, and polynomial functions for different orders are illustrated in Figures 2–9.

These results demonstrated that the new definition obeys velocities of change of functions in extremum points, inflection points, *etc.*

The comments and descriptions in Section 2 can be supported with analytical results. The velocities of change of any functions can be expressed by derivatives. The point of this study is to analyze the powers of derivative with respect to all real powers.

Theorem 4.1. *Assume that f(x) is any function such that f(x): R → R, and* $\alpha_1, \alpha_2, \ldots, \alpha_n \in R^+$. *Let f(x) be a positive, monotonically increasing/decreasing function in which the following conditions hold:*

(a) *If* $\alpha_1, \alpha_2, \ldots, \alpha_n \in R^+$, $\alpha_1 \leqslant \alpha_2 \leqslant \ldots \leqslant \alpha_n$ *and f(x) is positive and monotonically increasing, then* $f^{(\alpha_1)}(x) \leqslant f^{(\alpha_2)}(x) \leqslant \ldots\ldots \leqslant f^{(\alpha_{n-1})}(x) \leqslant f^{(\alpha_n)}(x)$.

(b) *If* $\alpha_1, \alpha_2, \ldots, \alpha_n \in R^+$, $\alpha_1 \leqslant \alpha_2 \leqslant \ldots \leqslant \alpha_n$, *and f(x) is positive and monotonically decreasing, then* $f^{(\alpha_1)}(x) \geqslant f^{(\alpha_2)}(x) \geqslant \ldots\ldots \geqslant f^{(\alpha_{n-1})}(x) \geqslant f^{(\alpha_n)}(x)$.

Proof.

(a) $f(x)$ is a monotonically increasing function, so $f(x_i) \leqslant f(x_j)$ for $x_i \leqslant x_j$, $\forall x_i, x_j \in R$. Then $f^{\alpha_i}(x_k+h) \leqslant f^{\alpha_j}(x_k+h)$ and $f^{\alpha_i}(x_k) \leqslant f^{\alpha_j}(x_k)$ where $1 \leqslant i < j \leqslant n$, k is any index, α_i, α_j are two constants and $\alpha_i \leqslant \alpha_j$. This case implies that $f^{(\alpha_i)}(x_k+h) - f^{(\alpha_i)}(x_k) \leqslant f^{\alpha_j}(x_k+h) - f^{(\alpha_j)}(x_k)$ and $(x_k+h)^{\alpha_i} - x_k^{\alpha_i} \leqslant (x_k+h)^{\alpha_j} - x_k^{\alpha_j}$. So, this implies the following quotient:

$$\lim_{h \to 0} L\left(\frac{f^{\alpha_i}(x_k+h) - f^{\alpha_i}(x_k)}{(x_k+h)^{\alpha_i} - x_k^{\alpha_i}}\right) \leqslant \lim_{h \to 0} L\left(\frac{f^{\alpha_j}(x_k+h) - f^{\alpha_j}(x_k)}{(x_k+h)^{\alpha_j} - x_k^{\alpha_j}}\right)$$

This implies the following important relations.

Step 1. Assuming that $n = 2$, there are two rational/irrational powers of functions such as α_1, $\alpha_2 \in R^+$, $\alpha_1 \leqslant \alpha_2$. Then:

$$\lim_{h \to 0} L\left(\frac{f^{\alpha_1}(x_k+h) - f^{\alpha_1}(x_k)}{(x_k+h)^{\alpha_1} - x_k^{\alpha_1}}\right) \leqslant \lim_{h \to 0} L\left(\frac{f^{\alpha_2}(x_k+h) - f^{\alpha_2}(x_k)}{(x_k+h)^{\alpha_2} - x_k^{\alpha_2}}\right)$$

Step 2. Assuming that there are $n-1$ rational/irrational powers of rates of change, there are $n-1$ rational/irrational powers of functions such as $\alpha_1, \alpha_2, \ldots, \alpha_{n-1} \in R^+$, $\alpha_1 \leqslant \alpha_2 \ldots \leqslant \alpha_{n-1}$. Then:

$$\lim_{h \to 0} L\left(\frac{f^{\alpha_1}(x_k+h) - f^{\alpha_1}(x_k)}{(x_k+h)^{\alpha_1} - x_k^{\alpha_1}}\right) \leqslant \lim_{h \to 0} L\left(\frac{f^{\alpha_2}(x_k+h) - f^{\alpha_2}(x_k)}{(x_k+h)^{\alpha_2} - x_k^{\alpha_2}}\right) \leqslant \ldots \leqslant \lim_{h \to 0} L\left(\frac{f^{\alpha_{n-1}}(x_k+h) - f^{\alpha_{n-1}}(x_k)}{(x_k+h)^{\alpha_{n-1}} - x_k^{\alpha_{n-1}}}\right)$$

Step 3. Assuming that $\alpha_1, \alpha_2, \ldots, \alpha_n \in R^+$, $\alpha_1 \leqslant \alpha_2 \leqslant \ldots \leqslant \alpha_n$, the following inequality holds:

$$f^{(n-1)}(x) = \lim_{h \to 0} L\left(\frac{f^{\alpha_{n-1}}(x_k+h) - f^{\alpha_{n-1}}(x_k)}{(x_k+h)^{\alpha_{n-1}} - x_k^{\alpha_{n-1}}}\right) \leqslant \lim_{h \to 0} L\left(\frac{f^{\alpha_n}(x_k+h) - f^{\alpha_n}(x_k)}{(x_k+h)^{\alpha_n} - x_k^{\alpha_n}}\right) = f^{(n)}(x).$$

This means that $f^{(\alpha_{n-1})}(x) \leqslant f^{(\alpha_n)}(x)$ and $f^{(\alpha_i)}(x) \leqslant f^{(\alpha_n)}(x)$ for $1 \leqslant i \leqslant n-1$.

(b) $f(x)$ is a monotonically decreasing function, so $f(x_i) \geqslant f(x_j)$ for $x_i \leqslant x_j$, $\forall x_i, x_j \in R$. Then $f^{\alpha_i}(x_k+h) \geqslant f^{\alpha_j}(x_k+h)$ and $f^{\alpha_i}(x_k) \geqslant f^{\alpha_j}(x_k)$ where $1 \leqslant i < j \leqslant n$, k is any index, α_i, α_j are two constants and $\alpha_i \leqslant \alpha_j$. This case implies that $f^{(\alpha_i)}(x_k+h) - f^{(\alpha_i)}(x_k) \geqslant f^{\alpha_j}(x_k+h) - f^{(\alpha_j)}(x_k)$ and $(x_k+h)^{\alpha_i} - x_k^{\alpha_i} \leqslant (x_k+h)^{\alpha_j} - x_k^{\alpha_j}$. So, this implies the following quotient:

$$\lim_{h \to 0} L\left(\frac{f^{\alpha_i}(x_k+h) - f^{\alpha_i}(x_k)}{(x_k+h)^{\alpha_i} - x_k^{\alpha_i}}\right) \geqslant \lim_{h \to 0} L\left(\frac{f^{\alpha_j}(x_k+h) - f^{\alpha_j}(x_k)}{(x_k+h)^{\alpha_j} - x_k^{\alpha_j}}\right)$$

This implies the following important relations.

Step 1. Assuming that $n = 2$, there are two rational/irrational powers of functions such as α_1, $\alpha_2 \in R^+$, $\alpha_1 \leqslant \alpha_2$. Then:

$$\lim_{h \to 0} L\left(\frac{f^{\alpha_1}(x_k+h) - f^{\alpha_1}(x_k)}{(x_k+h)^{\alpha_1} - x_k^{\alpha_1}}\right) \geqslant \lim_{h \to 0} L\left(\frac{f^{\alpha_2}(x_k+h) - f^{\alpha_2}(x_k)}{(x_k+h)^{\alpha_2} - x_k^{\alpha_2}}\right)$$

Step 2. Assuming that there are $n - 1$ rational/irrational powers of rates of changes, then there are $n - 1$ rational/irrational powers of functions such as $\alpha_1, \alpha_2, \ldots, \alpha_{n-1} \in R^+$, $\alpha_1 \leqslant \alpha_2 \ldots \leqslant \alpha_{n-1}$. Then:

$$\lim_{h \to 0} L\left(\frac{f^{\alpha_1}(x_k+h) - f^{\alpha_1}(x_k)}{(x_k+h)^{\alpha_1} - x_k^{\alpha_1}}\right) \leqslant \lim_{h \to 0} L\left(\frac{f^{\alpha_2}(x_k+h) - f^{\alpha_2}(x_k)}{(x_k+h)^{\alpha_2} - x_k^{\alpha_2}}\right) \leqslant \ldots \leqslant \lim_{h \to 0} L\left(\frac{f^{\alpha_{n-1}}(x_k+h) - f^{\alpha_{n-1}}(x_k)}{(x_k+h)^{\alpha_{n-1}} - x_k^{\alpha_{n-1}}}\right)$$

Step 3. Assuming that $\alpha_1, \alpha_2, \ldots, \alpha_n \in R^+$, $\alpha_1 \leqslant \alpha_2 \leqslant \ldots \leqslant \alpha_n$, the following inequality holds:

$$f^{(n-1)}(x) = \lim_{h \to 0} L\left(\frac{f^{\alpha_{n-1}}(x_k+h) - f^{\alpha_{n-1}}(x_k)}{(x_k+h)^{\alpha_{n-1}} - x_k^{\alpha_{n-1}}}\right) \leqslant \lim_{h \to 0} L\left(\frac{f^{\alpha_n}(x_k+h) - f^{\alpha_n}(x_k)}{(x_k+h)^{\alpha_n} - x_k^{\alpha_n}}\right) = f^{(n)}(x)$$

This means that $f^{(\alpha_{n-1})}(x) \leqslant f^{(\alpha_n)}(x)$ and $f^{(\alpha_i)}(x) \leqslant f^{(\alpha_n)}(x)$for $1 \leqslant i \leqslant n-1$.

Theorem 4.2. *Assume that f(x) is any function such that f(x): R → R, and $\alpha_1, \alpha_2, \ldots, \alpha_n \in R^-$. Let f(x) be a positive, monotonically increasing/decreasing function in which the following conditions hold:*

(a) If $\alpha_1, \alpha_2, \ldots, \alpha_n \in R^-$, $\alpha_1 \geqslant \alpha_2 \geqslant \ldots \geqslant \alpha_n$ and f(x) is positive and monotonically increasing, then $f^{(\alpha_1)}(x) \geqslant f^{(\alpha_2)}(x) \geqslant \ldots \geqslant f^{(\alpha_{n-1})}(x) \geqslant f^{(\alpha_n)}(x)$.

(b) If $\alpha_1, \alpha_2, \ldots, \alpha_n \in R^-$, $\alpha_1 \geqslant \alpha_2 \geqslant \ldots \geqslant \alpha_n$, and f(x) is positive and monotonically decreasing, then $f^{(\alpha_1)}(x) \geqslant f^{(\alpha_2)}(x) \geqslant \ldots\ldots \geqslant f^{(\alpha_{n-1})}(x) \geqslant f^{(\alpha_n)}(x)$.

Proof. The proof can be handled in two steps: increasing function and decreasing function.

(a) $f(x)$ is a positive, monotonically increasing function, so $f(x_i) \leqslant f(x_j)$ for $x_i \leqslant x_j$, $\forall x_i, x_j \in R$, and α_1, $\alpha_2 \in R^-$ and $\alpha_1 \leqslant \alpha_2$. Then $f^{\alpha_1}(x+h) \geqslant f^{\alpha_2}(x+h)$ and $f^{\alpha_1}(x) \geqslant f^{\alpha_2}(x)$, $|\alpha_1| \geqslant |\alpha_2|$.

$$A = \lim_{h \to 0} L\left(\frac{f^{\alpha_1}(x_i+h) - f^{\alpha_1}(x_i)}{(x_i+h)^{\alpha_1} - x_i^{\alpha_1}}\right) = \frac{\frac{d}{dh}\frac{(f^{|\alpha_1|}(x) - f^{|\alpha_1|}(x+h))}{f^{|\alpha_1|}(x) f^{|\alpha_1|}(x+h)}}{\frac{d}{dh}\frac{x^{|\alpha_1|} - (x+h)^{|\alpha_1|}}{x^{|\alpha_1|}(x+h)^{|\alpha_1|}}} = \frac{f'(x)\,(x)^{|\alpha_1|+1}}{f^{|\alpha_1|+1}(x)}$$

and,

$$B = \lim_{h \to 0} L\left(\frac{f^{\alpha_2}(x_i+h) - f^{\alpha_2}(x_i)}{(x_i+h)^{\alpha_2} - x_i^{\alpha_2}}\right) = \frac{\frac{d}{dh}\frac{(f^{|\alpha_2|}(x) - f^{|\alpha_2|}(x+h))}{f^{|\alpha_2|}(x) f^{|\alpha_2|}(x+h)}}{\frac{d}{dh}\frac{x^{|\alpha_2|} - (x+h)^{|\alpha_2|}}{x^{|\alpha_2|}(x+h)^{|\alpha_2|}}} = \frac{f'(x)\,(x)^{|\alpha_2|+1}}{f^{|\alpha_2|+1}(x)}$$

where $f^{|\alpha_1|+1}(x) \leqslant f^{|\alpha_2|+1}(x)$ and $x^{|\alpha_1|+1} \leqslant x^{|\alpha_2|+1}$. This case implies the following inequality:

$$A \geqslant B \Rightarrow \frac{f'(x)\,(x)^{|\alpha_1|+1}}{f^{|\alpha_1|+1}(x)} \geqslant \frac{f'(x)\,(x)^{|\alpha_2|+1}}{f^{|\alpha_2|+1}(x)} \Rightarrow \frac{(x)^{|\alpha_1|+1}}{f^{|\alpha_1|+1}(x)} \geqslant \frac{(x)^{|\alpha_2|+1}}{f^{|\alpha_2|+1}(x)}$$

This case is for two orders, and it can be enlarged to other orders. For $\alpha_1 \geqslant \alpha_2 \geqslant \ldots \geqslant \alpha_n$ and $|\alpha_1| \leqslant |\alpha_2| \leqslant \ldots \leqslant |\alpha_n|$,

$$\frac{f'(x)\,(x)^{|\alpha_1|+1}}{f^{|\alpha_1|+1}(x)} \geqslant \frac{f'(x)\,(x)^{|\alpha_2|+1}}{f^{|\alpha_2|+1}(x)} \geqslant \ldots \geqslant \frac{(x)^{|\alpha_{n-1}|+1}}{f^{|\alpha_{n-1}|+1}(x)} \geqslant \frac{(x)^{|\alpha_n|+1}}{f^{|\alpha_n|+1}(x)}$$

(b) $g(x)$ is a positive, monotonically decreasing function, so $g(x_i) \geqslant g(x_j)$ for $x_i \leqslant x_j$, $\forall x_i, x_j \in R$, and $\alpha_1, \alpha_2 \in R^-$ and $\alpha_1 \leqslant \alpha_2$. Then $g^{\alpha_1}(x+h) \leqslant g^{\alpha_2}(x+h)$ and $g^{\alpha_1}(x) \leqslant g^{\alpha_2}(x)$, $|\alpha_1| \geqslant |\alpha_2|$. Assuming that $f(x)$ is a positive, monotonically increasing function, $g(x) = \frac{1}{f(x)}$ is a positive, monotonically decreasing function:

$$A = \lim_{h\to 0} L\left(\frac{g^{\alpha_1}(x_i+h) - g^{\alpha_1}(x_i)}{(x_i+h)^{\alpha_1} - x_i^{\alpha_1}}\right) = \frac{\frac{d}{dh}\left(f^{|\alpha_1|}(x+h) - f^{|\alpha_1|}(x)\right)}{\frac{d}{dh}\frac{x^{|\alpha_1|} - (x+h)^{|\alpha_1|}}{x^{|\alpha_1|}(x+h)^{|\alpha_1|}}} = -\frac{f'(x) f^{|\alpha_1|-1}(x)}{x^{3|\alpha_1|-1}}$$

and,

$$B = \lim_{h\to 0} L\left(\frac{g^{\alpha_2}(x_i+h) - g^{\alpha_2}(x_i)}{(x_i+h)^{\alpha_2} - x_i^{\alpha_2}}\right) = \frac{\frac{d}{dh}\left(f^{|\alpha_2|}(x+h) - f^{|\alpha_2|}(x)\right)}{\frac{d}{dh}\frac{x^{|\alpha_2|} - (x+h)^{|\alpha_2|}}{x^{|\alpha_2|}(x+h)^{|\alpha_2|}}} = -\frac{f'(x) f^{|\alpha_2|-1}(x)}{x^{3|\alpha_2|-1}}$$

where $f^{|\alpha_2|-1}(x) \leqslant f^{|\alpha_1|-1}(x)$ and $x^{3|\alpha_2|-1} \leqslant x^{3|\alpha_1|-1}$. This case implies the following inequality.

$$A \leqslant B \Rightarrow -\frac{f'(x) f^{|\alpha_1|-1}(x)}{x^{3|\alpha_1|-1}} \geqslant -\frac{f'(x) f^{|\alpha_2|-1}(x)}{x^{3|\alpha_2|-1}} \Rightarrow -\frac{f'(x) f^{|\alpha_1|-1}(x)}{x^{3|\alpha_1|-1}} \geqslant -\frac{f'(x) f^{|\alpha_2|-1}(x)}{x^{3|\alpha_2|-1}}$$

This case is for two orders, and it can be enlarged to other orders. For $\alpha_1 \geqslant \alpha_2 \geqslant \ldots \geqslant \alpha_n$ and $|\alpha_1| \leqslant |\alpha_2| \leqslant \ldots \leqslant |\alpha_n|$,

$$-\frac{f'(x)\,(x)^{|\alpha_1|-1}}{f^{|\alpha_1|-1}(x)} \geqslant \frac{f'(x)\,(x)^{|\alpha_2|-1}}{f^{|\alpha_2|-1}(x)} \geqslant \ldots \geqslant \frac{(x)^{|\alpha_{n-1}|-1}}{f^{|\alpha_{n-1}|-1}(x)} \geqslant \frac{(x)^{|\alpha_n|-1}}{f^{|\alpha_n|-1}(x)}$$

Any complex number has direction and magnitude. It is known that derivative has direction and magnitude. So the relationship between complex numbers and derivatives must be verified.

Theorem 4.3. *Assuming that f(x) is a function such as f: R → R and α ∈ R, $f^{(\alpha)}(x)$ is a function of complex variables.*

Proof. Assume that $\alpha = \frac{\beta}{\delta}$ and $\delta \neq 0$. If $f(x) \geqslant 0$, the obtained results are positive, and they constitute the real part of complex numbers. The rational/irrational order derivative of $f(x)$ is:

$$f^{(\alpha)} = \frac{f'(x) f^{\alpha-1}(x)}{x^{\alpha-1}} = f'(x)\sqrt[\delta]{\frac{f^{\beta-\delta}(x)}{x^{\beta-\delta}}} = f'(x)\sqrt[\delta]{\left(\frac{f(x)}{x}\right)^{\beta-\delta}}$$

If the rational/irrational derivative is a function of complex variables, then $f^{(\alpha)}(x) = g(x) + ih(x)$, where $i = \sqrt{-1}$.

If $f(x) < 0$, there will be two cases:

Case 1: Assume that δ is odd.

If $\left(\frac{f(x)}{x}\right)^{\beta-\delta} \geqslant 0$ or $\left(\frac{f(x)}{x}\right)^{\beta-\delta} < 0$,

then the obtained function $f^{(\alpha)}(x)$ is a real function and $h(x) = 0$ for both cases since the multiplication of any negative number in odd steps yields a negative number.

Case 2: Assume that δ is even.

If $\left(\frac{f(x)}{x}\right)^{\beta-\delta} \geqslant 0$, then $h(x) = 0$ and $f^{(\alpha)}(x)$ is a real function.

If $\left(\frac{f(x)}{x}\right)^{\beta-\delta} < 0$, then the multiplication of any number in even steps yields a positive number for real numbers. However, it yields a negative result for complex numbers, so, $h(x) \neq 0$. This means that $f^{(\alpha)}(x)$ is a complex function.

In fact, $f^{(\alpha)}(x)$ is a complex function for both cases because $h(x) = 0$ for some situations.

Problem 1. The derivative operator D is a linear operator. The theoretical and application results presented in the previous sections demonstrated that the derivative operator does not have to be linear. The ordinary derivative operator D is linear; however, the rate of change of any function *versus* its independent variables does not have to be linear. This is in need of theoretical verification.

Problem 2. A new approach for derivatives was expressed in this paper. This new approach has the potential to change the integration rules in the case of orders of derivatives different from 1. The new rules for integration should be defined.

Problem 3. The geometrical meanings of ordinary derivatives are known. The geometrical meanings of this new approach for derivatives in the case of rational and irrational orders of derivatives are still unknown.

5. Conclusions

The ordinary definition of a derivative from Newton and l'Hôpital can be considered as order = 1; in this case, the derivative operator D is linear. This paper includes a new approach for the derivative concept in which the derivative operator is not linear. The following consequences can be put forth:

(a) All functions have maximum rates of changes at inflection points in the positive direction.
(b) All functions have minimum rates of changes at inflection points in the negative direction.
(c) The derivative operator does not have to be linear, since the rates of changes of functions are not linear.
(d) This new approach needs to prove geometrical meaning for derivative orders different from 1.
(e) This new approach brought out a new problem: how to handle the integration in cases of derivative orders different from 1.
(f) This new approach reveals that derivative and complex numbers have relationships.

Conflicts of Interest: The author declares no conflict of interest.

References

1. Newton, I. *Philosophiæ Naturalis Principia Mathematica*; Jussu Societatis Regiae ac Typis Joseph Streater. Prostat apud plures bibliopolas: London, UK, 1687.
2. L'Hôpital, G. *Analyse des Infiniment Petits pour l'Intelligence des Lignes Courbes (Infinitesimal Calculus with Applications to Curved Lines)*; François Montalant: Paris, France, 1696.
3. L'Hôpital, G. *Analyse des Infinement Petits*; Relnk Books: Paris, France, 1715.
4. Goldenbaum, U.; Jesseph, D. *Infinitesimal Differences: Controversies between Leibniz and His Contemporaries*; Walter de Gruyter: New York, NY, USA, 2008.
5. Baron, M.E. *The Origin of the Infinitesimal Calculus*; Dover Publications: New York, NY, USA, 1969.
6. Wren, F.L.; Garrett, J.A. The development of the fundamental concepts of infinitesimal analysis. *Am. Math. Mon.* **1933**, *40*, 269–291. [CrossRef]
7. Bliss, G.A. The evolution of problems of the calculus of variations. *Am. Math. Mon.* **1936**, *43*, 598–609. [CrossRef]

8. Taylor, A.E. L'Hôpital rule. *Am. Math. Mon.* **1952**, *59*, 20–24. [CrossRef]
9. Stewart, J.K. Another variation of Newton's method. *Am. Math. Mon.* **1951**, *58*, 331–334. [CrossRef]
10. Leibniz, G.F. Correspondence with l'Hôpital. Personal letter, 1695.
11. Das, S. *Functional Fractional Calculus*; Springer-Verlag Berlin Heidelberg: Berlin, Germany, 2011.
12. Herrmann, R. *Fractional Calculus: An Introduction for Physicists*; World Scientific: GigaHedron, Germany, 2011.
13. Oldham, K.B.; Spanier, J. *The Fractional Calculus*; Academic Press: New York, NY, USA, 1974.
14. Samko, S.G.; Ross, B. Integration and differentiation to a variable fractional order. *Integral Transform. Spec. Funct.* **1993**, *1*, 277–300. [CrossRef]
15. Kiryakova, V.S. *Generalized Fractional Calculus and Applications*; Wiley and Sons: New York, NY, USA, 1994.
16. Samko, S.G.; Kilbas, A.A.; Marichev, O.I. *Fractional Integrals and Derivatives Translated from the 1987 Russian Original*; Gordon and Breach: Yverdon, Switzerland, 1993.
17. Rubin, B. Fractional integrals and potentials. In *Pitman Monographs and Surveys in Pure and Applied Mathematics*; Longman: Harlow, UK, 1996; Volume 82.
18. Gorenflo, R.; Mainardi, F. Fractional oscillations and Mittag-Leffler functions. In Proceedings of the RAAM 1996, Kuwait University, Kuwait, Kuwait, 4–7 May 1996; pp. 193–208.
19. Mainardi, F.; Gorenflo, R. On Mittag-Leffler-type functions in fractional evolution processes. *J. Comput. Appl. Math.* **2000**, *118*, 283–299. [CrossRef]
20. Mainardi, F. *Fractional Calculus and Waves in Linear Viscoelasticity: An Introduction to Mathematical Models*; World Scientific: Singapore, 2010.
21. Podlubny, I. *Fractional Differential Equations*; Academic Press: New York, NY, USA, 1999.
22. Podlubny, I. Geometric and physical interpretation of fractional integration and fractional differentiation. *Fract. Calc. Appl. Anal.* **2002**, *5*, 367–386.
23. Karcı, A. Kesirli Türev için Yapılan Tanımlamaların Eksiklikleri ve Yeni Yaklaşım. In Proceeedings of the TOK-2013 Turkish Automatic Control National Meeting and Exhibition, Malatya, Turkey, 26–28 September 2003; pp. 1040–1045.
24. Karcı, A.; Karadoğan, A. Fractional order derivative and relationship between derivative and complex functions. In Proceeedings of the IECMSA-2013: 2nd International Eurasian Conference on Mathematical Sciences and Applications, Sarajevo, Bosnia and Herzogovina, 26–29 August 2013; pp. 55–56.
25. Karcı, A. A New approach for fractional order derivative and its applications. *Univers. J. Eng. Sci.* **2013**, *1*, 110–117.

6

Microtubules Nonlinear Models Dynamics Investigations through the $\exp(-\Phi(\xi))$-Expansion Method Implementation

Nur Alam [1] and Fethi Bin Muhammad Belgacem [2,*]

[1] Department of Mathematics, Pabna University of Science & Technology, Pabna 6600, Bangladesh; nuralam.pstu23@gmail.com
[2] Department of Mathematics, Faculty of Basic Education, PAAET, Al-Ardhiya 92400, Kuwait
* Correspondence: fbmbelgacem@gmail.com

Academic Editor: Reza Abedi

Abstract: In this research article, we present exact solutions with parameters for two nonlinear model partial differential equations(PDEs) describing microtubules, by implementing the $\exp(-\Phi(\xi))$-Expansion Method. The considered models, describing highly nonlinear dynamics of microtubules, can be reduced to nonlinear ordinary differential equations. While the first PDE describes the longitudinal model of nonlinear dynamics of microtubules, the second one describes the nonlinear model of dynamics of radial dislocations in microtubules. The acquired solutions are then graphically presented, and their distinct properties are enumerated in respect to the corresponding dynamic behavior of the microtubules they model. Various patterns, including but not limited to regular, singular kink-like, as well as periodicity exhibiting ones, are detected. Being the method of choice herein, the $\exp(-\Phi(\xi))$-Expansion Method not disappointing in the least, is found and declared highly efficient.

Keywords: The $\exp(-\Phi(\xi))$-Expansion Method; models of microtubules; exact solutions; periodic solutions; rational solutions; solitary solutions; trigonometric solutions

---

1. Introduction

Microtubules (MTs) are major cytoskeletal proteins. MTs are cytoskeletal biopolymers shaped as nanotubes. They are hollow cylinders formed by Proto-Filaments (PFs) representing a series of proteins known as tubulin dimers. Each dimer is an electric dipole. These dimers are in a straight position within the PFs or placed in radial positions pointing out of the cylindrical surface. MTs compriseaninteresting type of protein structure that may be a good candidate for designing and manufacturing electronic nano-devices. MTs dynamical behavior is modeled by nonlinear partial differential equations (NPDEs). These equations are mathematical models of physical circumstances that emerge in various fields of engineering, plasma physics, solid state physics, optical fibers, chemistry, hydrodynamics, biology, fluid mechanics and geochemistry. To date solving NPDEs exactly or approximately, a plethora of methods have been in use. These include, but are not limited to, (G'/G)-expansion [1–6], Frobenius decomposition [7], local fractional variation iteration [8], local fractional series expansion [9], multiple exp-function algorithm [10,11], transformed rational function [12], exp-function method [13,14], trigonometric series function [15], inverse scattering [16], homogeneous balance [17,18], first integral [19–22], F-expansion [23–25], Jacobi function [26–29], Sumudu transform [30–32], solitary wave ansatz [33–36], novel (G'/G) -expansion [37–42], modified direct algebraic method [43,44], and last but not least, the $\exp(-\Phi(\xi))$-Expansion Method [45–50].

The objective of this paper is to apply the latter method, namely the $\exp(-\Phi(\xi))$-Expansion Method, to construct the exact solutions for the following two NPDEs modeling MT dynamics, [51–59]. In particular, in presenting the questions to be solved, for comparison purposes, we follow the initial set up established by Zayed and Alurrfi [56], solving the extended Riccatti equations (see Equations (1) and (2)). We then depart generically from their development by using an entirely distinct method, albeit we compare our final results with theirs in [56], keeping in focus the developments in [57–59], as well.

(i) The model of nonlinear dynamics of microtubules assuming a single longitudinal degree of freedom per tubulin dimer is described by the nonlinear PDE (see [59]),

$$m\frac{\partial^2 z(x,t)}{\partial t^2} - kl^2\frac{\partial^2 z(x,t)}{\partial x^2} - qE - Az(x,t) + Bz^3(x,t) + \gamma\frac{\partial z(x,t)}{\partial t} = 0 \tag{1}$$

where A, and B are positive parameters, m, is the mass of the dimer, $z(x,t)$, is the traveling wave, E is the magnitude of intrinsic electric field, l, is the MT length, $q > 0$, is the excess charge within the dipole, γ, is the viscosity coefficient and, k, is a harmonic constant describing the nearest-neighbor interaction between the dimers belonging to the same PFs. In [48], authors have used the Jacobi elliptic function method to find the exact solutions of Equation (1), the physical details and derivations of which were discussed there, although omitted here for obvious reasons.

(ii) The nonlinear PDE describing the nonlinear dynamics of radially dislocated MTs:

$$I\frac{\partial^2 z(x,t)}{\partial t^2} - kl^2\frac{\partial^2 z(x,t)}{\partial x^2} + pEz(x,t) - \frac{pE}{6}z^3(x,t) + \Gamma\frac{\partial z(x,t)}{\partial t} = 0 \tag{2}$$

Here, $z(x,t)$, is the corresponding angular displacement when the whole dimer rotates and, l, is the MT length, p is the magnitude of intrinsic electric field, k, stands for inter-dimer bonding interaction within the same PFs, I, is the moment of inertia of the single dimer and Γ is the viscosity coefficient. In [57], authors have used the simple equation method to find the exact solutions of Equation (2), after relating physical aspects and equation derivation being omitted here.

This paper is organized as follows: In Section 2, we give the description of the $\exp(-\Phi(\xi))$-Expansion Method, while in Section 3, we apply the said method to solve the given NPDEs, Equations (1) and (2). In Section 4, physical explanations are given, followed by the conclusion in Section 5. The paper ends with relevant acknowledgments, and a rich list of references for interested readers.

2. Description of the $\exp(-\Phi(\xi))$-Expansion Method

Following th initial setup in [56], we consider the nonlinear evolution equation in the form,

$$F(u, u_t, u_x, u_{tt}, u_{xt}, u_{xx}, \cdots\cdots) = 0 \tag{3}$$

where, F, is a polynomial in, $u(x,t)$, and its partial derivatives, involving nonlinear terms and highest order derivatives. The focal steps of the method are as follows:

Step 1. It is well known that, for a given wave equation, a travelling wave, $u(\xi)$, is a solution which depends upon, x, and, t, only through a unified variable, ξ, such that,

$$u(x,t) = u(\xi), \xi = k_1 x + \omega t \tag{4}$$

where, k_1 and ω, are constants. Based on this we have,

$$\frac{\delta}{\delta t} = \omega\frac{\delta}{\delta\xi}, \frac{\delta^2}{\delta t^2} = \omega^2\frac{\delta^2}{\delta\xi^2}, \frac{\delta}{\delta x} = k_1\frac{\delta}{\delta\xi}, \text{ and, } \frac{\delta^2}{\delta x^2} = {k_1}^2\frac{\delta^2}{\delta\xi^2} \tag{5}$$

and so on, for other derivatives.

We reduce Equation (3) to the following ODE:

$$Q(u, u', u'', \cdots\cdots) = 0 \tag{6}$$

Here, Q is a polynomial in, $u(\xi)$, and its total derivatives, such that $' = \frac{d}{d\xi}$.

Step 2. We assume that Equation (6) has the formal solution:

$$u(\xi) = \sum_{i=0}^{N} A_i(\exp(-\Phi(\xi)))^i \tag{7}$$

where, the A_i's are constants to be determined, such that $A_N \neq 0$ and $\Phi = \Phi(\xi)$ satisfies the following ODE:

$$\Phi'(\xi) = \exp(-\Phi(\xi)) + \mu \exp(\Phi(\xi)) + \lambda \tag{8}$$

Consequently, we get the following possibilities for Equation (8):

Cluster 1: When $\mu \neq 0, \lambda^2 - 4\mu > 0$, we get,

$$\Phi(\xi) = \ln\left(\frac{-\sqrt{(\lambda^2 - 4\mu)}\tanh\left(\frac{\sqrt{(\lambda^2 - 4\mu)}}{2}(\xi + E)\right) - \lambda}{2\mu}\right) \tag{9}$$

Cluster 2: When $\mu \neq 0, \lambda^2 - 4\mu < 0$, we get,

$$\Phi(\xi) = \ln\left(\frac{\sqrt{(4\mu - \lambda^2)}\tan\left(\frac{\sqrt{(4\mu - \lambda^2)}}{2}(\xi + E)\right) - \lambda}{2\mu}\right) \tag{10}$$

Cluster 3: When $\mu = 0, \lambda \neq 0$, and $\lambda^2 - 4\mu > 0$, we obtain,

$$\Phi(\xi) = -\ln\left(\frac{\lambda}{\exp(\lambda(\xi + E)) - 1}\right) \tag{11}$$

Cluster 4: When $\mu \neq 0, \lambda \neq 0$, and $\lambda^2 - 4\mu = 0$, we obtain

$$\Phi(\xi) = \ln\left(-\frac{2(\lambda(\xi + E) + 2)}{\lambda^2(\xi + E)}\right) \tag{12}$$

Cluster 5: When $\mu = 0, \lambda = 0$, and $\lambda^2 - 4\mu = 0$, we then have,

$$\Phi(\xi) = \ln(\xi + E) \tag{13}$$

where $A_N, \cdots\cdots, V, \lambda, \mu$, are constants to be determined, such that $A_N \neq 0$. The positive integer, m, can be determined by considering the homogeneous balance between nonlinear terms and the highest order derivatives occurring in the ODE in Equation (6), after using Equation (7).

Step 3. We interchange Equation (7) into Equation (6) and then we expand the function $\exp(-\Phi(\xi))$. As a result of this interchange, we get a polynomial of $\exp(-\Phi(\xi))$. We equate all the coefficients of same power of $\exp(-\Phi(\xi))$ to zero. This procedure yields a system of algebraic equations which could be solved to obtain the values of $A_N, \cdots\cdots, V, \lambda, \mu$ which after substitution into Equation (7) along with general solutions of Equation (8) completes the setup for getting the traveling wave solutions of the NPDE in Equation (3).

3. Applications

In this section, we will apply the $\exp(-\Phi(\xi))$-Expansion Method described in Section 2 to find the exact solutions of the NPDE Equations (1) and (2).

3.1. Exact Solutions of the NPDE Equation(1)

In this subsection, we find the exact wave solutions of Equation (1). To this end, we use the transformation (4) to reduce Equation (1) into the nonlinear ordinary differential equation (NODE),

$$P\psi''(\xi) - Q\psi'(\xi) - \psi(\xi) + \psi^3(\xi) - R = 0 \tag{14}$$

where,

$$P = \frac{m\omega^2 - kl^2{k_1}^2}{A}, Q = \frac{\gamma\omega}{A}, R = \frac{qE}{A\sqrt{A/B}} \tag{15}$$

and,

$$z(\xi) = \sqrt{\frac{A}{B}}\psi(\xi) \tag{16}$$

Balancing, $\psi''(\xi)$, with, $\psi^3(\xi)$, in Equation (14), we get $N = 1$. Consequently, we have,

$$\psi(\eta) = A_0 + A_1(\exp(-\Phi(\xi))) \tag{17}$$

where A_0, A_1 are constants to be determined such that $A_N \neq 0$, while λ, μ, are arbitrary.

Substituting Equation (17) into Equation (14) and equating the coefficients of $\exp(-\Phi(\xi))^3$, $\exp(-\Phi(\xi))^2$, $\exp(-\Phi(\xi))^1$, $\exp(-\Phi(\xi))^0$, to zero, we respectively obtain,

$$\exp(-\Phi(\xi))^3 : 2PA_1 + {A_1}^3 = 0 \tag{18}$$

$$\exp(-\Phi(\xi))^2 : 3A_0{A_1}^2 + QA_1 + 3PA_1\lambda = 0 \tag{19}$$

$$\exp(-\Phi(\xi))^1 : 2PA_1\mu + P\lambda^2 A_1 - A_1 + QA_1\lambda + 3{A_0}^2A_1 = 0 \tag{20}$$

and,

$$\exp(-\Phi(\xi))^0 : A_0 - R + PA_1\mu\lambda + QA_1\mu + {A_0}^3 = 0 \tag{21}$$

Now, solving Equations (18)–(21) yields,

$$\begin{gathered} A_0 = A_0,\ A_1 = \alpha,\ \lambda = -\frac{1}{3P}(3A_0\alpha + Q), \text{ and,} \\ \mu = \frac{1}{18P^2}(3A_0\alpha Q + 2Q^2 + 9P - 9{A_0}^2P),\ R = \frac{1}{27P^2}\{Q\alpha(2Q^2 + 9P)\} \end{gathered} \tag{22}$$

where, $\alpha = \pm\sqrt{-2P}$, and A_0, P, and, Q, are arbitrary constants.

Substituting Equation (22) into Equation (17), we obtain

$$\psi(\xi) = A_0 + \alpha(\exp(-\Phi(\xi))) \tag{23}$$

Now, substituting Equations (9)–(13) into Equation (23) respectively, we get the following five traveling wave solutions of the NPDE Equation (1).

When $\mu \neq 0$, $\lambda^2 - 4\mu > 0$,

$$z_1(\xi) = \sqrt{\frac{A}{B}}\{A_0 - \alpha(\frac{2\mu}{\sqrt{\lambda^2 - 4\mu}\tanh(\frac{\sqrt{\lambda^2 - 4\mu}}{2}(\xi + E)) + \lambda}\} \tag{24}$$

where E is an arbitrary constant.

When $\mu \neq 0, \lambda^2 - 4\mu < 0$,

$$z_2(\xi) = \sqrt{\frac{A}{B}}\{A_0 + \alpha(\frac{2\mu}{\sqrt{4\mu - \lambda^2}\tan(\frac{\sqrt{4\mu - \lambda^2}}{2}(\xi + E)) - \lambda}\} \tag{25}$$

where, E, is an arbitrary constant.

When $\mu = 0, \lambda \neq 0$, and $\lambda^2 - 4\mu > 0$,

$$z_3(\xi) = \sqrt{\frac{A}{B}}\{A_0 + \alpha(\frac{\lambda}{\exp(\lambda(\xi + E)) - 1})\} \tag{26}$$

where, E, is an arbitrary constant.

When $\mu \neq 0, \lambda \neq 0$, and $\lambda^2 - 4\mu = 0$,

$$z_4(\xi) = \sqrt{\frac{A}{B}}\{A_0 - \alpha(\frac{\lambda^2(\xi + E)}{2(\lambda(\xi + E)) + 2)})\} \tag{27}$$

where, E, is an arbitrary constant.

When $\mu = 0, \lambda = 0$, and $\lambda^2 - 4\mu = 0$,

$$z_5(\xi) = \sqrt{\frac{A}{B}}\{A_0 + \alpha(\frac{1}{\xi + E})\} \tag{28}$$

where, E, is an arbitrary constant.

3.2. Exact Solutions of the NPDE Equation (2)

In this subsection, we find the exact solutions of Equation (2). To this end, we use the transformation Equation (4) to reduce Equation (2) into the following NODE,

$$S\psi''(\xi) - T\psi'(\xi) + \psi(\xi) - \psi^3(\xi) = 0 \tag{29}$$

where,

$$S = \frac{I\omega^2 - kl^2{k_1}^2}{pE}, T = \frac{\Gamma\omega}{pE} \tag{30}$$

and,

$$z(\xi) = \sqrt{6}\psi(\xi) \tag{31}$$

Balancing $\psi''(\xi)$ with $\psi^3(\xi)$ in Equation (29), we get $N = 1$. Consequently, we have the formal solution of Equation (29), as follows:

$$\psi(\xi) = A_0 + A_1(\exp(-\Phi(\xi))) \tag{32}$$

where A_0, A_1 are constants to be determined such that $A_N \neq 0$, while λ, μ, are arbitrary. Substituting Equation (32) into Equation (29) and equating the coefficients of $\exp(-\Phi(\xi))^3$, $\exp(-\Phi(\xi))^2$, $\exp(-\Phi(\xi))^1$, $\exp(-\Phi(\xi))^0$ to zero, we respectively obtain

$$\exp(-\Phi(\xi))^3 : \ 2SA_1 - {A_1}^3 = 0 \tag{33}$$

$$\exp(-\Phi(\xi))^2 : 3SA_1\lambda - 3A_0{A_1}^2 + TA_1 = 0 \tag{34}$$

$$\exp(-\Phi(\xi))^1 : \ A_1 + 2SA_1\mu + S\lambda^2 A_1 + TA_1\lambda - 3{A_0}^2 A_1 = 0 \tag{35}$$

and,

$$\exp(-\Phi(\xi))^0 : A_0 + SA_1\mu\lambda + TA_1\mu - {A_0}^3 = 0 \tag{36}$$

Solving the Equation (33)–(36) yields:

Cluster 1: We have,

$$A_0 = A_0, A_1 = \frac{2}{3}T, \lambda = \frac{3}{2T}(2A_0 - 1), \mu = \frac{9}{4T^2}(A_0{}^2 - A_0), S = \frac{2}{9}T^2 \tag{37}$$

Of course, A_0, T, are arbitrary constants.

Cluster 2 : We have,

$$A_0 = A_0,\ A_1 = -\frac{2}{3}T, \lambda = -\frac{3}{2T}(2A_0 + 1), \mu = \frac{9}{4T^2}(A_0{}^2 + A_0), S = \frac{2}{9}T^2 \tag{38}$$

where A_0, T are arbitrary constants.

For cluster 1, substituting Equation (37) into Equation (32), we obtain

$$u(\xi) = A_0 + \frac{2T}{3}(\exp(-\Phi(\xi))) \tag{39}$$

while, for cluster 2, substituting Equation (38) into Equation (32), we obtain

$$u(\xi) = A_0 - \frac{2T}{3}(\exp(-\Phi(\xi))) \tag{40}$$

Now, substituting Equations (9)–(13) into Equation (39), respectively, we get the following five traveling wave solutions of the NPDE Equation (2).

When, $\mu \neq 0, \lambda^2 - 4\mu > 0$,

$$z_1(\xi) = \sqrt{6}\{A_0 - \frac{2T}{3}(\frac{2\mu}{\sqrt{\lambda^2 - 4\mu}\tanh(\frac{\sqrt{\lambda^2 - 4\mu}}{2}(\xi + E)) + \lambda}\} \tag{41}$$

where E is an arbitrary constant.

When $\mu \neq 0, \lambda^2 - 4\mu < 0$,

$$z_2(\xi) = \sqrt{6}\{A_0 + \frac{2T}{3}(\frac{2\mu}{\sqrt{4\mu - \lambda^2}\tan(\frac{\sqrt{4\mu - \lambda^2}}{2}(\xi + E)) - \lambda})\} \tag{42}$$

where E is an arbitrary constant.

When, $\mu = 0, \lambda \neq 0$, and $\lambda^2 - 4\mu > 0$,

$$z_3(\xi) = \sqrt{6}\{A_0 + \frac{2T}{3}(\frac{\lambda}{\exp(\lambda(\xi + E)) - 1})\} \tag{43}$$

where E is an arbitrary constant.

When $\mu \neq 0, \lambda \neq 0$, and $\lambda^2 - 4\mu = 0$,

$$z_4(\xi) = \sqrt{6}\{A_0 - \frac{2T}{3}(\frac{\lambda^2(\xi + E)}{2(\lambda(\xi + E)) + 2})\} \tag{44}$$

where E is an arbitrary constant.

When $\mu = 0, \lambda = 0$, and $\lambda^2 - 4\mu = 0$,

$$z_5(\xi) = \sqrt{6}\{A_0 + \frac{2T}{3}(\frac{1}{\xi + E})\} \tag{45}$$

where E is an arbitrary constant.

At this point, inserting Equations (9)–(13) into Equation (40), respectively, we get the following other five traveling wave solutions of the NPDE Equation (2).

When, $\mu \neq 0, \lambda^2 - 4\mu > 0$,

$$z_6(\xi) = \sqrt{6}\{A_0 + \frac{2T}{3}(\frac{2\mu}{\sqrt{\lambda^2 - 4\mu}\tanh(\frac{\sqrt{\lambda^2 - 4\mu}}{2}(\xi + E)) + \lambda})\} \tag{46}$$

where, E, is an arbitrary constant.

When $\mu \neq 0, \lambda^2 - 4\mu < 0$,

$$z_7(\xi) = \sqrt{6}\{A_0 - \frac{2T}{3}(\frac{2\mu}{\sqrt{4\mu - \lambda^2}\tan(\frac{\sqrt{4\mu - \lambda^2}}{2}(\xi + E)) - \lambda})\} \tag{47}$$

where, E, is an arbitrary constant.

When, $\mu = 0, \lambda \neq 0$, and $\lambda^2 - 4\mu > 0$,

$$z_8(\xi) = \sqrt{6}\{A_0 - \frac{2T}{3}(\frac{\lambda}{\exp(\lambda(\xi + E)) - 1})\} \tag{48}$$

where, E, is an arbitrary constant.

When, $\mu \neq 0, \lambda \neq 0$, and $\lambda^2 - 4\mu = 0$,

$$z_9(\xi) = \sqrt{6}\{A_0 + \frac{2T}{3}(\frac{\lambda^2(\xi + E)}{2(\lambda(\xi + E)) + 2)})\} \tag{49}$$

where, E, is an arbitrary constant.

When, $\mu = 0, \lambda = 0$, and $\lambda^2 - 4\mu = 0$,

$$z_{10}(\xi) = \sqrt{6}\{A_0 - \frac{2T}{3}(\frac{1}{\xi + E})\} \tag{50}$$

where, E, is an arbitrary constant.

4. Comparison

The papers [58,59] by Zdravkovic *et al.* are key to our present work. They collectively considered solutions of the nonlinear PDE describing the nonlinear dynamics of radially dislocated MTs using the simplest equation method. The solutions of the nonlinear PDE describing the nonlinear dynamics of radially dislocated MTs obtained by the $\exp(-\Phi(\xi))$-Expansion Method are different from those of the simplest equation method. It is n oteworthy to point out that some of our solutions coincide with already published results, if parameters taken particular values which authenticate our solutions. Moreover, Zdravkovic *et al.* [58] investigated the nonlinear PDE describing the nonlinear dynamics of radially dislocated MTs using the simplest equation method to obtain exact solutions via the simplest equation method and achieved only two solutions (see Appendix). Furthermore, ten solutions of the nonlinear PDE describing the nonlinear dynamics of radially dislocated MTs are constructed by applying the $\exp(-\Phi(\xi))$-Expansion Method. Zdravkovic *et al.* [58] (see also [59]) apply the simplest equation method to the nonlinear PDE describing the nonlinear dynamics of radially dislocated MTs, and they only solve kink type solutions, but we apply the $\exp(-\Phi(\xi))$-Expansion Method to the nonlinear PDE describing the nonlinear dynamics of radially dislocated MTs and solve kink type solutions, singular kink type solutions and plane periodic type solutions. On the other hand, the auxiliary equation used in this paper is different, so obtained solutions are also different. Similarly, for any nonlinear evolution equation, it can be shown that the $\exp(-\Phi(\xi))$-Expansion Method is much more direct and user-friendly than other methods.

5. Physical Interpretations of Some Obtained Solutions

In this section, attempting to shed lights on the corresponding physical behavior, we to discuss nonlinear dynamics of MTs whether as nano-bioelectronics transmission lines like or radially dislocated MTs, based on the obtained traveling wave solutions, from Equations (24)–(28), and (41)–(50), respectively. We examine the nature of some obtained solutions of Equations (1) and (2) by selecting particular values of the parameters and graphing the resulting exact solutions using mathematical software Maple 13, represented in Figures 1–6.

From our obtained solutions, we observe that Equations (24)–(28), and (41)–(50), exude kink type solitons, singular kink shape solitons, and periodic solutions. Equation (24) shows kink shaped soliton profile for, $A_0 = 1, m = 1, \omega = -1, k_1 = 1, k = 2, l = 2, A = 2, B = 3, \mu = 1, \lambda = 3, E = 1$, within the interval $-10 \leqslant x, t \leqslant 10$ which is represented in Figures 1 and 2. Equation (25) provides a periodic solution profile for, $A_0 = 1, m = 1, \omega = -1, k_1 = 1, k = 2, l = 2, A = 2, B = 3, \mu = 3, \lambda = 1, E = 5$ within the interval $-1 \leqslant x, t \leqslant 1$, which is represented in Figures 3 and 4. Equation (26) provides a singular kink soliton profile for, $A_0 = 1, m = 1, \omega = -1, k_1 = 1, k = 2, l = 2, A = 2, B = 3, \mu = 0, \lambda = 2, E = 1$, within the interval $-10 \leqslant x, t \leqslant 10$, which is represented in Figures 5 and 6. Equations (27) and (28) also represent singular kink type wave solutions which are similar to Figures 5 and 6. Equations (41) and (46) provide kink soliton profile, for $A_0 = 2, T = \frac{3}{2}, \omega = -1, k_1 = 1, \mu = 1, \lambda = 3$, and $E = 1$, within the interval, $-10 \leqslant x, t \leqslant 10$, as in Figures 1 and 2. Equations (42) and (47) provide periodic solutions for, $A_0 = 2, T = \frac{3}{2}, \omega = -1, k_1 = 1, \mu = 3, \lambda = 1, E = 5$, within the interval, $-1 \leqslant x, t \leqslant 1$, as in Figures 3 and 4. Equations (43) and (48), provide singular kink soliton profiles for, $A_0 = 2, T = \frac{3}{2}, \omega = -1, k_1 = 1, \mu = 0, \lambda = 2$, and, $E = 1$, within the interval $-10 \leqslant x, t \leqslant 10$, as in Figures 5 and 6. Equations (44) and (45), as well as Equations (49) and (50), also represent singular Kink type wave solutions which are similar to Figures 5 and 6.

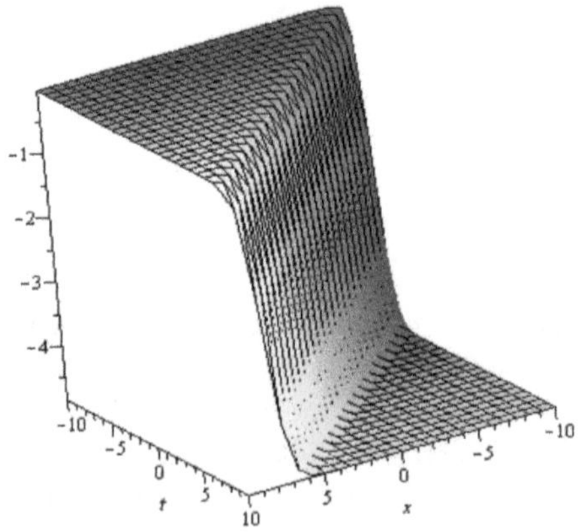

Figure 1. The solitary wave 3D graphics of Equation (24) shows a kink shaped soliton profile for, $A_0 = 1, m = 1, \omega = -1, k_1 = 1, k = 2, l = 2, A = 2, B = 3, \mu = 1, \lambda = 3, E = 1$ within the interval $-10 \leqslant x, t \leqslant 10$.

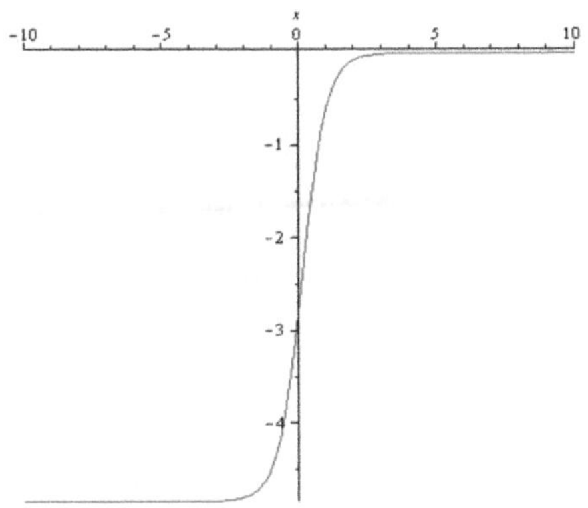

Figure 2. The solitary wave 2D graphics of Equation (24) shows a kink shaped soliton profile for, $A_0 = 1, m = 1, \omega = -1, k_1 = 1, k = 2, l = 2, A = 2, B = 3, \mu = 1, \lambda = 3, E = 1, t = 2$.

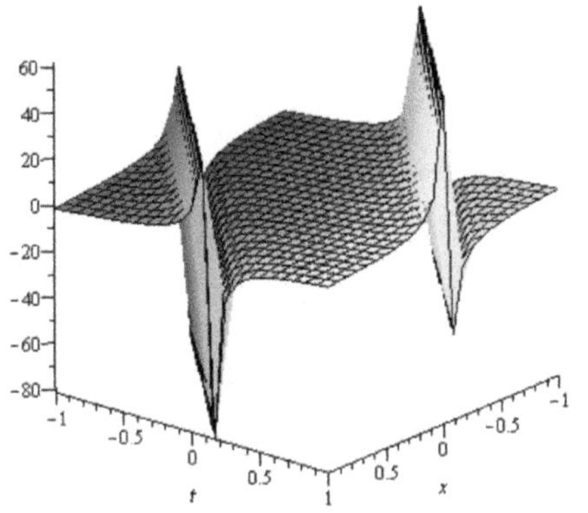

Figure 3. The solitary wave 3D graphics of Equiation (25) provides a periodic solution profile for, $A_0 = 1, m = 1, \omega = -1, k_1 = 1, k = 2, l = 2, A = 2, B = 3, \mu = 3, \lambda = 1, E = 5$ within the interval $-1 \leqslant x, t \leqslant 1$.

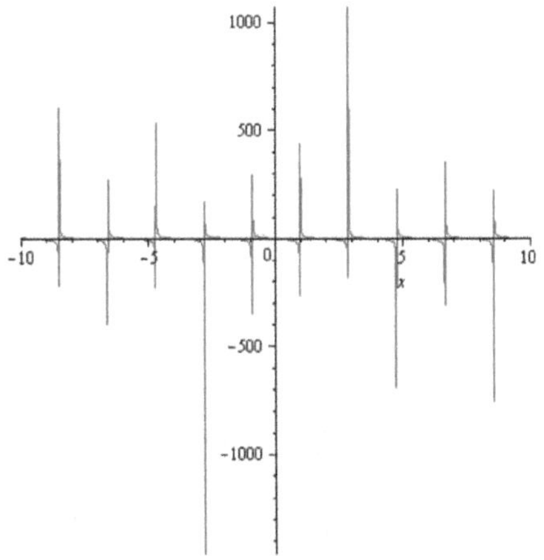

Figure 4. The solitary wave 2D graphics of Equation (25) provides a periodic solution profile for, $A_0 = 1, m = 1, \omega = -1, k_1 = 1, k = 2, l = 2, A = 2, B = 3, \mu = 3, \lambda = 1, E = 5, t = 2$.

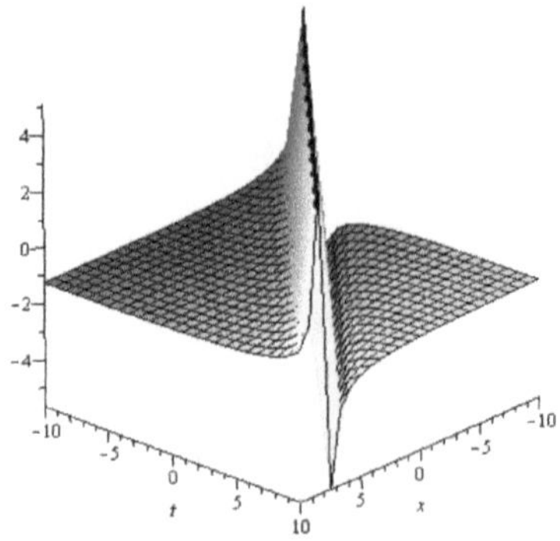

Figure 5. The solitary wave 3D graphics of Equation (26) provides a singular kink soliton profile for, $A_0 = 1, m = 1, \omega = -1, k_1 = 1, k = 2, l = 2, A = 2, B = 3, \mu = 0, \lambda = 2, E = 1$ within the interval $-10 \leqslant x, t \leqslant 10$.

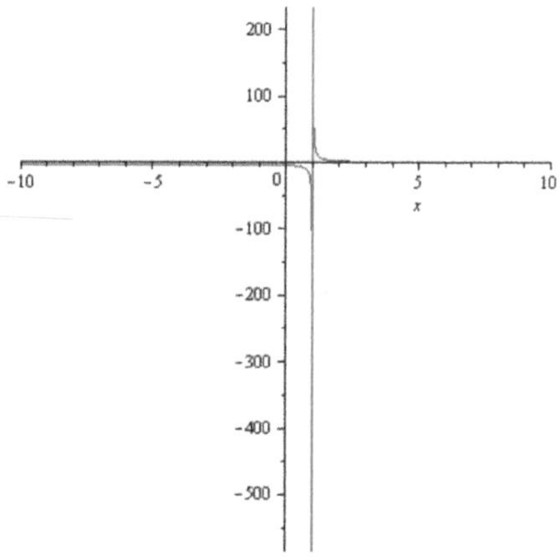

Figure 6. The solitary wave 2D graphics of Equation (26) provides a singular kink soliton profile for, $A_0 = 1, m = 1, \omega = -1, k_1 = 1, k = 2, l = 2, A = 2, B = 3, \mu = 0, \lambda = 2, E = 1, t = 2$.

6. Conclusions

The $\exp(-\Phi(\xi))$-Expansion Method has been appliedto Equations (1) and (2), which describe the nonlinear dynamics of microtubules assuming a single longitudinal degree of freedom per tubulin dimer [59] and the dynamics of radial dislocations in MTs, respectively. The said method was instrumental in the provision of new analytical solutions such as kink type solutions, singular kink type solutions and plane periodic type solutions which are shown in Figures 1–3. On comparing our results in this paper with the well-known results obtained in [50,58,59], we deduce that our results are new and not published elsewhere. All analytical solutions obtained by The $\exp(-\Phi(\xi))$-Expansion Method in the paper have been controlled, whether they are verified to Equation (1) and Equation (2) with the aid of commercial software Maple, and all new solutions have been verified to the original equations Equations (1) and (2). Zayed and Alurrfi [56] recently solved the two equations but used the alternative generalized Ricatti projective method. There, they also obtained trigonometric, hyperbolic and rational solutions but failed to obtain the exponential ones that we got. Our distinction resides mostly in obtaining extra solution types using our method. Of course, the choice of parameters yields different facets of the solutions and their graphic presentation so as to be type representative, without rendering the paper so voluminous, should more realizations be expected.

Acknowledgments: The authors acknowledge and salute the mathematics editorial board management, and thank the consequent anonymous referees' diligent efforts and critiques that helped improve the flow, style and scientific veracity of this paper. Furthermore, Fethi Bin Muhammad Belgacem wishes to acknowledge the support of the Public Authority for Applied Education and Training, Kuwait, through the Research Department grant No. PAAET RDBE-13-09.

Author Contributions: In light of the recent scholarly interest in nano-bioelctronics Microtubules dynamic behavior, both authors agreed to investigate this phenomenon but with the advent of an original method not alreay used in the literature, which gave the authors the opportunity to compare with existing solutions. The contribution is therefore equally and evenly shared between both authors. Due to the subject importance and the desire that common understanding of these physical phenomena be widely spread, the authors invite interested readers to communicate, request, share relevant pieces of information, and collaborate, if so wished!

Conflicts of Interest: The authors declare no conflict of interest.

Appendix

Zdravkovic *et al.* [56] studied solutions of of the nonlinear PDE describing the nonlinear dynamics of radially dislocated MTs using the simplest equation method and achieved the following exact solutions:

$$\psi_1(x,t) = \pm\frac{1}{2}\left[1 + \tanh y + \frac{1}{\cosh^2 y(d + \tanh y)}\right]$$

$$\psi_2(x,t) = \pm\frac{1}{2}\left[1 + \tanh(\frac{y}{2}) + \frac{1}{\sinh y)}\right]$$

References

1. Alam, M.A.; Akbar, M.A.; Mohyud-Din, S.T. General traveling wave solutions of the strain wave equation in microstructured solids via the new approach of generalized (G'/G)-Expansion method. *Alex. Eng. J.* **2014**, *53*, 233–241. [CrossRef]
2. Alam, M.N.; Akbar, M.A.; Hoque, M.F. Exact traveling wave solutions of the (3+1)-dimensional mKdV-ZK equation and the (1+1)-dimensional compound KdVB equation using new approach of the generalized (G'/G)-expansion method. *PramanJ. Phys.* **2014**, *83*, 317–329. [CrossRef]
3. Alam, M.N.; Akbar, M.A. A new (G'/G)-expansion method and its application to the Burgers equation. *Walailak J. Sci. Technol.* **2014**, *11*, 643–658.
4. Hafez, M.G.; Alam, M.N.; Akbar, M.A. Exact traveling wave solutions to the Klein-Gordon equation using the novel (G'/G)-expansion method. *Results Phys.* **2014**, *4*, 177–184. [CrossRef]
5. Alam, M.N.; Akbar, M.A. The new approach of generalized (G'/G)-expansion method for nonlinear evolution equations. *Ain Shams Eng.* **2014**, *5*, 595–603. [CrossRef]
6. Younis, M.; Rizvi, S.T.R. Dispersive dark optical soliton in (2+1)-dimensions by G'/G-expansion with dual-power law nonlinearity. *Opt. Int. J. Light Electron Opt.* **2015**, *126*, 5812–5814. [CrossRef]
7. Ma, W.X.; Wu, H.Y.; He, J.S. Partial differential equations possessing Frobeniusintegrable decomposition technique. *Phys. Lett. A* **2007**, *364*, 29–32. [CrossRef]
8. Yang, Y.J.; Baleanu, D.; Yang, X.J. A Local fractional variational iteration method for Laplace equation within local fractional operators. *Abstr. Appl. Anal.* **2013**, *2013*, 202650. [CrossRef]
9. Yang, A.M.; Yang, X.J.; Li, Z.B. Local fractional series expansion method for solving wave and diffusion equations on cantor sets. *Abstr. Appl. Anal.* **2013**, *2013*, 351057. [CrossRef]
10. Ma, W.X.; Zhu, Z. Solving the (3+1)-dimensional generalized KP and BKP equations by the multiple exp-function algorithm. *Appl. Math. Comput.* **2012**, *218*, 11871–11879. [CrossRef]
11. Ma, W.X.; Huang, T.; Zhang, Y. A multiple exp-function method for nonlinear differential equations and its application. *Phys. Scr.* **2010**, *82*, 065003. [CrossRef]
12. Ma, W.X.; Lee, J.H. A transformed rational function method and exact solutions to the (3+1) dimensional Jimbo-Miwa equation. *Chaos Solitons Fractals* **2009**, *42*, 1356–1363. [CrossRef]
13. He, J.H.; Wu, X.H. Exp-function method for nonlinear wave equations. *Chaos Solitons Fractals* **2006**, *30*, 700–708. [CrossRef]
14. Younis, M.; Rizvi, S.T.R.; Ali, S. Analytical and soliton solutions: Nonlinear model of an obioelectronics transmission lines. *Appl. Math. Comput.* **2015**, *265*, 994–1002. [CrossRef]

15. Zhang, Z.Y. New exact traveling wave solutions for the nonlinear Klein-Gordonequation. *Turk. J. Phys.* **2008**, *32*, 235–240.
16. Ablowitz, M.J.; Segur, H. *Solitions and Inverse Scattering Transform*; SIAM: Philadelphia, PA, USA, 1981.
17. Fan, E.; Zhang, H. A note on the homogeneous balance method. *Phys. Lett. A* **1998**, *246*, 403–406. [CrossRef]
18. Wang, M.L. Exact solutions for a compound KdV-Burgers equation. *Phys. Lett. A* **1996**, *213*, 279–287. [CrossRef]
19. Moosaei, H.; Mirzazadeh, M.; Yildirim, A. Exact solutions to the perturbed nonlinear Schrodinger equation with Kerr law nonlinearity by using the first integral method. *Nonlinear Anal. Model. Control* **2011**, *16*, 332–339.
20. Bekir, A.; Unsal, O. Analytic treatment of nonlinear evolution equations using the first integral method. *Pramana J. Phys.* **2012**, *79*, 3–17. [CrossRef]
21. Lu, B.H.Q.; Zhang, H.Q.; Xie, F.D. Traveling wave solutions of nonlinear partial differential equations by using the first integral method. *Appl. Math. Comput.* **2010**, *216*, 1329–1336. [CrossRef]
22. Feng, Z.S. The first integral method to study the Burgers–KdVequation. *J. Phys. A Math. Gen.* **2002**, *35*, 343–349. [CrossRef]
23. Abdou, M.A. The extended F-expansion method and its application for a class of nonlinear evolution equations. *Chaos Solitons Fractals* **2007**, *31*, 95–104. [CrossRef]
24. Ren, Y.J.; Zhang, H.Q. A generalized F-expansion method to find abundant families of Jacobi elliptic function solutions of the (2+1)-dimensional Nizhnik-Novikov-Veselovequation. *Chaos Solitons Fractals* **2006**, *27*, 959–979. [CrossRef]
25. Zhang, J.L.; Wang, M.L.; Wang, Y.M.; Fang, Z.D. The improved F-expansion method and its applications. *Phys. Lett. A* **2006**, *350*, 103–109. [CrossRef]
26. Dai, C.Q.; Zhang, J.F. Jacobian elliptic function method for nonlinear differential-difference equations. *Chaos Solitons Fractals* **2006**, *27*, 1042–1049. [CrossRef]
27. Fan, E.; Zhang, J. Applications of the Jacobi elliptic function method to special-type nonlinear equations. *Phys. Lett. A* **2002**, *305*, 383–392. [CrossRef]
28. Liu, S.; Fu, Z.; Liu, S.; Zhao, Q. Jacobi elliptic function expansion method and periodic wave solutions of nonlinear wave equations. *Phys. Lett. A* **2001**, *289*, 69–74. [CrossRef]
29. Zhao, X.Q.; Zhi, H.Y.; Zhang, H.Q. Improved Jacobi elliptic function method with symbolic computation to construct new double-periodic solutions for the generalized Ito system. *Chaos Solitons Fractals* **2006**, *28*, 112–126. [CrossRef]
30. Belgacem, F.B.M. Sumudu Transform Applications to Bessel Functions and Equations. *Appl. Math. Sci.* **2010**, *4*, 3665–3686.
31. Belgacem, F.B.M.; Karaballi, A.A. Sumudu transform Fundamental Properties investigations and applications. *J. Appl. Math. Stoch. Anal.* **2006**, *2006*, 1–23. [CrossRef]
32. Belgacem, F.B.M. Sumudu Applications to Maxwell's Equations. *PIERS Online* **2009**, *5*, 1–6. [CrossRef]
33. Younis, M.; Ali, S. Bright, dark and singular solitons in magneto-electro-elastic circular rod. *Waves Random Complex Media* **2015**, *25*, 549–555. [CrossRef]
34. Younis, M.; Ali, S. Solitary wave and shock wave solutions to the transmission line model for nano-ionic currents along microtubules. *Appl. Math. Comput.* **2014**, *246*, 460–463. [CrossRef]
35. Ali, S.; Rizvi, S.T.R.; Younis, M. Traveling wave solutions for nonlinear dispersive water wave systems with time dependent coefficients. *Nonlinear Dyn.* **2015**, *82*, 1755–1762. [CrossRef]
36. Younis, M.; Ali, S.; Mahmood, S.A. Solitons for compound KdV-Burgers' equation with variable coefficients and power law nonlinearity. *Nonlinear Dyn.* **2015**, *81*, 1191–1196. [CrossRef]
37. Alam, M.N.; Belgacem, F.B.M. Application of the Novel (G'/G)-Expansion Method to the Regularized Long Wave Equation. *Waves Wavelets Fractals Adv. Anal.* **2015**, *1*, 51–67. [CrossRef]
38. Alam, M.N.; Hafez, M.G.; Belgacem, F.B.M.; Akbar, M.A. Applications of the novel (G'/G)-expansion method to find new exact traveling wave solutions of the nonlinear coupled Higgs field equation. *Nonlinear Stud.* **2015**, *22*, 613–633.
39. Alam, M.N.; Belgacem, F.B.M.; Akbar, M.A. Analytical treatment of the evolutionary (1+1) dimensional combined KdV-mKdV equation via novel (G'/G)-expansion method. *J. Appl. Math. Phys.* **2015**, *3*, 61765. [CrossRef]

40. Alam, M.N.; Belgacem, F.B.M. New generalized (G'/G)-expansion method applications to coupled Konno-Oono and right-handed noncommutative Burgers equations. *Adv. Pure Math* **2016**, in press.
41. Alam, M.N.; Hafez, M.G.; Akbar, M.A.; Belgacem, F.B.M. Application of new generalized $(G'G)$-expansion method to the (3+1)-dimensional Kadomtsev-Petviashvili equation. *Ital. J. Pure Appl. Math* **2016**, in press.
42. Alam, M.N.; Belgacem, F.B.M. Traveling Wave Solutions for the (1+1) Dim Compound KdVB Equation by the Novel (G'/G)-Expansion Method. *Int. J. Mod. Nonlinear Theory Appl.* **2016**, in press.
43. Younis, M.; Rehman, H.; Iftikhar, M. Travelling wave solutions to some nonlinear evolution equations. *Appl. Math. Comput.* **2014**, *249*, 81–88. [CrossRef]
44. Hereman, W.; Banerjee, P.P.; Korpel, A.; Assanto, G.; van Immerzeele, A.; Meerpoel, A. Exact solitary wave solutions of nonlinear evolution and wave equations using a direct algebraic method. *J. Phys. A: Math. Gen.* **1986**, *19*, 607. [CrossRef]
45. Hafez, M.G.; Alam, M.N.; Akbar, M.A. Traveling wave solutions for some important coupled nonlinear physical models via the coupled Higgs equation and the Maccarisystem. *J. King Saud Univ.Sci.* **2015**, *27*, 105–112. [CrossRef]
46. Hafez, M.G.; Alam, M.N.; Akbar, M.A. Application of the $\exp(-\Phi(\xi))$-expansion method to find exact solutions for the solitary wave equation in an un-magnetized dusty plasma. *World Appl. Sci. J.* **2014**, *32*, 2150–2155.
47. Alam, M.N.; Hafez, M.G.; Akbar, M.A.; Roshid, H.O. Exact traveling wave solutions to the (3+1)-dimensionalmKdV-ZK and the (2+1)-dimensional Burgers equations via $\exp(-\Phi(\xi))$-expansion method. *Alex. Eng. J.* **2015**, *54*, 635–644.
48. Alam, M.N.; Hafez, M.G.; Akbar, M.A.; Roshid, H.O. Exact Solutions to the (2+1)-Dimensional Boussinesq Equation via $\exp(-\Phi(\xi))$-Expansion Method. *J. Sci. Res.* **2015**, *7*, 1–10. [CrossRef]
49. Zayed, E.M.E.; Amer, Y.A.; Shohid, R.M.A. The (G'/G)-expansion method and the $\exp(-\Phi(\xi))$-Expansion Method with applications to a higher order dispersive nonlinear schrodingerequation. *Sci. Res. Essays* **2015**, *10*, 218–231.
50. Zekovic, S.; Muniyappan, A.; Zdravkovic, S.; Kavitha, L. Employment of Jacobian elliptic functions for solving problems in nonlinear dynamics of microtubules. *Chin. Phys. B* **2014**, 23, 020504. [CrossRef]
51. Sataric, M.V.; Sekulic, D.L.; Sataric, B.M.; Zdravkovic, S. Role of nonlinear localized Ca^{2+} pulses along microtubules in tuning the mechano-Sensitivity of hair cells. *Prog. Biophys. Mol. Biol.* **2015**, *119*, 162–174. [CrossRef] [PubMed]
52. Sekulic, D.; Sataric, M.V. An improved nanoscale transmission line model of microtubules: The effect of nonlinearity on the propagation of electrical signals. *Facta Univ. Ser. Electron. Energ.* **2015**, *28*, 133–142. [CrossRef]
53. Sekulic, D.L.; Sataric, B.M.; Tuszynski, J.A.; Sataric, M.V. Nonlinear ionic pulses along microtubules. *Eur. Phys. J. E Soft Matter* **2011**, *34*, 1–11. [CrossRef] [PubMed]
54. Sekulic, D.; Sataric, M.V.; Zivanov, M.B. Symbolic computation of some new nonlinear partial differential equations of nanobiosciences using modified extended tanh-function method. *Appl. Math. Comput.* **2011**, *218*, 3499–3506. [CrossRef]
55. Sataric, M.V.; Sekulic, D.; Zivanov, M.B. Solitoniclonic currents along microtubules. *J. Comput. Theor. Nanosci.* **2010**, *7*, 2281–2290. [CrossRef]
56. Zayed, E.M.E.; Alurrfi, K.A.E. The generalized projective riccati equations method and its applications for solving two nonlinear PDEs describing microtubules. *Int. J. Phys. Sci.* **2015**, *10*, 391–402.
57. Sekulic, D.L.; Sataric, M.V. Microtubule as Nanobioelectronic nonlinear circuit. *Serbian J. Electr. Eng.* **2012**, *9*, 107–119. [CrossRef]
58. Zdravkovic, S.; Sataric, M.V.; Maluckov, A.; Balaz, A. A nonlinear model of the dynamics of radial dislocations in microtubules. *Appl. Math. Comput.* **2014**, *237*, 227–237. [CrossRef]
59. Zdravkovic, S.; Sataric, M.V.; Zekovic, S. Nonlinear Dynamics of Microtibules—A longitudinal Model. *Europhys. Lett.* **2013**, *102*, 38002. [CrossRef]

7

Chaos Control in Three Dimensional Cancer Model by State Space Exact Linearization Based on Lie Algebra

Mohammad Shahzad

Nizwa College of Applied Sciences, Ministry of Higher Education, Nizwa 611, Oman; dmsinfinite@gmail.com

Academic Editor: J. Alberto Conejero

Abstract: This study deals with the control of chaotic dynamics of tumor cells, healthy host cells, and effector immune cells in a chaotic Three Dimensional Cancer Model (TDCM) by State Space Exact Linearization (SSEL) technique based on Lie algebra. A non-linear feedback control law is designed which induces a coordinate transformation thereby changing the original chaotic TDCM system into a controlled one linear system. Numerical simulation has been carried using *Mathematica* that witness the robustness of the technique implemented on the chosen chaotic system.

Keywords: SSEL; Lie bracket; Lie derivative; nonlinear feedback; TDCM

---

1. Introduction

Cancer pertains to a class of diseases characterized by out-of-control cell growth. Unsatisfactory performance of immune system against cancerous abnormal cells and consequently their chaotic growth leads to serious health damages and even death. So this chaotic nature of cancer needs to be controlled. The chaotic dynamics of cancer growth have been extensively studied in the literature to understand the mechanism of the disease and to predict its future behavior. Interactions of cancer cells with healthy host cells and immune system cells are the main components of these models and these interactions may yield different outcomes. Some important phenomena of cancer progression such as cancer dormancy, creeping through, and escape from immune surveillance have been investigated [1]. De Pillis and Radunskaya [2] included a normal tissue cell population in this model, performed phase space analysis, and investigated the effect of chemotherapy treatment by using optimal control theory whereas Kirschner and Panetta [3] examined the cancer cell growth in the presence of the effector immune cells and the cytokine IL-2 which has an essential role in the activation and stimulation of the immune system. They implied that antigenicity of the cancer cells plays an essential role in the recognition of cancer cells by the immune system. They observed oscillations in the cancer cell populations which is also demonstrated in the Kuznetsov's model and in addition, they obtained a stable limit cycle for some parameter range of the antigenicity. One can find many other models of the cancer-immune interactions with their dynamical analysis as well as investigations of optimal therapy effects. Although all these models include different cell populations, they share basic common characteristics such as existence of cancer free equilibria which is the main attention of investigating the therapy effects, coexisting equilibria where cancer and other cells are present in the body and in competition, and finally cancer escape and uncontrolled growth [4]. It has been observed that most of the interesting dynamics occur around coexisting equilibria which may yield oscillations in the cell populations, and converge to a stable limit cycle, as we mentioned before.

During the last decade, a lot of work based on different approaches has been done to control the chaos through the mathematical model representing cancer dynamics. Gohary and Alwasel [5] have studied the chaos and optimal control of cancer model with completely unknown parameters together with the asymptotic stability analysis of biologically feasible steady-states whereas Gohary [6] studied the problem of optimal control of cancer self-remission and cancer unstable steady-states together with the stability analysis of biologically feasible equilibrium states using a local stability approach. Baghernia *et al.* [7] considered the nonlinear prey-predator model to show the natural interaction between cancer and immune cells. Furthermore, a controlling strategy is proposed based on sliding mode control to convert the unstable states to the desired chaotic status.

Motivated by the aforementioned studies, the author aims to control the chaotic dynamics of TDCM using SSEL technique based on Lie algebra. The main advantages of this approach are that it is not only robust but also in this approach the control is injected only on healthy host cells (*i.e.*, $H(t)$) and effect of control can be seen on the rest of state variables (*i.e.*, $T(t)$ & $E(t)$).

Rest of the study has been organized as follows: In Section 2, a brief about the methodology has been described whereas Section 3 is on problem formulation. Section 4, contains the implementations of the technique on the considered system (TDCM). Section 5, is based on numerical simulations and lastly the whole study has been concluded in Section 6.

2. State Space Exact Linearization

It is a technique of controlling a chaotic system that involves transformation of a given non-linear system into a linear system by injecting a suitable control input [8]. Let us consider a non-linear dynamical system as:

$$\dot{x} = f(x) \tag{1}$$

After injecting control term it can be written as:

$$\dot{x} = f(x) = g(x)u(x) \tag{2}$$

where $x \in \mathbb{R}^n$, is the state vector; and $u \in \mathbb{R}$, is the control parameter, $f : \mathbb{R}^n \to \mathbb{R}^n$ and $g : \mathbb{R}^n \to \mathbb{R}^n$ are both smooth vector fields on $\mathbb{R}^n$ and Equation (2) is called feedback linearizable in the domain $\Omega \in \mathbb{R}^n$ if there exists a smooth reversible change of coordinates $z \in T(x)$, $x \in \Omega$ and a smooth transformation feedback $v = \alpha(x) + \beta(x)u$, $x \in \Omega$, where $v \in \mathbb{R}$ is the new control if the closed loop system is linear [9].

For two vector fields, $f(x)$ and $g(x)$, the different order Lie brackets are denoted by the symbols as follows:

$ad_f^k g(x) = [f, ad_f^{k-1} g](x)$ for $k = 1,2,3,...$ with $ad_f^0 g(x) = g(x)$, and each of $ad_f^k g(x) \in \mathbb{R}^n$ for $k = 1,2,3,....$ If we write $ad_f^k g(x) = \begin{bmatrix} \left(ad_f^k g(x)\right)_1 & \left(ad_f^k g(x)\right)_2 & \cdots & \left(ad_f^k g(x)\right)_n \end{bmatrix}^T$; then $\left(ad_f^k g(x)\right)_j$ is computed by the formula:

$$\left(ad_f^k g(x)\right)_j = \sum_{j=1}^{n} \left[f_i \frac{\partial}{\partial x_i} \left(ad_f^{k-1} g(x)\right)_j - \left(ad_f^{k-1} g(x)\right)_i \frac{\partial}{\partial x_i} f_j \right]$$

In some neighborhood $N(x_0)$ of a point x_0, if the matrix,

$$M = \begin{bmatrix} g(x_0) & ad_f g(x_0) & ad_f^2 g(x_0)... & ad_f^{n-1} g(x_0) \end{bmatrix}$$

has a rank n and $S = \text{span}\left\{g, ad_f g, ad_f^2 g, ..., ad_f^{n-2} g\right\}$ is involutive, then there exists a real valued function $\lambda(x) \in N(x_0)$, such that:

$L_g\lambda(x) = L_{ad_f g}\lambda(x) = L_{ad_f^2 g}\lambda(x) = ... = L_{ad_f^{n-2} g}\lambda(x) = 0$, and $L_{ad_f^{n-1} g}\lambda(x) \neq 0$ where $L_F\lambda(x)$ denotes the Lie derivative of the real valued function $\lambda(x)$ with respect to the vector field F. If that happens there exists transformation in $N(x_0)$, given by:

$$\begin{aligned} z &= \begin{bmatrix} z_1 & z_2 & z_3 & \dots & z_n \end{bmatrix}^T = T(x) \\ &= \begin{bmatrix} T_1(x) & T_2(x) & \dots & T_n(x) \end{bmatrix}^T \\ &= \begin{bmatrix} \lambda(x) & L_f\lambda(x) & \dots & L_f^{n-1}\lambda(x) \end{bmatrix}^T \end{aligned} \tag{3}$$

and $v = L_f^n\lambda(x) + L_g L_f^{n-1}\lambda(x)u$ that transforms the non-liner system into the linear controllable system [10–14]:

$$\begin{aligned} \dot{z}_1 &= z_2 \\ \dot{z}_2 &= z_3 \\ \dot{z}_{n-1} &= z_n \\ \dot{z}_n &= v \end{aligned} \tag{4}$$

3. Problem Formulation

The simplicity and elusiveness of the TDCM in its various forms have attracted the attention of mathematicians for decades. The equations of motion of TDCM [1] are given by:

$$\begin{aligned} \frac{dT}{dt} &= r_1 T\left(1 - \frac{T}{k_1}\right) - a_{12}TH - a_{13}TE \\ \frac{dH}{dt} &= r_2 H\left(1 - \frac{H}{k_2}\right) - a_{21}TH \\ \frac{dE}{dt} &= \frac{r_3 TE}{T+k_3} - a_{31}TE - d_3 E \end{aligned} \tag{5}$$

where $T(t)$ denotes the number of cancer cells; $H(t)$ denotes the healthy host cells and $E(t)$ denotes effecter immune cells at the time t; r_1 is the growth rate of cancer cells in the absence of any effect from other cell populations with maximum carrying capacity k_1, a_{12}, and a_{13} refers to the cancer cells killing rate by the healthy host cells and effecter cells respectively; r_2 is the growth rate of healthy host cells with maximum carrying capacity k_2; a_{21} is the rate of inactivation of the healthy cells by cancer cells. The rate of recognition of the cancer cells by the immune system depends on the antigenicity of the cancer cells. Since this recognition process is very complex, in order to keep the model simple, assume the stimulation of the immune system depends directly on the number of cancer cells with positive constants r_3 and k_3. The effecter cells are inactivated by the cancer cells at the rate a_{31} as well as they die naturally at the rate d_3. We assume that the cancer cells proliferate faster than the healthy cells (*i.e.*, $r_1 > r_2$) and all system parameters are being kept positive.

In order to make Equation (5) dimensionless, let us introduce: $x_1 = \frac{T}{k_1}$, $x_2 = \frac{H}{k_2}$, $x_3 = \frac{E}{k_3}$, $\tau = r_1 t$, $A_{12} = \frac{a_{12}k_2}{r_1}$, $A_{13} = \frac{a_{13}k_3}{r_1}$, $R_2 = \frac{r_2}{r_1}$, $A_{21} = \frac{a_{21}k_1}{r_1}$, $R_3 = \frac{r_3}{r_1}$, $K_3 = \frac{k_3}{k_1}$, $A_{31} = \frac{a_{31}k_1}{r_1}$, $D_3 = \frac{d_3}{r_1}$, the non-dimensional form of the TDCM Equation (5), can be written as:

$$\text{TDCM}: \begin{cases} \dot{x}_1 = x_1(1-x_1) - A_{12}x_1x_2 - A_{13}x_1x_3 \\ \dot{x}_2 = R_2x_2(1-x_2) - A_{21}x_1x_2 \\ \dot{x}_3 = \frac{R_3x_1x_3}{x_1+k_3} - A_{31}x_1x_3 - D_3x_3 \end{cases} \tag{6}$$

4. Control of the Chaotic System

In order to apply the control technique, the above system of Equation (6) can be written as $\dot{x} = f(x)$ where, $x = \begin{bmatrix} x_1 & x_2 & x_3 \end{bmatrix}^T$, and,

$$f(x) = \begin{bmatrix} x_1(1-x_1) - A_{12}x_1x_2 - A_{13}x_1x_3 & R_2x_2(1-x_2) - A_{21}x_1x_2 & \frac{R_3x_1x_3}{x_1+k_3} - A_{31}x_1x_3 - D_3x_3 \end{bmatrix}^T$$

Parametric entrainment control $u(x_1, x_2, x_3)$ is applied to the parameter R_2 in the second equation of the Equation (6) and one gets,

$$\dot{x} = f(x) + g(x)u \tag{7}$$

where $g(x) = \begin{bmatrix} 0 & x_2 & 0 \end{bmatrix}^T$ and $u \in \mathbb{R}$.

Using Equations (6) and (7), and using Lie bracket,

$$ad_f g(x) = \begin{bmatrix} A_{12}x_1x_2 & 2R_2x_2^2 & 0 \end{bmatrix}^T \tag{8}$$

$$ad_f^2 g(x) = \begin{bmatrix} A_{12}x_1^2x_2 & A_{21}^2x_1x_2^2 + 2R_2x_2^2(R_2 - A_{21}x_1) & -A_{21}x_1x_2\left\{\frac{R_3K_3x_3}{(x_1+K_3)^2} - A_{31}x_3\right\} \end{bmatrix}^T \tag{9}$$

Lemma 1. *For any $x_0 \in \mathbb{R}^3 \backslash span\{(1,0,0),(0,0,1)\} \cup span\{(0,1,0),(0,0,1)\}$, there exists an open set $N(x_0)$ containing x_0 where the matrix $M = \begin{bmatrix} g(x_0) & ad_f g(x_0) & ad_f^2 g(x_0) \end{bmatrix}$ has rank 3 and $S = span\left\{g, ad_f g\right\}$ is involutive.*

Proof. Let

$$|M| = \begin{vmatrix} 0 & A_{12}x_1x_2 & A_{12}x_1^2x_2 \\ x_2 & 2R_2x_2^2 & A_{21}^2x_1x_2^2 + 2R_2x_2^2(R_2 - A_{21}x_1) \\ 0 & 0 & -A_{21}x_1x_2\left\{\frac{R_3K_3x_3}{(x_1+K_3)^2} - A_{31}x_3\right\} \end{vmatrix} = A_{12}^2x_1^2x_2^3x_3\left(\frac{R_3K_3}{(x_1+K_3)^2} - A_{31}\right) \neq 0, \text{ for}$$

non-zero values of state variables involved in $|M|$. It is also confirmed by Figure 1.

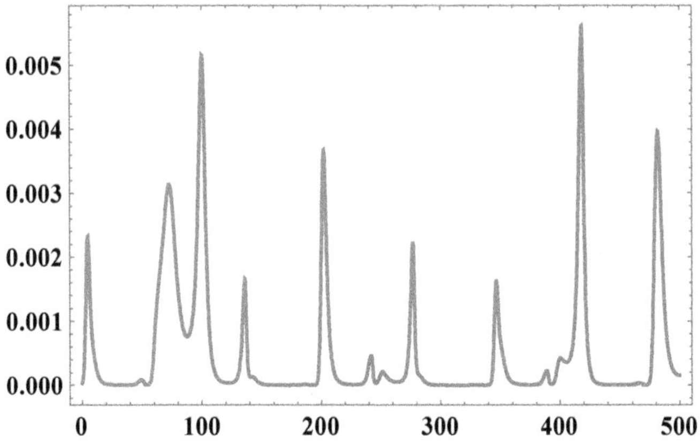

Figure 1. Time series of Det (M).

Therefore $\rho(M) = 3$ or equal to the order of the system. With the help of Equations (7) and (8) one can show that:

$$\left[g, ad_f g\right] = \begin{pmatrix} A_{12}x_1x_2 \\ 2R_2x_2^2 \\ 0 \end{pmatrix} = ad_f g(x) \tag{10}$$

which shows that $\left[g, ad_f g\right]$ belongs to $S = \text{span}\left[g, ad_f g\right]$. Hence, S is involutive.

Lemma 2. *For any thrice differentiable function $\Psi(x_3)$, there exists, a smooth transformation $z = \begin{bmatrix} \Psi & L_f\Psi & L_f^2\Psi \end{bmatrix}^T$ with a smooth inverse, defined on an open set $N(x_0)$ where $x_0 \in \mathbb{R}^3 \backslash span\{(1,0,0),(0,0,1)\} \cup span\{(0,1,0),(0,0,1)\}$ that reduces Equation (6) to a linear controllable form.*

Proof. Since Lemma 1 holds, there exists a real valued function $\lambda(x)$ such that $L_g\lambda(x) = 0$ and $L_{ad_f g}\lambda(x) = 0$ but, $L_{ad_f^2 g}\lambda(x) \neq 0$. Now, $L_g\lambda(x) = 0$ implies $\frac{\partial \lambda}{\partial x_2} = 0$ and $L_{ad_f g}\lambda(x) = 0$ implies $\frac{\partial \lambda}{\partial x_1} = 0$. Hence $\lambda(x)$ is independent of x_1 and x_2 but depends on x_3. Thus,

$$\lambda(x) = \Psi(x_3) \tag{11}$$

where simple calculations yield, $L_{ad_f^2 g}\lambda(x) = -A_{12}x_1x_2x_3\left\{\frac{R_3K_3}{(x_1+K_3)^2} - A_{31}\right\}\Psi'(x_3) \neq 0$ as x_1, x_2 & $x_3 \neq 0$.

With the help Equation (11) one can easily calculate the following Lie derivatives given as

$$\begin{aligned}
L_f\lambda(x) &= x_3\left(\frac{R_3x_1}{x_1+K_3} - A_{31}x_1 - D_3\right)\Psi'(x_3) \\
L_f^2\lambda(x) &= x_3A^2\Psi'(x_3) + x_3x_1BC\Psi'(x_3) + x_3^2A^2\Psi''(x_3) \\
L_f^3\lambda(x) &= x_3\left[A^3 + 3x_1ABC + x_3A^3 + x_3x_1ABC^2 - \frac{2R_3K_3x_1^2C^2}{(x_1+K_3)^3}\right]\Psi'(x_3) \\
&\quad -x_3x_1B\left[x_1C + A_{12}R_2x_2(1-x_2) + A_{12}A_{21}x_1x_2 + A_{13}x_3A\right]\Psi'(x_3) \\
&\quad +x_3^2A\left[3x_1BC + 2A^2\right]\Psi''(x_3) + x_3^3A^3\Psi'''(x_3) \\
L_gL_f^2\lambda(x) &= -x_1x_2x_3A_{12}B\Psi'(x_3)
\end{aligned} \tag{12}$$

where, $A = \frac{R_3x_1}{x_1+K_3} - A_{31}x_1 - D_3$, $B = \frac{R_3K_3}{(x_1+K_3)^2} - A_{31}$, $C = 1 - x_1 - A_{12}x_2 - A_{13}x_3$.

With the help of Equation (12), the transformation Equation (3) takes the form

$$z = \begin{bmatrix} z_1 \\ z_2 \\ z_3 \end{bmatrix} = T(x) = \begin{bmatrix} T_1(x) \\ T_2(x) \\ T_3(x) \end{bmatrix} = \begin{bmatrix} \lambda(x) \\ L_f\lambda(x) \\ L_f^2\lambda(x) \end{bmatrix} = \begin{bmatrix} \Psi(x_3) \\ x_3A\Psi'(x_3) \\ x_3\left(A^2 + x_1BC\right)\Psi'(x_3) + x_3^2A^2\Psi''(x_3) \end{bmatrix} \tag{13}$$

Inverse transformation can be calculated from Equation (13) as

$$x = \begin{bmatrix} x_1 \\ x_2 \\ x_3 \end{bmatrix} = T^{-1}(z) = \begin{bmatrix} T_1^{-1}(z) \\ T_2^{-1}(z) \\ T_3^{-1}(z) \end{bmatrix} \tag{14}$$

Controller u is obtained from the Equation (3) as

$$u = \frac{1}{L_gL_f^2\lambda(x)}\left[v - L_f^3\lambda(x)\right] \tag{15}$$

These will change the TDCM given by Equation (6) into a linear controllable system:

$$\begin{aligned}
\dot{z}_1 &= z_2 \\
\dot{z}_2 &= z_3 \\
\dot{z}_3 &= v
\end{aligned} \tag{16}$$

where v is considered linear in z_1, z_2, and z_3. Without loss of generality, one may choose the linear form of v as $v = a_1z_1 + a_2z_2 + a_3z_3$ where, $a_1, a_2, a_3 \in \mathbb{R}$.

Theorem 1 [8]. *(Stabilization at a point): The controller* $v = a_1z_1 + a_2z_2 + a_3z_3$ *stabilizes the equilibrium point* $z = (0,0,0)$ *iff* $a_1, a_2, a_3 < 0$ *and* $a_1 + a_2a_3 > 0$.

Theorem 2 [8]. *(Stabilization onto a limit cycle—Hopf bifurcation): The controller* $v = a_1z_1 + a_2z_2 + a_3z_3$ *stabilizes the system on to a stable limit cycle if* $a1, a2, a3 < 0$ *and* $a_1 + a_2a_3 > 0$.

Now, using Equations (12) and (15), the controller u can be written as:

$$u = \frac{-1}{x_1 x_2 x_3 A_{12} B}\left[\begin{array}{c} a_1\Psi(x_3) + \left(a_2 x_3 A + a_3 x_3 A^2 + a_3 x_3 x_1 BC\right)\Psi'(x_3) + a_3 x_3^2 A^2 \Psi''(x_3) \\ -x_3\left[A^3 + 3x_1 ABC + x_3 x_1 ABC^2 - \frac{2R_3K_3x_1^2C^2}{(x_1+K_3)^3}\right]\Psi'(x_3) \\ -x_3 x_1 B\left[x_1 C + A_{12}R_2 x_2(1-x_2) + A_{12}A_{21}x_1 x_2 + A_{13}x_3 A\right]\Psi'(x_3) \\ +x_3^2 A\left[3x_1 BC + 2A^2\right]\Psi''(x_3) + x_3^3 A^3 \Psi'''(x_3). \end{array}\right] \tag{17}$$

Equation (6) with an output function $\Psi(x_3)$ becomes a controlled one when the control loop is closed with control input given by Equation (17).

Now the problem is studied for the particular forms of $\Psi(x_3)$. However, there are many choices for $\Psi(x_3)$ (*i.e.*, linear, quadratic, *etc.* in x_3). Here we choose only linear because the other forms will give more complicated forms of u after tedious calculations.

Let, a linear output $\Psi(x_3) = x_3 - x_g$ will give control law:

$$u = \frac{-1}{x_1 x_2 x_3 A_{12} B}\left[\begin{array}{c} a_1(x_3 - x_g) + \left(a_2 x_3 A + a_3 x_3 A^2 + a_3 x_3 x_1 BC\right) \\ -x_3\left[A^3 + 3x_1 ABC + x_3 x_1 ABC^2 - \frac{2R_3K_3x_1^2C^2}{(x_1+K_3)^3}\right] \\ -x_3 x_1 B\left[x_1 C + A_{12}R_2 x_2(1-x_2) + A_{12}A_{21}x_1 x_2 + A_{13}x_3 A\right] \end{array}\right] \tag{18}$$

stabilizes Equation (6) to the control goal $\vec{x}_{g1} = \begin{pmatrix} 18.87 & -17.87 + 2.5x_g & x_g \end{pmatrix}^T$ or $\vec{x}_{g2} = \begin{pmatrix} 0.13 & 0.87 + 2.5x_g & x_g \end{pmatrix}^T$ where x_g is the parameter that determines the control goal.

Let $\Psi(x_3) = x_3 + K$, where K is arbitrary constant to be determined later. Controlling Equation (17) to the origin and changing the values of K, one can control x to the control goal x_g. In this case control law takes the form:

$$u = \frac{-1}{x_1 x_2 x_3 A_{12} B}\left[\begin{array}{c} a_1(x_3 + K) + \left(a_2 x_3 A + a_3 x_3 A^2 + a_3 x_3 x_1 BC\right) \\ -x_3\left[A^3 + 3x_1 ABC + x_3 x_1 ABC^2 - \frac{2R_3K_3x_1^2C^2}{(x_1+K_3)^3}\right] \\ -x_3 x_1 B\left[x_1 C + A_{12}R_2 x_2(1-x_2) + A_{12}A_{21}x_1 x_2 + A_{13}x_3 A\right] \end{array}\right] \tag{19}$$

for the transformations, from linear to nonlinear and *vice versa*, we have the following relations respectively,

$$\begin{bmatrix} z_1 \\ z_2 \\ z_3 \end{bmatrix} = \begin{bmatrix} x_3 + K \\ x_3 A \\ x_3 A^2 + x_3 x_1 BC \end{bmatrix} \tag{20}$$

Similarly, using reverse transformation, someone can find x in terms of z. Below the same has been written using *Mathematica*:

$$\begin{bmatrix} x_1 \\ x_2 \\ x_3 \end{bmatrix} = \begin{bmatrix} -2.5(D \pm E) \\ \dfrac{2\left(K - z_1 + z_3 + \frac{z_1^2}{z_1 - K}\right)}{5(z_1 - K)(D \pm E)\left(\frac{18}{25(D\pm E)^2} - \frac{1}{5}\right)} + 2.5\left(D \pm E - z_1 + K + 0.4\right) \\ (z_1 - K) \end{bmatrix} \tag{21}$$

where $D = -3.8 - \frac{z_2}{z_1 - K}$ and $E = \sqrt{14.04 + \frac{8.4z_2}{K - z_1} + \frac{z_2^2}{(K - z_1)^2}}$.

Since the feedback control law stabilizes the equilibrium point of Equation (16), then using Equation (21) and changing the value of K, one can control x_3 to the goal x_g and the variation of K is given by the formula:

$$K = -x_g \tag{22}$$

As x_3 goes to the goal x_g, the state vector x goes to $\vec{x}_{g1} = \begin{pmatrix} 18.87 & -17.87 + 2.5x_g & x_g \end{pmatrix}^T$ or $\vec{x}_{g2} = \begin{pmatrix} 0.13 & 0.87 + 2.5x_g & x_g \end{pmatrix}^T$.

5. Numerical Simulations

The system parameters involved in Equation (6) have been chosen as: $A_{12} = 1$; $A_{13} = 2.5$; $A_{21} = 1.5$; $R_2 = 0.6$; $A_{31} = 0.2$; $R_3 = 4.5$; $k_3 = 1$; $D_3 = 0.5$ with the initial conditions: $T(0) = 0.1$; $H(0) = 0.1$; $E(0) = 0.1$, and for $K = -1$; $a_1 = -0.1$; $a_2 = -0.9$; $a_3 = -0.6$, a controller given by Equation (19) is evaluated. Using *Mathematica*, the following different graphs have been plotted to show the robustness as well as effectiveness of the implemented technique. Figures 2 and 3 depict the uncontrolled and controlled time series of the state variables of the system Equation (6), respectively whereas the phase plots (a graph between state vector & its derivative) of three state vectors are shown in Figures 4–6 for controlled and uncontrolled one, respectively. Someone can observe that chaotic attractors are being replaced by regular ones as the controller is injected. Comparative parametric plots of controlled and uncontrolled state vectors pairwise and 3D have been sketched (Figures 7–11).

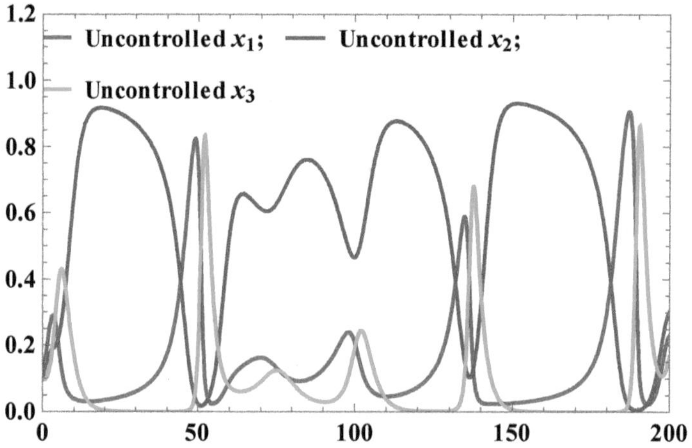

Figure 2. Time series of uncontrolled x_1, x_2, and x_3.

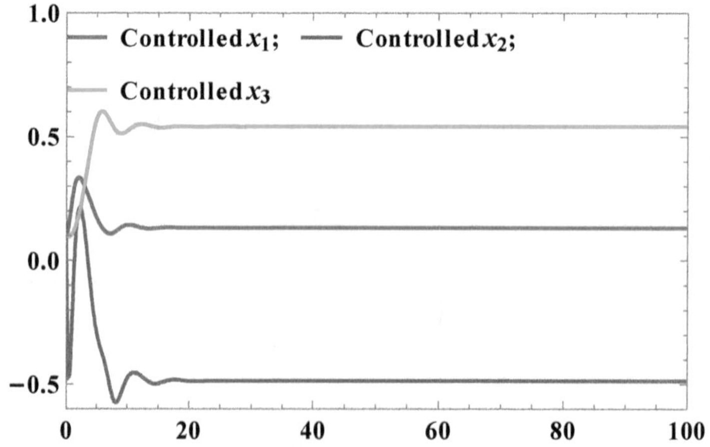

Figure 3. Time series of uncontrolled x_1, x_2, and x_3.

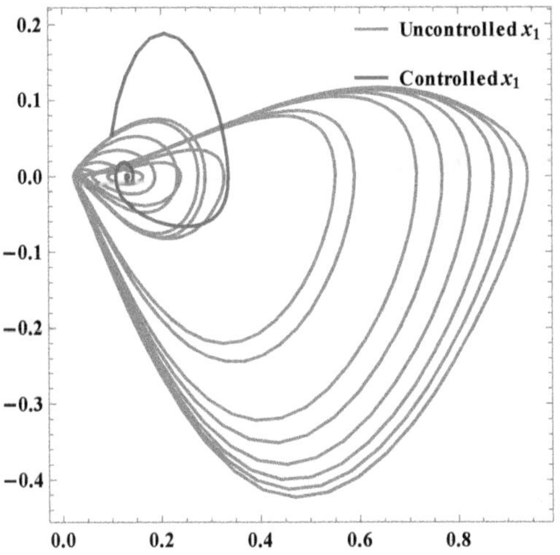

Figure 4. Phase plots of x_1.

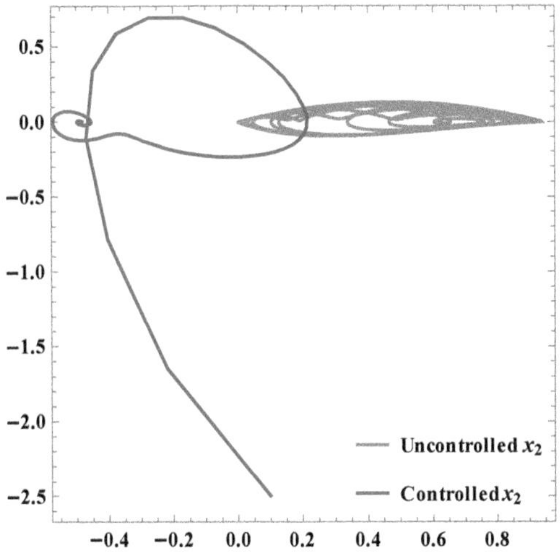

Figure 5. Phase plots of x_2.

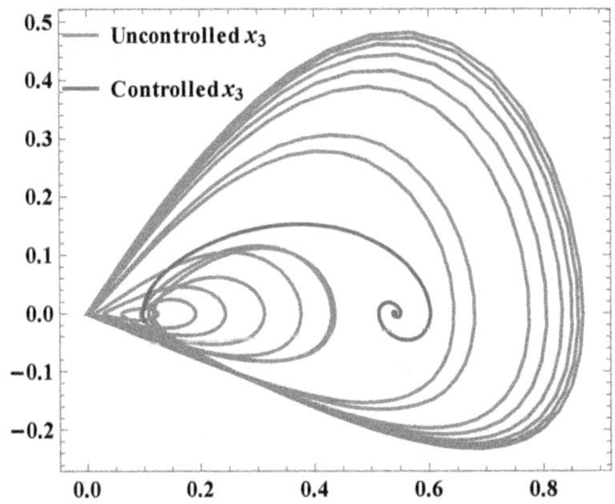

Figure 6. Phase plots of x_3.

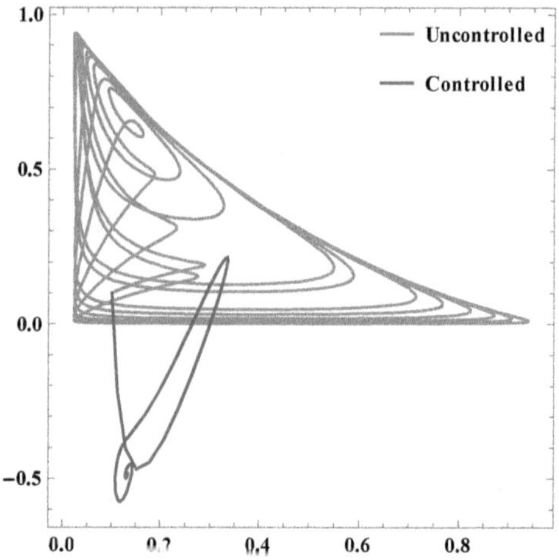

Figure 7. Parametric plots between x_1 & x_2.

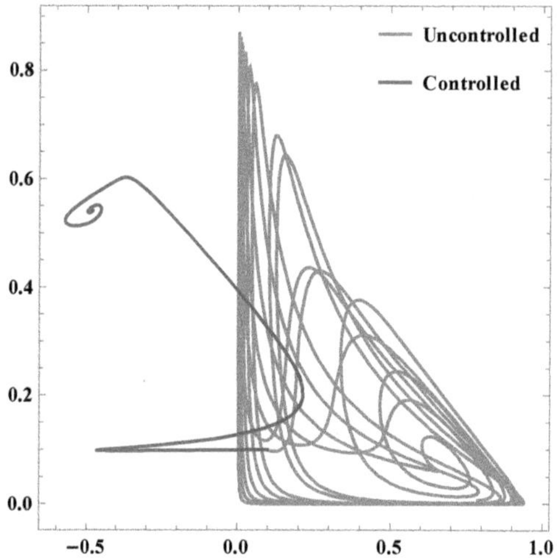

Figure 8. Parametric plots between x_2 & x_3.

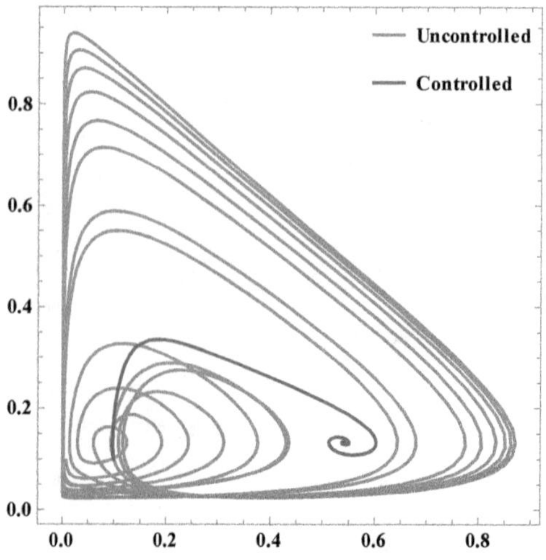

Figure 9. Parametric plots between x_3 & x_1.

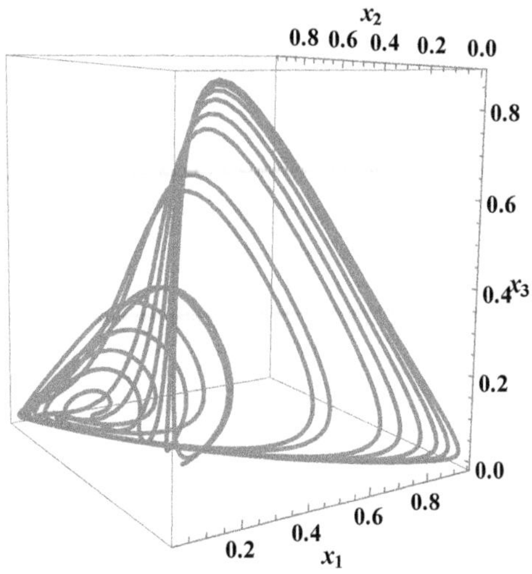

Figure 10. Parametric 3D plot of uncontrolled state vectors.

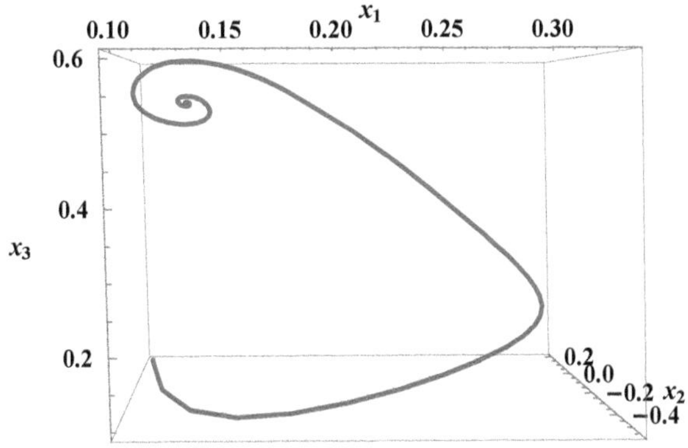

Figure 11. Parametric 3D plot of controlled state vectors.

6. Conclusions

In this paper, we have controlled the chaotic dynamics of TDCM using the SSEL method based on Lie algebra. Without the loss of generality, an equivalent linear system to the considered chaotic system has been obtained using Lie algebra. Also, a single control term has been injected to the chaotic system and the control has been observed in all three state vectors representing the number of cancer cells $T(t)$; the healthy host cells $H(t)$; and the effecter immune cells $E(t)$ in a very short time. The robustness of the technique in controlling the chaotic behavior can be observed through the presented plots.

Conflicts of Interest: The author declares no conflict of interest.

References

1. Itik, M.; Banks, S.P. Chaos in a three dimensional cancer model. *IJBC* **2010**, *20*, 71–79. [CrossRef]
2. De Pillis, L.G.; Radunskaya, A. The dynamics of an optimally controlled tumor model: A case study. *Math. Comput. Model.* **2003**, *37*, 1221–1244. [CrossRef]
3. Kirschner, D.; Panetta, J.C. Modeling immunotherapy of the tumor-immune interaction. *J. Math. Biol.* **1998**, *37*, 235–252. [CrossRef] [PubMed]
4. D'Onofrio, A. A general framework for modeling tumor-immune system competition and immunotherapy: Mathematical analysis and biomedical inferences. *Phys. D Nonlinear Phenom.* **2005**, *208*, 220–235. [CrossRef]

5. El-Gohary, A.; Alwasel, I.A. The chaos and optimal control of cancer model with complete unknown parameters. *Chaos Solitons Fractals* **2009**, *42*, 2865–2874. [CrossRef]
6. El-Gohary, A. Chaos and optimal control of cancer self-remission and tumor system steady states. *Chaos Solitons Fractals* **2008**, *37*, 1305–1316. [CrossRef]
7. Baghernia, P.; Moghaddam, R.K.; Kobravi, H. Cancer sliding mode control considering to chaotic manners of system. *J. Med. Imaging Health Inform.* **2015**, *5*, 448–457. [CrossRef]
8. Islam, M.; Islam, B.; Islam, N. Chaos control in Shimizu Morioka system by Lie algebraic exact linearization. *Int. J. Dyn. Control* **2014**, *2*, 386–394. [CrossRef]
9. Andrievskii, B.R.; Fradkov, A.L. Control of chaos: Methods and applications. *Autom. Remote Control* **2003**, *64*, 673–713. [CrossRef]
10. Chen, L.Q.; Liu, Y.Z. A modified exact linearization control for chaotic oscillators. *Nonlinear Dyn.* **1999**, *20*, 309–317. [CrossRef]
11. Liqun, C.; Yanzhu, L. Control of Lorenz chaos by the exact linearization. *Appl. Math. Mech.* **1998**, *19*, 67–73. [CrossRef]
12. Tsagas, G.R.; Mazumdar, H.P. On the control of a dynamical system by a linearization method via Lie Algebra. *Rev. Bull. Calcutta Math. Soc.* **2000**, *8*, 25–32.
13. Alvarez-Gallegos, J. Non-linear regulation of a Lorenz system by feedback linearization techniques. *Dyn. Control* **1994**, *4*, 272–289.
14. Shinbrot, T.; Grebogi, C.; Ott, E.; Yorkee, J.A. Using small perturbation to control chaos. *Nature* **1993**, *363*, 411–474. [CrossRef]

8

Birkhoff Normal Forms, KAM Theory and Time Reversal Symmetry for Certain Rational Map

Erin Denette [1], Mustafa R. S. Kulenović [1,*] and Esmir Pilav [2]

[1] Department of Mathematics, University of Rhode Island, Kingston, RI 02881-0816, USA; edenette@uri.edu
[2] Department of Mathematics, University of Sarajevo, 71000 Sarajevo, Bosnia and Herzegovina; esmir.pilav@pmf.unsa.ba
[*] Correspondence: mkulenovic@uri.edu

Academic Editor: Palle E.T. Jorgensen

Abstract: By using the KAM(Kolmogorov-Arnold-Moser) theory and time reversal symmetries, we investigate the stability of the equilibrium solutions of the system:

$$x_{n+1} = \frac{1}{y_n}, \quad y_{n+1} = \frac{\beta x_n}{1+y_n}, \qquad n = 0, 1, 2, \ldots,$$

where the parameter $\beta > 0$, and initial conditions x_0 and y_0 are positive numbers. We obtain the Birkhoff normal form for this system and prove the existence of periodic points with arbitrarily large periods in every neighborhood of the unique positive equilibrium. We use invariants to find a Lyapunov function and Morse's lemma to prove closedness of invariants. We also use the time reversal symmetry method to effectively find some feasible periods and the corresponding periodic orbits.

Keywords: area preserving map; Birkhoff normal form; difference equation; KAM theory; periodic solutions; symmetry; time reversal

MSC: 37E40, 37J40, 37N25, 39A28, 39A30

---

1. Introduction

The following rational system of difference equations:

$$\begin{cases} x_{n+1} &= \frac{1}{y_n} \\ y_{n+1} &= \frac{\beta x_n}{1+y_n} \end{cases}, \quad n = 0, 1, \ldots \tag{1}$$

and the corresponding equation:

$$y_{n+1} = \frac{\beta}{y_{n-1}(1+y_n)} \qquad n = 0, 1, \ldots, \tag{2}$$

where the parameter $\beta > 0$ and initial conditions x_0, y_0 are positive numbers were considered in [1] and [2]. The authors established the boundedness of all solutions of system (1) by using the invariant:

$$I(x_n, y_n) = \beta x_n + y_n + \frac{1}{x_n} + \frac{\beta}{y_n} + \frac{y_n}{x_n}. \tag{3}$$

Equation (2) and its invariant (3), where $x_n = 1/y_{n-1}$ were obtained in [3,4] and the stability of the equilibrium by means of Lyapunov function generated by invariant (3) was derived in [5,6], (pp. 247–250). Equation (2) is also a special case of equation:

$$y_{n+1} = \frac{By_n y_{n-1} + Ey_{n-1} + F}{by_n y_{n-1} + ey_{n-1} + f} \quad n = 0, 1, \ldots \tag{4}$$

with all nonnegative coefficients and initial conditions. Equation (4) is a rational difference equation with quadratic terms which is a subject of recent research, see [7–10].

In this paper, we will show that the corresponding map can be transformed into an area preserving map for which we will find the Birkhoff Normal form, and, using it, we will apply the KAM theorem to prove the stability of the unique positive equilibrium and the existence of periodic points with an arbitrarily large period in every neighborhood of the unique positive equilibrium. In addition, we prove that the corresponding map is conjugate to its inverse map through the involution map. Then, we will use this conjugacy to find some feasible periods of this map. The KAM theory will be enough to prove the stability of the equilibrium for $\beta \neq 2$, and then we use the invariant (3) and Morse's lemma to prove the stability in the remaining case $\beta = 2$, see [5]. In addition, Morse's lemma implies that all invariants are locally simple closed curves. A very recent paper [11] gives some effective tests for difference equation to have a continuous invariant. The method of invariants for the construction of a Lyapunov function and proving stability of the equilibrium points was used successfully in [5,6,12], and the KAM theory was used for the same objective in [12–16]. The class of difference equations which admit an invariant is not a large class even in the case of rational difference equations, see [11]. In the case when a difference equation's corresponding map is area preserving and does not possess an invariant, the only tool left seems to be KAM theory, see [17] for such an example. Furthermore, the corresponding Equation (4) can be embedded by iteration into a fourth order difference equation:

$$y_{n+1} = \frac{y_{n-2}y_{n-3}(1 + y_{n-1})(1 + y_{n-2})}{\beta + y_{n-2}(1 + y_{n-1})}, \quad n = 0, 1, \ldots,$$

which is increasing in all its arguments and yet exhibits the chaos.

Let T be the map associated to the system (1), *i.e.*,

$$T\begin{pmatrix} x \\ y \end{pmatrix} = \begin{pmatrix} \frac{1}{y} \\ \frac{\beta x}{1+y} \end{pmatrix}. \tag{5}$$

The map (5) has the unique fixed point $(1/\bar{y}, \bar{y})$ in the positive quadrant, where

$$\bar{y}^2(1 + \bar{y}) = \beta.$$

An invertible map $T : \mathbb{R}^2 \to \mathbb{R}^2$ is *area preserving* if the area of $T(A)$ equals the area of A for all measurable subsets A [6,18,19]. As is known, a differentiable map T is area preserving if the determinant of its Jacobian matrix is equal ± 1, that is $\det J_T = \pm 1$ at every point of domain of T, see [18,19]. We claim that in logarithmic coordinates (u, v) where $u = \ln(\bar{y}x)$, and $v = \ln(y/\bar{y})$, the map (5) is area preserving.

Lemma 1. *The map* (5) *is an area preserving map in the logarithmic coordinates.*

Proof. The Jacobian matrix of the map T is

$$J_T(x, y) = \begin{pmatrix} 0 & -\frac{1}{y^2} \\ \frac{\beta}{y+1} & -\frac{\beta x}{(y+1)^2} \end{pmatrix} \tag{6}$$

with

$$detJ_T(x,y) = \frac{\beta}{y^2(y+1)}.$$

We substitute $u = \ln(\bar{y}x)$, $v = \ln(y/\bar{y})$ and rewrite the map in (u,v) coordinates to obtain the map

$$\begin{pmatrix} u \\ v \end{pmatrix} \to \begin{pmatrix} -v \\ \ln\beta + u - \ln(e^v\bar{y}+1) - 2\ln\bar{y} \end{pmatrix}.$$

The Jacobian matrix of this map is

$$J_T(u,v) = \begin{pmatrix} 0 & -1 \\ 1 & \frac{1}{e^v\bar{y}+1} - 1 \end{pmatrix}, \tag{7}$$

and so $detJ_T(u,v) = 1$. □

A fixed point $(\bar{x},\bar{y})$ is an *elliptic point* of an area preserving map if the eigenvalues of $J_T(\bar{x},\bar{y})$ form a purely imaginary, complex conjugate pair $\lambda, \bar{\lambda}$, see [6,18].

Lemma 2. *The map T in (x,y) coordinates has an elliptic fixed point $(1/\bar{y},\bar{y})$. In the logarithmic coordinates, the corresponding fixed point is $(0,0)$.*

Proof. For the fixed points in (x,y) coordinates, solving $1/y = x$ and $\beta x/(1+y) = y$ yields the fixed point $(1/\bar{y},\bar{y})$ where $\bar{y}$ is the unique positive solution of $\bar{y}^2(1+\bar{y}) = \beta$. Evaluating the Jacobian matrix (6) of T at $(1/\bar{y},\bar{y})$ gives

$$J_T(1/\bar{y},\bar{y}) = \begin{pmatrix} 0 & -\frac{1}{\bar{y}^2} \\ \frac{\beta}{\bar{y}+1} & -\frac{\beta}{\bar{y}(\bar{y}+1)^2} \end{pmatrix}.$$

By using $\beta = \bar{y}^3 + \bar{y}^2$, we obtain that the eigenvalues of $J_T(1/\bar{y},\bar{y})$ are $\lambda, \bar{\lambda}$ where

$$\lambda = \frac{-\bar{y} + i\sqrt{(\bar{y}+2)(3\bar{y}+2)}}{2(\bar{y}+1)}. \tag{8}$$

Since $|\lambda| = 1$, we have that $(1/\bar{y},\bar{y})$ is an elliptic fixed point.

Under the logarithmic coordinate change $(x,y) \to (u,v)$, the fixed point $(1/\bar{y},\bar{y})$ becomes $(0,0)$. Evaluating the Jacobian matrix (7) of T at $(0,0)$ gives

$$J_T(0,0) = \begin{pmatrix} 0 & -1 \\ 1 & \frac{1}{\bar{y}+1} - 1 \end{pmatrix}$$

with eigenvalues which are given by (8). □

The rest of the paper is organized into three sections. The second section contains a derivation of the Birkhoff normal form for map T and an application of the KAM theory, which proves stability of the equilibrium and the existence of an infinite number of periodic solutions for $\beta \neq 2$. The third section makes use of the invariant (3) in proving stability for $\beta = 2$ and the construction of a Lyapunov function. The fourth section uses the symmetries for the map T showing that this map is conjugate to its inverse through an involution. Then, we use time reversal symmetry method [13,20] based on the symmetries to effectively find some feasible periods and corresponding orbits of the map T.

2. The KAM Theory and Birkhoff Normal Form

The KAM Theorem asserts that, in any sufficiently small neighborhood of a non degenerate elliptic fixed point of a smooth area-preserving map, there exists many invariant closed curves. We explain this theorem in some detail. Consider a smooth, area-preserving map $(x,y) \to T(x,y)$ of the plane that has $(0,0)$ as an elliptic fixed point. After a linear transformation, one can represent the map in the form

$$z \to \lambda z + g(z,\bar{z}),$$

where λ is the eigenvalue of the elliptic fixed point, $z = x + iy$ and $\bar{z} = x - iy$ are complex variables, and g vanishes with its derivative at $z = 0$. Assume that the eigenvalue λ of the elliptic fixed point satisfies the non-resonance condition $\lambda^k \neq 1$ for $k = 1, \ldots, q$, for some $q \geq 4$. Then, Birkhoff showed that there exists new, canonical complex coordinates $(\zeta,\bar{\zeta})$ relative to which the mapping takes the normal form

$$\zeta \to \lambda\zeta e^{i\tau(\zeta\bar{\zeta})} + h(\zeta,\bar{\zeta})$$

in a neighborhood of the elliptic fixed point, where $\tau(\zeta\bar{\zeta}) = \tau_1|\zeta|^2 + \ldots + \tau_s|\zeta|^{2s}$ is a real polynomial, $s = [(q-2)/2]$ and h vanishes with its derivatives up to order $q-1$. The numbers $\tau_1, \ldots, \tau_s$ are called twist coefficients. Consider an invariant annulus $\epsilon < |\zeta| < 2\epsilon$ in a neighborhood of an elliptic fixed point, for ϵ, a very small positive number. Note that, if we neglect the remainder h, the normal form approximation $\zeta \to \lambda\zeta e^{i\tau(\zeta\bar{\zeta})}$ leaves invariant all circles $|\zeta|^2 = const$. The motion restricted to each of these circles is a rotation by some angle. In addition, please note that, if at least one of the twist coefficients τ_j is nonzero, the angle of rotation will vary from circle to circle. A radial line through the fixed point will undergo twisting under the map. The KAM theorem (Moser's twist theorem) says that, under the addition of the remainder term, most of these invariant circles will survive as invariant closed curves under the full map.

Theorem 3. *Assume that $\tau(\zeta\bar{\zeta})$ is not identically zero and ϵ is sufficiently small, then the map T has a set of invariant closed curves of positive Lebesque measure close to the original invariant circles. Moreover, the relative measure of the set of surviving invariant curves approaches full measure as ϵ approaches 0. The surviving invariant closed curves are filled with dense irrational orbits.*

The KAM theorem requires that the elliptic fixed point be non-resonant and non degenerate. Note that for $q = 4$ the non-resonance condition $\lambda^k \neq 1$ requires that $\lambda \neq \pm 1$ or $\pm i$. The above normal form yields the approximation

$$\zeta \to \lambda\zeta + c_1\zeta^2\bar{\zeta} + O(|\zeta|^4)$$

with $c_1 = i\lambda\tau_1$ and τ_1 being the first twist coefficient. We will call an elliptic fixed point non-degenerate if $\tau_1 \neq 0$.

Consider a general map T that has a fixed point at the origin with complex eigenvalues λ and $\bar{\lambda}$ satisfying $|\lambda| = 1$ and $Im(\lambda) \neq 0$. By putting the linear part of such a map into Jordan Normal form, we may assume that T has the following form near the origin

$$T\begin{pmatrix} x_1 \\ x_2 \end{pmatrix} = \begin{pmatrix} Re(\lambda) & -Im(\lambda) \\ Im(\lambda) & Re(\lambda) \end{pmatrix}\begin{pmatrix} x_1 \\ x_2 \end{pmatrix} + \begin{pmatrix} g_1(x_1,x_2) \\ g_2(x_1,x_2) \end{pmatrix}.$$

One can now pass to the complex coordinates $z = x_1 + ix_2$ to obtain the complex form of the system

$$z \to \lambda z + \xi_{20}z^2 + \xi_{11}z\bar{z} + \xi_{02}\bar{z}^2 + \xi_{30}z^3 + \xi_{21}z^2\bar{z} + \xi_{12}z\bar{z}^2 + \xi_{03}\bar{z}^3 + O(|z|^4).$$

The coefficient c_1 can be computed directly using the formula below derived by Wan in the context of Hopf bifurcation theory [21]. In [22], it is shown that, when one uses area-preserving coordinate changes, Wan's formula yields the twist coefficient τ_1 that is used to verify the non-degeneracy condition necessary to apply the KAM theorem. We use the formula:

$$c_1 = \frac{\xi_{20}\xi_{11}(\bar{\lambda}+2\lambda-3)}{(\lambda^2-\lambda)(\bar{\lambda}-1)} + \frac{|\xi_{11}|^2}{1-\bar{\lambda}} + \frac{2|\xi_{02}|^2}{\lambda^2-\bar{\lambda}} + \xi_{21},$$

where

$$\xi_{20} = \frac{1}{8}\{(g_1)_{x_1x_1} - (g_1)_{x_2x_2} + 2(g_2)_{x_1x_2} + i[(g_2)_{x_1x_1} - (g_2)_{x_2x_2} - 2(g_1)_{x_1x_2}]\},$$

$$\xi_{11} = \frac{1}{4}\{(g_1)_{x_1x_1} + (g_1)_{x_2x_2} + i[(g_2)_{x_1x_1} + (g_2)_{x_2x_2}]\},$$

$$\xi_{02} = \frac{1}{8}\{(g_1)_{x_1x_1} - (g_1)_{x_2x_2} - 2(g_2)_{x_1x_2} + i[(g_2)_{x_1x_1} - (g_2)_{x_2x_2} + 2(g_1)_{x_1x_2}]\},$$

$$\xi_{21} = \frac{1}{16}((g_1)_{x_1x_1x_1} + (g_1)_{x_1x_2x_2} + (g_2)_{x_1x_1x_2} + (g_2)_{x_2x_2x_2}),$$

$$+\frac{i}{16}((g_2)_{x_1x_1x_1} + (g_2)_{x_1x_2x_2} - (g_1)_{x_1x_1x_2} - (g_1)_{x_2x_2x_2}).$$

Theorem 4. *The elliptic fixed point $(0,0)$, in the (u,v) coordinates, is non-degenerate for $\beta \neq 2$ and non-resonant for $\beta > 0$.*

Proof. Let F be the function defined by

$$F\begin{pmatrix} u \\ v \end{pmatrix} = \begin{pmatrix} -v \\ \ln\beta + u - \ln(e^v\bar{y}+1) - 2\ln\bar{y} \end{pmatrix}.$$

Then, F has the unique elliptic fixed point $(0,0)$. The Jacobian matrix of F at (u,v) is given by

$$J_F(u,v) = \begin{pmatrix} 0 & -1 \\ 1 & \frac{1}{e^v\bar{y}+1} - 1 \end{pmatrix}.$$

At $(0,0)$, $J_F(u,v)$ has the form

$$J_0 = J_F(0,0) = \begin{pmatrix} 0 & -1 \\ 1 & \frac{1}{\bar{y}+1} - 1 \end{pmatrix}. \quad (9)$$

The eigenvalues of (9) are λ and $\bar{\lambda}$ where

$$\lambda = \frac{-\bar{y} + i\sqrt{(\bar{y}+2)(3\bar{y}+2)}}{2(\bar{y}+1)}.$$

One can prove that

$$|\lambda| = 1$$

$$\lambda^2 = -\frac{\bar{y}(\bar{y}+4)+2}{2(\bar{y}+1)^2} - \frac{i\sqrt{(\bar{y}+2)(3\bar{y}+2)}\bar{y}}{(\bar{y}+1)(2\bar{y}+2)}$$

$$\lambda^3 = \frac{\bar{y}(2\bar{y}(\bar{y}+3)+3)}{2(\bar{y}+1)^3} - \frac{i(2\bar{y}+1)\sqrt{(\bar{y}+2)(3\bar{y}+2)}}{2(\bar{y}+1)^3},$$

$$\lambda^4 = \frac{\bar{y}(-\bar{y}^3+8\bar{y}+8)+2}{2(\bar{y}+1)^4} + \frac{i\bar{y}\sqrt{(\bar{y}+2)(3\bar{y}+2)}(\bar{y}(\bar{y}+4)+2)}{2(\bar{y}+1)^4}$$

from which follows that $\lambda^k \neq 1$ for $k = 1, 2, 3, 4$ and $\beta > 0$.

Now, we have that

$$F\begin{pmatrix} u \\ v \end{pmatrix} = \begin{pmatrix} 0 & -1 \\ 1 & \frac{1}{\bar{y}+1} - 1 \end{pmatrix}\begin{pmatrix} u \\ v \end{pmatrix} + \begin{pmatrix} f_1(\delta, u, v) \\ f_2(\delta, u, v) \end{pmatrix},$$

where

$$\begin{aligned} f_1(\delta, u, v) &= 0, \\ f_2(\delta, u, v) &= \frac{v\bar{y}}{\bar{y}+1} - \ln\left(e^v \bar{y} + 1\right) - 2\ln\bar{y} + \ln\beta \end{aligned}$$

Then, the system $(u_{n+1}, v_{n+1}) = F(u_n, v_n)$ is equivalent to

$$\begin{pmatrix} u_{n+1} \\ v_{n+1} \end{pmatrix} = \begin{pmatrix} 0 & -1 \\ 1 & \frac{1}{\bar{y}+1} - 1 \end{pmatrix}\begin{pmatrix} u_n \\ v_n \end{pmatrix} + \begin{pmatrix} f_1(u_n, v_n) \\ f_2(u_n, v_n) \end{pmatrix}.$$

Let

$$\begin{pmatrix} u_n \\ v_n \end{pmatrix} = P\begin{pmatrix} \tilde{u}_n \\ \tilde{v}_n \end{pmatrix},$$

where

$$P = \frac{1}{\sqrt{D}}\begin{pmatrix} \frac{\bar{y}}{2(\bar{y}+1)} & -\frac{\sqrt{(\bar{y}+2)(3\bar{y}+2)}}{2(\bar{y}+1)} \\ 1 & 0 \end{pmatrix}, \quad P^{-1} = \sqrt{D}\begin{pmatrix} 0 & 1 \\ -\frac{2(\bar{y}+1)}{\sqrt{(\bar{y}+2)(3\bar{y}+2)}} & \frac{\bar{y}}{\sqrt{(\bar{y}+2)(3\bar{y}+2)}} \end{pmatrix}$$

and

$$D = \frac{\sqrt{(\bar{y}+2)(3\bar{y}+2)}}{2(\bar{y}+1)}.$$

Thus, the system $(u_{n+1}, v_{n+1}) = F(u_n, v_n)$ is transformed into its Birkhoff normal form

$$\begin{pmatrix} \tilde{u}_{n+1} \\ \tilde{v}_{n+1} \end{pmatrix} = \begin{pmatrix} -\frac{\bar{y}}{2\bar{y}+2} & -\frac{\sqrt{(\bar{y}+2)(3\bar{y}+2)}}{2(\bar{y}+1)} \\ \frac{\sqrt{(\bar{y}+2)(3\bar{y}+2)}}{2\bar{y}+2} & -\frac{\bar{y}}{2\bar{y}+2} \end{pmatrix}\begin{pmatrix} \tilde{u}_n \\ \tilde{v}_n \end{pmatrix} + P^{-1}H\left(P\begin{pmatrix} \tilde{u}_n \\ \tilde{v}_n \end{pmatrix}\right),$$

where

$$H\begin{pmatrix} u \\ v \end{pmatrix} := \begin{pmatrix} f_1(u, v) \\ f_2(u, v) \end{pmatrix}.$$

Let

$$G\begin{pmatrix} u \\ v \end{pmatrix} = \begin{pmatrix} g_1(u, v) \\ g_2(u, v) \end{pmatrix} = P^{-1}H\left(P\begin{pmatrix} u \\ v \end{pmatrix}\right).$$

By a straightforward calculation, we obtain that

$$\begin{aligned} g_1(u, v) &= \sqrt{D}\left(\frac{u\bar{y}}{\sqrt{D}(\bar{y}+1)} - \ln\left(\bar{y}e^{\frac{u}{\sqrt{D}}} + 1\right) - 2\ln\bar{y} + \ln\beta\right) \\ g_2(u, v) &= \frac{\bar{y}\left(\sqrt{D}(\bar{y}+1)\left(-\ln\left(\bar{y}e^{\frac{u}{\sqrt{D}}} + 1\right) - 2\ln\bar{y} + \ln\beta\right) + u\bar{y}\right)}{(\bar{y}+1)\sqrt{(\bar{y}+2)(3\bar{y}+2)}} \end{aligned}$$

Another calculation gives

$$\begin{aligned}
\xi_{20}|_{u=v=0} &= -\frac{\bar{y}\left(\sqrt{(\bar{y}+2)(3\bar{y}+2)}+i\bar{y}\right)}{8\sqrt{D}(\bar{y}+1)^2\sqrt{(\bar{y}+2)(3\bar{y}+2)}}\\
\xi_{11}|_{u=v=0} &= -\frac{\bar{y}\left(1+\frac{i\bar{y}}{\sqrt{(\bar{y}+2)(3\bar{y}+2)}}\right)}{4\sqrt{D}(\bar{y}+1)^2}\\
\xi_{02}|_{u=v=0} &= -\frac{\bar{y}\left(\sqrt{(\bar{y}+2)(3\bar{y}+2)}+i\bar{y}\right)}{8\sqrt{D}(\bar{y}+1)^2\sqrt{(\bar{y}+2)(3\bar{y}+2)}}\\
\xi_{21}|_{u=v=0} &= \frac{(\bar{y}-1)\bar{y}\left(\sqrt{(\bar{y}+2)(3\bar{y}+2)}+i\bar{y}\right)}{16D(\bar{y}+1)^3\sqrt{(\bar{y}+2)(3\bar{y}+2)}}
\end{aligned}.$$

By using

$$\begin{aligned}
\xi_{20}\xi_{11} &= \frac{\bar{y}^2\left(\sqrt{(\bar{y}+2)(3\bar{y}+2)}+i\bar{y}\right)^2}{32D(\bar{y}+1)^4(\bar{y}+2)(3\bar{y}+2)}\\
\xi_{11}\overline{\xi_{11}} &= \frac{\bar{y}^2}{4D(\bar{y}+1)^2(3\bar{y}^2+8\bar{y}+4)}\\
\xi_{02}\overline{\xi_{02}} &= \frac{\bar{y}^2}{16D(\bar{y}+1)^2(3\bar{y}^2+8\bar{y}+4)}
\end{aligned},$$

a straightforward calculation yields

$$\begin{aligned}
c_1 &= \frac{\xi_{20}\xi_{11}(\bar{\lambda}+2\lambda-3)}{(\lambda^2-\lambda)(\bar{\lambda}-1)}+\frac{|\xi_{11}|^2}{1-\bar{\lambda}}+\frac{2|\xi_{02}|^2}{\lambda^2-\bar{\lambda}}+\xi_{21}\\
&= \frac{(\bar{y}-1)\bar{y}(\bar{y}+1)}{(3\bar{y}+2)\sqrt{(\bar{y}+2)(3\bar{y}+2)}\left(4+\bar{y}\left(3\bar{y}-i\sqrt{(\bar{y}+2)(3\bar{y}+2)}+8\right)\right)}
\end{aligned}.$$

It can be proved that

$$\tau_1 = -i\bar{\lambda}c_1 = -\frac{(\bar{y}-1)\bar{y}}{2(\bar{y}+2)(3\bar{y}+2)^2},$$

which implies that $\tau_1 \neq 0$ for $\beta \neq 2$ since $\bar{y}^2(1+\bar{y}) = \beta$.
□

The following result is a consequence of Moser's twist map theorem [13,19,23,24].

Theorem 5. *Let T be a map* (5) *associated to the system* (1), *and $(\bar{x},\bar{y})$ be a non-degenerate elliptic fixed point. If $\beta \neq 2$, then there exist periodic points with arbitrarily large periods in every neighborhood of $(\bar{x},\bar{y})$. In addition, $(\bar{x},\bar{y})$ is a stable fixed point.*

3. Invariant

In this section, we use the invariant to find a Lyapunov function and prove stability of the equilibrium for all values of parameter $\beta > 0$, see [5,6] for similar results.

Lemma 6. *The unique equilibrium $(\frac{1}{\bar{y}},\bar{y})$ of Equation* (1) *is a critical point of the invariant* (3).

Proof. The system (1) possesses an invariant given by Equation (3). The function $I(x,y)$ associated with Equation (3) has partial derivatives

$$\frac{\partial I}{\partial x} = \beta - \frac{1}{x^2} - \frac{y}{x^2}, \quad \frac{\partial I}{\partial y} = 1 - \frac{\beta}{y^2} + \frac{1}{x}. \tag{10}$$

The unique equilibrium of Equation (1) satisfies that $\overline{x} = \frac{1}{\overline{y}}$ and $\overline{y}^2(1+\overline{y}) = \beta$. Hence $\overline{x}$ is the unique positive solution of the equation $\beta x^3 - x - 1 = 0$. Equation (10) implies that any critical point (x,y) of Equation (3) satisfies $y = \beta x^2 - 1$ and $(x+1)y^2 - \beta x = 0$. Substitution yields

$$\beta^2 x^5 + \beta^2 x^4 - 2\beta x^3 - 2\beta x^2 + (1-\beta)x + 1 = 0. \tag{11}$$

Equation (11) can be rewritten as

$$(\beta x^3 - x - 1)(\beta x^2 + \beta x - 1) = 0,$$

which has $\overline{x}$ as a solution. □

Lemma 7. *The graph of the function $I(x,y)$ associated with Equation (3) is a simple closed curve in a neighborhood of the equilibrium of Equation (1). The equilibrium point $(\bar{x},\bar{y})$ is stable.*

Proof. The Hessian matrix associated with $I(x,y)$ is

$$H(x,y) = \begin{bmatrix} \frac{2}{x^3} + \frac{2y}{x^3} & -\frac{1}{x^2} \\ -\frac{1}{x^2} & \frac{2\beta}{y^3} \end{bmatrix}$$

with determinant

$$\det(H(x,y)) = \frac{4\beta(1+y)}{x^3y^3} - \frac{1}{x^4}.$$

For (x,y) a critical point of $I(x,y)$,

$$\det(H(x,y)) = \frac{4\beta^2}{xy^3} - \frac{1}{x^4}. \tag{12}$$

For $(\overline{x},\overline{y})$ the unique equilibrium of Equation (1), we can further reduce Equation (12) to

$$\det(H(\overline{x},\overline{y})) = \frac{4\beta^2\overline{y}}{\overline{y}^3} - \overline{y}^4 = \frac{4\beta^2 - \overline{y}^6}{\overline{y}^2}.$$

Note that since the equation $\beta x^3 - x - 1 = 0$ has $\bar{x}$ as its unique positive solution, the equation $\beta/y^3 - 1/y - 1 = \frac{\beta - y^2 - y^3}{y^3} = 0$ has $\bar{y}$ as its unique positive solution. Let us define $f(y) = y^3 + y^2 - \beta$. We observe that $f(0) = -\beta < 0$ and $f(\sqrt[6]{4\beta^2}) = 2\beta + \sqrt[3]{4\beta^2} - \beta > 0$, which guarantees that $0 < \overline{y} < \sqrt[6]{4\beta^2}$. Now the Morse's lemma [18] guarantees the result provided $\det(H(\overline{x},\overline{y})) > 0$. However, $\frac{4\beta^2 - \overline{y}^6}{\overline{y}^2} > 0$ if and only if $\overline{y} < \sqrt[6]{4\beta^2}$, which is indeed the case. In view of the Morse's lemma [18], the level sets of the function $I(x,y)$ are diffeomorphic to circles in the neighborhood of $(\overline{x},\overline{y})$. In addition, the function

$$V(x,y) = I(x,y) - I(\bar{x},\bar{y})$$

is a Lyapunov function, and so the equilibrium point $(\bar{x},\bar{y})$ is stable, see [5]. □

See Figure 1 for the family of invariant curves around the equilibrium. See Figure 2 for the bifurcation diagrams which indicate the appearance of chaos.

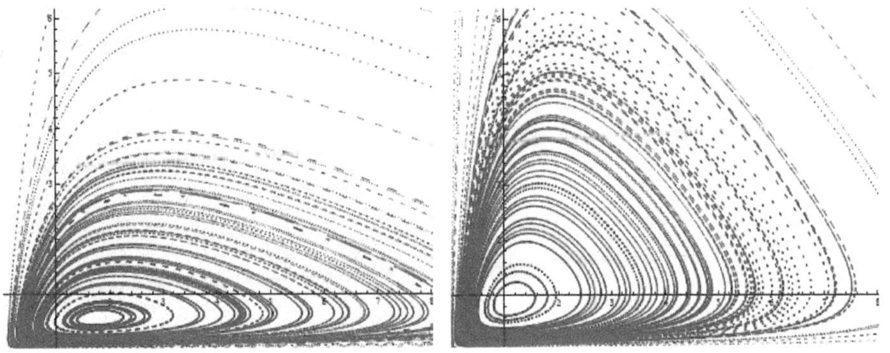

(a) (b)

Figure 1. Some orbits of the map T for (**a**) $a = 0.5$ and (**b**) $a = 1.5$. The plots are generated by Dynamica 3 [6].

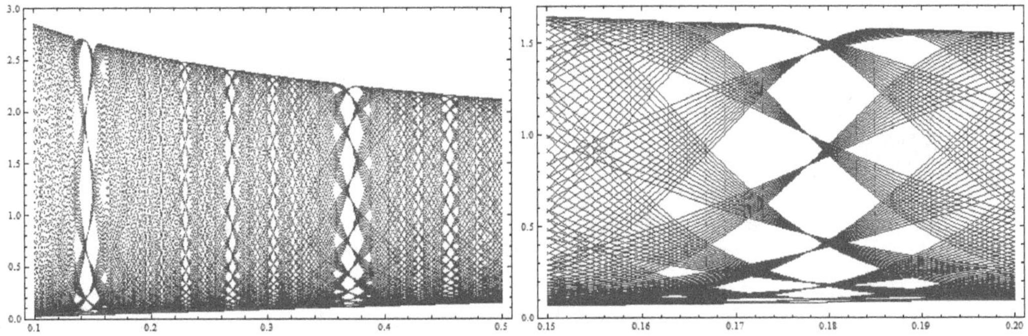

Figure 2. A bifurcation diagram in $(\beta - x)$-plane. The plots are generated by Dynamica 3 [6].

4. Symmetries

In the study of area-preserving maps, symmetries play an important role since they yield special dynamic behavior. A transformation R of the plane is said to be a *time reversal symmetry* for T if $R^{-1} \circ T \circ R = T^{-1}$, meaning that applying the transformation R to the map T is equivalent to iterating the map backwards in time, see [13,20]. If the time reversal symmetry R is an involution, *i.e.*, $R^2 = id$, then the time reversal symmetry condition is equivalent to $R \circ T \circ R = T^{-1}$, and T can be written as the composition of two involutions $T = I_1 \circ I_0$, with $I_0 = R$ and $I_1 = T \circ R$. Note that if $I_0 = R$ is a reversor, then so is $I_1 = T \circ R$. In addition, the jth involution, defined as $I_j := T^j \circ R$, is also a reversor.

The invariant sets of the involution maps,

$$S_{0,1} = \{(x,y) | I_{0,1}(x,y) = (x,y)\}$$

are one-dimensional sets called the symmetry lines of the map. Once the sets $S_{0,1}$ are known, the search for periodic orbits can be reduced to a one-dimensional root finding problem using the following result, see [13,20]:

Theorem 8. *If* $(x,y) \in S_{0,1}$ *then* $T^n(x,y) = (x,y)$ *if and only if*

$$\begin{cases} T^{n/2}(x,y) \in S_{0,1}, & \text{for } n \text{ even;} \\ T^{(n\pm1)/2}(x,y) \in S_{1,0}, & \text{for } n \text{ odd.} \end{cases}$$

That is, according to this result, periodic orbits can be found by searching in the one-dimensional sets $S_{0,1}$, rather than in the whole domain. Periodic orbits of different orders can then be found at the intersection of the symmetry lines S_j $j = 1, 2, \ldots$ associated to the jth involution; for example, if $(x, y) \in S_j \cap S_k$, then $T^{j-k}(x, y) = (x, y)$. In addition, the symmetry lines are related to each other by the following relations: $S_{2j+i} = T^j(S_i)$, $S_{2j-i} = I_j(S_i)$, for all i, j.

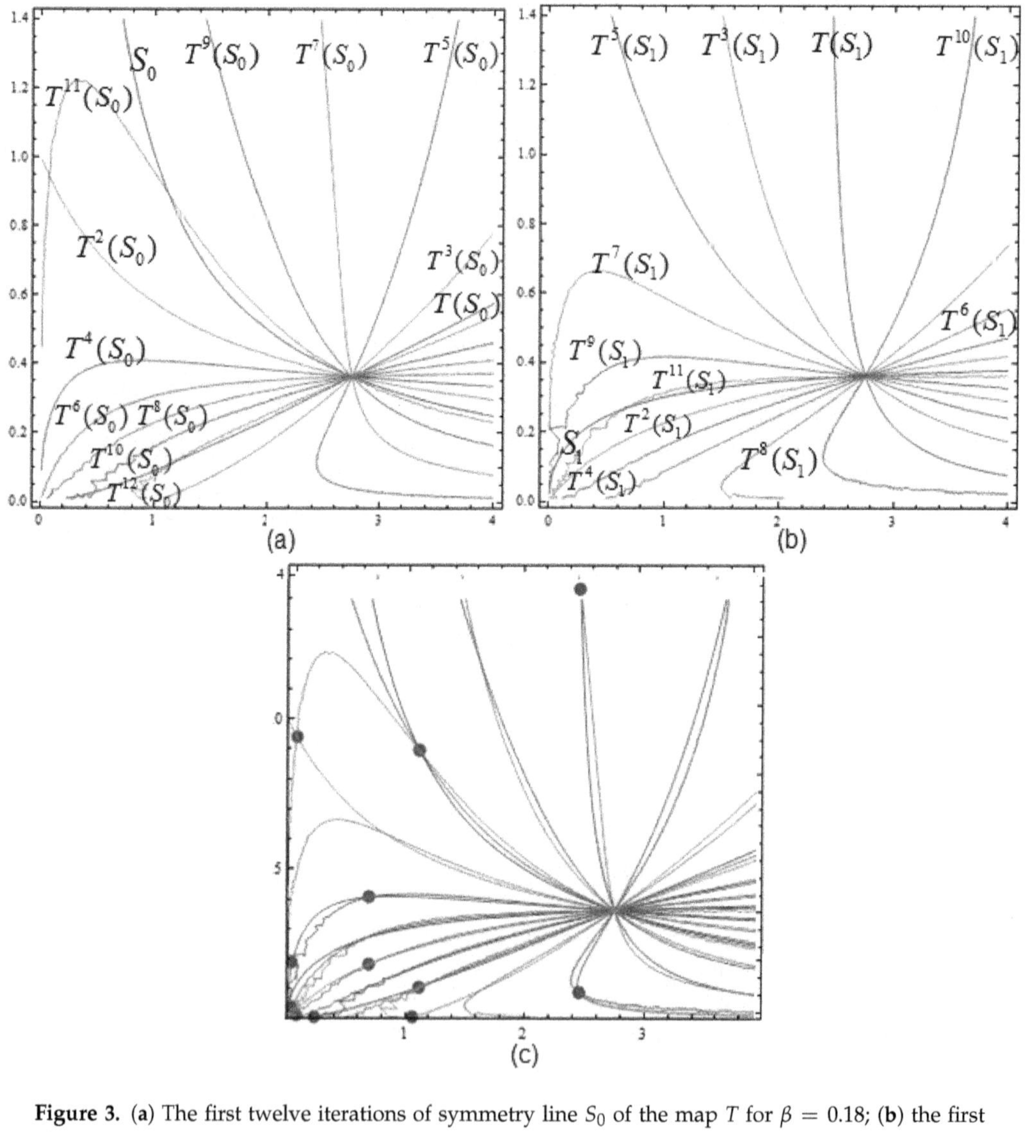

Figure 3. (**a**) The first twelve iterations of symmetry line S_0 of the map T for $\beta = 0.18$; (**b**) the first eleven iterations of symmetry line S_1 of the map T for $\beta = 0.18$; (**c**) the periodic orbits of period 22 (red) and 18 (blue).

The inverse of the map (5) is the map $T^{-1}(x, y) = \left(\frac{y(1+1/x)}{\beta}, \frac{1}{x}\right)$. The involution $R(x, y) = \left(\frac{1}{y}, \frac{1}{x}\right)$ is a reversor for (3). Indeed,

$$(R \circ T \circ R)(x, y) = (R \circ T)\left(\frac{1}{y}, \frac{1}{x}\right) = R\left(x, \frac{\beta/y}{1+1/x}\right) = \left(\frac{y\,(1+1/x)}{\beta}, \frac{1}{x}\right) = T^{-1}(x, y).$$

Thus, $T = I_1 \circ I_0$ where $I_0(x,y) = R(x,y)$ and $I_1(x,y) = T \circ R = \left(x, \frac{\beta}{y(1+1/x)}\right)$.

The symmetry lines corresponding to I_0 and I_1 are

$$S_0 = \{(x,y) : xy = 1\},\ S_1 = \{(x,y) : \beta x = y^2(x+1)\}.$$

Periodic orbits on the symmetry line S_0 with even period n are searched for by starting with points $(x_0, 1/x_0) \in S_0$ and imposing that $(x_{n/2}, y_{n/2}) \in S_0$, where $(x_{n/2}, y_{n/2}) = T^{n/2}(x_0, 1/x_0)$. This reduces to a one-dimensional root finding for the equation $x_{n/2}y_{n/2} = 1$, where the unknown is x_0. Furthermore, periodic orbits on S_0 with odd period n are obtained by solving for x_0 the equation $\beta x_{(n+1)/2} = y^2_{(n+1)/2}(1 + x_{(n+1)/2})$, where $(x_{(n+1)/2}, y_{(n+1)/2}) = T^{(n+1)/2}(x_0, 1/x_0)$.

For example, for $\beta = 1.8$, in Figure 3, we have an intersection between the symmetry lines S_0 and $S_{22} = T^{11}(S_0)$, $S_4 = T^2(S_0)$ and $S_{22} = T^{11}(S_0)$, and S_1 and $S_{23} = T^{11}(S_1)$ of the map T. The intersection points of this lines correspond to the periodic orbits of period 22, 18 and 22, respectively.

See Figure 3 for some examples of the periodic orbits of periods 18 and 22.

5. Conclusions

By using the KAM (Kolmogorov-Arnold-Moser) theory, invariants and corresponding Lyapunov function and time reversal symmetries, we proved the stability of the equilibrium solution of the system:

$$x_{n+1} = \frac{1}{y_n},\ y_{n+1} = \frac{\beta x_n}{1+y_n}, \quad n = 0,1,\ldots,$$

where the parameter $\beta > 0$, and initial conditions x_0 and y_0 are positive numbers. We obtain the Birkhoff normal form for this system and used them to prove the existence of periodic points with arbitrarily large periods in every neighborhood of the unique positive equilibrium.

Author Contributions: All three authors have written this paper and the final form of this paper is approved by all three authors.

Conflicts of Interest: The authors declare no conflict of interest.

References

1. Amleh, A.M.; Camouzis, E.; Ladas, G. On the Dynamics of a Rational Difference Equation, Part I. *Int. J. Differ. Equ.* **2008**, *3*, 1–35.
2. Drymmonis, E.; Camouzis, E.; Ladas, G.; Tikjha, G.W. Patterns of boundedness of the rational system $x_{n+1} = \frac{\alpha_1}{A_1+B_1x_n+C_1y_n}$ and $y_{n+1} = \frac{\alpha_2+\beta_2x_n+\gamma_2y_n}{A_2+B_2x_n+C_2y_n}$. *J. Differ. Equ. Appl.* **2012**, *18*, 89–110.
3. Feuer, J.; Janowski, E.; Ladas, G. Invariants for some rational recursive sequences with periodic coefficients. *J. Differ. Equ. Appl.* **1996**, *2*, 167–174.
4. Grove, E.A.; Janowski, E.J.; Kent, C.M.; Ladas, G. On the rational recursive sequence $x_{n+1} = \frac{\alpha x_n+\beta}{(\gamma x_n+\delta)x_{n-1}}$. *Comm. Appl. Nonlinear Anal.* **1994**, *1*, 61–72.
5. Kulenović, M.R.S. Invariants and related Liapunov functions for difference equations. *Appl. Math. Lett.* **2000**, *13*, 1–8.
6. Kulenović, M.R.S.; Merino, O. *Discrete Dynamical Systems and Difference Equations with Mathematica*; Chapman and Hall/CRC: Boca Raton, FL, USA; London, UK, 2002.
7. Chan, D.M.; Kent, C.M.; Ortiz-Robinson, N.L. Convergence results on a second-order rational difference equation with quadratic terms. *Adv. Differ. Equ.* **2009**, doi:10.1155/2009/985161.
8. Dehghan, M.; Kent, C.M.; Mazrooei-Sebdani, R.; Ortiz, N.L.; Sedaghat, H. Monotone and oscillatory solutions of a rational difference equation containing quadratic terms. *J. Differ. Equ. Appl.* **2008**, *14*, 1045–1058.
9. Dehghan, M.; Mazrooei-Sebdani, R.; Sedaghat, H. Global behaviour of the Riccati difference equation of order two. *J. Differ. Equ. Appl.* **2011**, *17*, 467–477.
10. Kulenović, M.R.S.; Moranjkić, S.; Nurkanović, Z. Naimark-Sacker bifurcation of second order rational difference equation with quadratic terms. *J. Nonlinear Sci. Appl.* **2016**, in press.

11. Cima, A.; Gasull, A.; Manosa, V. Non-integrability of measure preserving maps via Lie symmetries. *J. Differ. Equ.* **2015**, *259*, 5115–5136.
12. Kulenović, M.R.S.; Nurkanović, Z.; Pilav, E. Birkhoff Normal Forms and KAM theory for Gumowski-Mira Equation. *Sci. World J. Math. Anal.* **2014**, doi:10.1155/2014/819290.
13. Gidea, M.; Meiss, J.D.; Ugarcovici, I.; Weiss, H. Applications of KAM Theory to Population Dynamics. *J. Biol. Dyn.* **2011**, *5*, 44–63.
14. Kocic, V.L.; Ladas, G.; Tzanetopoulos, G.; Thomas, E. On the stability of Lyness' equation. *Dynam. Contin. Discrete Impuls. Syst.* **1995**, *1*, 245–254.
15. Kulenović, M.R.S.; Nurkanović, Z. Stability of Lyness' Equation with Period-Two Coefficient via KAM Theory. *J. Concr. Appl. Math.* **2008**, *6*, 229–245.
16. Ladas, G.; Tzanetopoulos, G.; Tovbis, A. On May's host parasitoid model. *J. Differ. Equ. Appl.* **1996**, *2*, 195–204.
17. Hrustić, S.J.; Kulenović, M.R.S.; Nurkanović, Z.; Pilav, E. Birkhoff Normal Forms, KAM theory and Symmetries for Certain Second Order Rational Difference Equation with Quadratic Term. *Int. J. Differ. Equ.* **2015**, *10*, 181–199.
18. Hale, J.K.; Kocak, H. *Dynamics and Bifurcation*; Springer-Verlag: New York, NY, USA, 1991.
19. Tabor, M. *Chaos and Integrability in Nonlinear Dynamics: An introduction*; A Wiley-Interscience Publication. John Wiley & Sons, Inc.: New York, NY, USA, 1989.
20. Del-Castillo-Negrete, D.; Greene, J.M.; Morrison, E.J. Area preserving nontwist maps: Periodic orbits and transition to chaos. *Phys. D* **1996**, *91*, 1–23.
21. Wan, Y.H. Computation of the stability condition for the Hopf bifurcation of diffeomorphisms on $\mathcal{R}^2$. *SIAM J. Appl. Math.* **1978**, *34*, 167–175.
22. Moeckel, R. Generic bifurcations of the twist coefficient. *Ergodic Theory Dyn. Syst.* **1990**, *10*, 185–195.
23. MacKay, R.S. *Renormalization in Area-Preserving Maps*; World Scientific: River Edge, NJ, USA, 1993.
24. Siegel, C.; Moser, J. *Lectures on Celestial Mechanics*; Springer-Varlag: New York, NY, USA, 1971.

9

Coefficient Inequalities of Second Hankel Determinants for Some Classes of Bi-Univalent Functions

Rayaprolu Bharavi Sharma [1] and Kalikota Rajya Laxmi [2,*]

1 Department of Mathematics, Kakatiya University, Warangal, Telangana-506009, India; rbsharma005@gmail.com
2 Department of Mathematics, SRIIT, Hyderabad, Telangana-501301, India
* Correspondence: rajyalaxmi2206@gmail.com

Academic Editor: Hari M. Srivastava

Abstract: In this paper, we investigate two sub-classes $S^*(\theta, \beta)$ and $K^*(\theta, \beta)$ of bi-univalent functions in the open unit disc Δ that are subordinate to certain analytic functions. For functions belonging to these classes, we obtain an upper bound for the second Hankel determinant $H_2(2)$.

Keywords: analytic functions; univalent functions; bi-univalent functions; second Hankel determinants

---

1. Introduction

Let A be the class of the functions of the form

$$f(z) = z + \sum_{k=2}^{\infty} a_k z^k, \tag{1}$$

which are analytic in the open unit disc $\Delta = \{z : |z| < 1\}$. Further, by S we shall denote the class of all functions in A that are univalent in Δ.

Let P denote the family of functions $p(z)$, which are analytic in Δ such that $p(0) = 1$, and $\Re p(z) > 0$ $(z \in \Delta)$ of the form

$$P(z) = 1 + \sum_{n=1}^{\infty} c_n z^n. \tag{2}$$

For two functions f and g, analytic in Δ, we say that the function f is subordinate to g in Δ, and we write it as $f(z) \prec g(z)$ if there exists a Schwarz function ω, which is analytic in Δ with $\omega(0) = 0$, $|\omega(z)| < 1$ $(z \in \Delta)$ such that

$$f(z) = g(\omega(z)). \tag{3}$$

Indeed, it is known that

$$f(z) \prec g(z) \Rightarrow f(0) = g(0) \text{ and } f(\Delta) \subset g(\Delta). \tag{4}$$

Every function $f \in S$ has an inverse f^{-1}, which is defined by $f^{-1}(f(z)) = z$, $(z \in \Delta)$

$$\text{and } f\left(f^{-1}(w)\right) = w, \quad \left(|w| < r_0(f); r_0(f) \geqslant \frac{1}{4}\right). \tag{5}$$

In fact, the inverse function is given by

$$f^{-1}(w) = w - a_2 w^2 + \left(2a_2^2 - a_3\right) w^3 - \left(5a_2^3 - 5a_2 a_3 + a_4\right) w^4 + ...$$

A function $f \in A$ is said to be bi-univalent in Δ if both f and f^{-1} are univalent in Δ.

Let $\sum$ denote the class of bi-univalent functions defined in the unit disc Δ.

We notice that $\sum$ is non empty .One of the best examples of bi-univalent functions is $f(z) = \log\left(\frac{1+z}{1-z}\right)$, which maps the unit disc univalently onto a strip $|\mathrm{Im}w| < \frac{\pi}{2}$, which in turn contains the unit disc. Other examples are z, $\frac{z}{1-z}$, $-\log(1-z)$.

However, the Koebe function is not a member of $\sum$ because it maps unit disc univalent onto the entire complex plane minus a slit along $-\frac{1}{4}$ to $-\infty$. Hence, the image domain does not contain the unit disc.

Other examples of univalent function that are not in the class $\sum$ are $z - \frac{z^2}{2}$, $\frac{z}{1-z^2}$.

In 1967, Lewin [1] first introduced class $\sum$ of bi-univalent function and showed that $|a_2| \leqslant 1.51$ for every $f \in \sum$. Subsequently, in 1967, Branan and Clunie [2] conjectured that $|a_2| \leqslant \sqrt{2}$ for bi-star like functions and $|a_2| \leqslant 1$ for bi-convex functions. Only the last estimate is sharp; equality occurs only for $f(z) = \frac{z}{1-z}$ or its rotation.

Later, Netanyahu [3] proved that $\max_{f \in \sum} |a_2| = \frac{4}{3}$. In 1985, Kedzierawski [4] proved Brannan and Clunie's conjecture for bi-starlike functions. In 1985, Tan [5] obtained that $|a_2| < 1.485$, which is the best known estimate for bi-univalent functions. Since then, various subclasses of the bi-univalent function classes $\sum$ were introduced, and non-sharp estimates on the first two coefficients $|a_2|$ and $|a_3|$ in the Taylor Maclaurin's series expansion were found in several investigations. The coefficient estimate problem for each of $|a_n|$ $(n \in N\{2, 3\})$ is still an open problem.

In 1976, Noonan and Thomas [6] defined q^{th}Hankel determinant of f for $q \geqslant 1$ and $n \geqslant 1$, which is stated by

$$H_q(n) = \begin{vmatrix} a_n & a_{n+1} & \cdots & a_{n+q-1} \\ a_{n+1} & a_{n+2} & \cdots & a_{n+q} \\ \vdots & \vdots & \vdots & \vdots \\ a_{n+q-1} & a_{n+q} & \cdots & a_{n+2q-2} \end{vmatrix}.$$

These determinants are useful, for example, in showing that a function of bounded characteristic in Δ, *i.e.*, a function that is a ratio of two bounded analytic functions with its Laurent series around the origin having integral coefficient is rational.

The Hankel determinant plays an important role in the study of singularities (for instance, see [7] Denies, p.329 and Edrei [8]).A Hankel determinant plays an important role in the study of power series with integral coefficients. In 1966, Pommerenke [9] investigated the Hankel determinants of areally mean p-valent functions, univalent functions as well as of starlike functions, and, in 1967 [10], he proved that the Hankel determinants of univalent functions satisfy $H_q(n) < Kn^{-(\frac{1}{2}+\beta)q+\frac{3}{2}}$ $(n = 1,2,...,q = 2, 3, ...)$ where $\beta > \frac{1}{4000}$ and K depend only on q.

Later, Hayman [11] proved that $H_2(n) < An^{\frac{1}{2}}$ $(n = 1,2,...; A$ an absolute constant) for areally mean univalent functions. The estimates for the Hankel determinant of areally mean p-valent functions have been investigated [12–14]. Elhosh [15,16] obtained bounds for Hankel determinants of univalent functions with a positive Hayman index α and k-fold symmetric and close to convex functions. Noor [9] determined the rate of growth of $H_q(n)$ as $n \to \infty$ for the functions in S with bounded boundary.

Ehrenborg [17] studied the Hankel determinant of exponential polynomials. The Hankel transform of an integer sequence and some of its properties were discussed by Layman [18].

One can easily observe that the Fekete-Szego functional $\left|a_3 - a_2^2\right| = H_2(1)$. This function was further generalized with μ real as well as complex. Fekete-Szego gave a sharp estimate of $\left|a_3 - \mu a_2^2\right|$ for μ real. The well-known results due to them is

$$\left|a_3 - \mu a_2^2\right| \leqslant \begin{cases} 4\mu - 3 & \mu \leqslant 1 \\ 1 + 2\exp\left(\dfrac{-2\mu}{1-\mu}\right) & 0 \leqslant \mu \leqslant 1 \\ 3 - 4\mu & \mu \geqslant 0 \end{cases}.$$

On the other hand, Zaprawa [19,20] extended the study on Fekete-Szego problem to some classes of bi-univalent functions. Ali [21] found sharp bounds on the first four coefficients and a sharp estimate for the Fekete-Szego functional $\left|\gamma_3 - t\gamma_2^2\right|$, where t is real, for the inverse function of f defined as $f^{-1}(w) = w + \sum\limits_{k=2}^{\infty} \gamma_k w^k$ to the class of strongly starlike functions of order α $(0 < \alpha \leqslant 1)$.

Recently S.K. Lee *et al.* [22] obtained the second Hankel determinant $H_2(2) = a_2 a_4 - a_3^2$ for functions belonging to subclasses of Ma-Minda starlike and convex functions. T. Ram Reddy [23] obtained the Hankel determinants for starlike and convex functions with respect to symmetric points. T. Ram Reddy *et al.* [24,25] also obtained the second Hankel determinant for subclasses of p-valent functions and p-valent starlike and convex function of order α.

Janteng [26] has obtained sharp estimates for the second Hankel determinant for functions whose derivative has a positive real part.Afaf Abubaker [27] studied sharp upper bound of the second Hankel determinant of subclasses of analytic functions involving a generalized linear differential operator. In 2015, the second Hankel determinant for bi-starlike and bi-convex function of order β was obtained by Erhan Deniz [28].

2. Preliminaries

Motivated by above work, in this paper, we introduce certain subclasses of bi-univalent functions and obtained an upper bound to the coefficient functional $a_2 a_4 - a_3^2$ for the function f in these classes defined as follows:

Definition 2.1.: A function $f \in A$ is said to be in the class $S^*(\theta, \beta)$ if it satisfies the following conditions:

$$\Re\left\{e^{i\theta}\left\{\frac{zf'(z)}{f(z)}\right\}\right\} > \beta\cos\theta \quad (\forall z \in \Delta) \tag{6}$$

$$\Re\left\{e^{i\theta}\left\{\frac{wg'(w)}{g(w)}\right\}\right\} > \beta\cos\theta \quad (\forall w \in \Delta) \tag{7}$$

where g is an extension of f^{-1} to Δ.

Note: 1. For $\theta = 0$, the class $S^*(\theta, \beta)$ reduces to the class $S_\sigma^*(\beta)$, and, for this class, coefficient inequalities of the second Hankel determinant were studied by Deniz *et al* [28].

2. For $\theta = 0$ and $\beta = 0$, the class $S^*(\theta, \beta)$ reduces to the class S_σ^*, and, for this class, coefficient inequalities of the second Hankel determinant were studied by Deniz *et al* [28].

Definition 2.2.: A function $f \in A$ is said to be in the class $K^*(\theta, \beta)$ if it satisfies the following conditions:

$$\Re\left\{e^{i\theta}\left\{1 + \frac{zf''(z)}{f'(z)}\right\}\right\} > \beta\cos\theta \quad (\forall z \in \Delta) \tag{8}$$

$$\Re\left\{e^{i\theta}\left\{1 + \frac{wg''(w)}{g'(w)}\right\}\right\} > \beta\cos\theta \quad (\forall w \in \Delta) \tag{9}$$

where g is an extension of f^{-1} in Δ.

Note: 1. For $\theta = 0$, the class $K^*(\theta, \beta)$ reduces to the class $K^*_\sigma(\beta)$, and, for this class, coefficient inequalities ofthe second Hankel determinantwere studied by Deniz *et al* [28].

2. For $\theta = 0$ and $\beta = 0$ the class $K^*(\theta, \beta)$ reduces to the class K^*_σ, and, for this class, coefficient inequalities ofthe second Hankel determinantwere studied by Deniz *et al* [28].

To prove our results, we require the following Lemmas:

Lemma 2.1. [14] Let the function $p \in P$ be given by the following series:

$$p(z) = 1 + c_1 z + c_2 z^2 + c_3 z^3 + ... \qquad (z \in \Delta). \tag{10}$$

Then the sharp estimate is given by .

Lemma 2.2. [29] The power series for the function $p \in P$ is given (10) converges in the unit disc Δ to a function in P if and only if Toeplitz determinants

$$D_n = \begin{vmatrix} 2 & c_1 & c_2 & \rightleftharpoons & c_n \\ c_{-1} & 2 & c_1 & \rightleftharpoons & c_{n-1} \\ \vdots & \vdots & \vdots & \vdots & \vdots \\ c_{-n} & c_{-n+1} & c_{-n+2} & \rightleftharpoons & 2 \end{vmatrix}, \quad n \in N$$

and $c_{-k} = \overline{c}_k$ are all non-negative. These are strictly positive except for $p(z) = \sum_{k=1}^{m} \rho_k P_0(e^{it_k} z)$, $\rho_k > 0$, t_k real and $t_k \neq t_j$ for k ≠ j, where $P_0(z) = \left(\frac{1+z}{1-z}\right)$; in this case, $D_n > 0$ for $n < (m-1)$ and $D_n = 0$ for $n \geqslant m$.

This necessary and sufficient condition found in the literature [29] is due to Caratheodary and Toeplitz. We may assume without any restriction that $c_1 > 0$. On using Lemma (2.2) for $n = 2$ and $n = 3$ respectively, we get

$$D_2 = \begin{vmatrix} 2 & c_1 & c_2 \\ \overline{c}_1 & 2 & c_1 \\ \overline{c}_2 & \overline{c}_1 & 2 \end{vmatrix} = \left[8 + 2\mathrm{Re}\left\{c_1^2 c_2\right\} - 2|c_2|^2 - 4c_1^2\right] \geqslant 0.$$

It is equivalent to

$$2c_2 = \left\{c_1^2 + x\left(4 - c_1^2\right)\right\}, \text{ for some } x, \ |x| \leqslant 1 \tag{11}$$

$$D_3 = \begin{vmatrix} 2 & c_1 & c_2 & c_3 \\ \overline{c}_1 & 2 & c_1 & c_2 \\ \overline{c}_2 & \overline{c}_1 & 2 & c_1 \\ \overline{c}_3 & \overline{c}_2 & \overline{c}_1 & 2 \end{vmatrix} \geqslant 0.$$

Then $D_3 \geqslant 0$ is equivalent to

$$\left|\left(4c_3 - 4c_1c_2 + c_1^3\right)\left(4 - c_1^2\right) + c_1\left(2c_2 - c_1^2\right)^2\right| \leqslant 2\left(4 - c_1^2\right)^2 - 2\left|\left(2c_2 - c_1^2\right)\right|^2. \tag{12}$$

From the relations (2.6) and (2.7), after simplifying, we get

$$4c_3 = c_1^3 + 2\left(4 - c_1^2\right)c_1 x - c_1\left(4 - c_1^2\right)x^2 + 2\left(4 - c_1^2\right)\left(1 - |x|^2\right)z,$$

for some $x, \ z$ with

$$|x| \leqslant 1 \text{and } |z| \leqslant 1. \tag{13}$$

3. Main Results

We now prove our main result for the function f in the class $S^*(\theta, \beta)$.

Theorem 3.1. Let the function f given by (1.1) be in the class $S^*(\theta, \beta)$. Then

$$\left|a_2a_4 - a_3^2\right| \leqslant \begin{cases} \dfrac{16(1-\beta)^4\cos^4\theta}{3} + \dfrac{4}{3}(1-\beta)^2\cos^2\theta, & \beta \in \left[0, 1 - \dfrac{1}{2\sqrt{2}\cos\theta}\right] \\ \dfrac{3(1-\beta)^2\cos^2\theta}{2\left[1-2(1-\beta)^2\cos^2\theta\right]}, & \beta \in \left(1 - \dfrac{1}{2\sqrt{2}\cos\theta}, 1\right) \end{cases}.$$

Proof: Let $f \in S(\theta, \beta; h)$ and $g = f^{-1}$. From (6) and (7) it follows that

$$e^{i\theta}\left\{\frac{zf'(z)}{f(z)}\right\} = [(1-\beta)p(z) + \beta]]\cos\theta + i\sin\theta \tag{14}$$

$$e^{i\theta}\left\{\frac{wg'(w)}{g(w)}\right\} = [(1-\beta)q(w) + \beta]]\cos\theta + i\sin\theta \tag{15}$$

where $p(z) = 1 + c_1z + c_2z^2 + c_3z^3 + ...\ \in P, (z \in \Delta)$ and $q(w) = 1 + d_1w + d_2w^2 + d_3w^3 + ...\ \in P, (w \in \Delta)$. Now, equating the coefficients in (14) and (15), we have

$$e^{i\theta}a_2 = c_1(1-\beta)\cos\theta \tag{16}$$

$$e^{i\theta}\left(2a_3 - a_2^2\right) = c_2(1-\beta)\cos\theta \tag{17}$$

$$e^{i\theta}\left(3a_4 - 3a_2a_3 + a_2^3\right) = c_3(1-\beta)\cos\theta \tag{18}$$

and

$$-e^{i\theta}a_2 = d_1(1-\beta)\cos\theta \tag{19}$$

$$e^{i\theta}\left(3a_2^2 - 2a_3\right) = d_2(1-\beta)\cos\theta \tag{20}$$

$$e^{i\theta}\left(-3a_4 + 12a_2a_3 - 10a_2^3\right) = d_3(1-\beta)\cos\theta \tag{21}$$

Now from (16) and (19) we get

$$c_1 = -d_1 \tag{22}$$

and

$$c_2 = e^{-i\theta}p_1(1-\beta)\cos\theta. \tag{23}$$

Now, from (17) and (20), we get

$$a_3 = e^{-2i\theta}c_1^2(1-\beta)^2\cos^2\theta + \frac{e^{-i\theta}(1-\beta)\cos\theta(c_2-d_2)}{4}. \tag{24}$$

Additionally, from (18) and (21), we get

$$a_4 = \frac{2}{3}e^{-3i\theta}c_1^3(1-\beta)^3\cos^3\theta + \frac{5}{8}e^{-2i\theta}c_1(c_2-d_2)(1-\beta)^2\cos^2\theta + \frac{1}{6}e^{-i\theta}(c_3-d_3)(1-\beta)\cos\theta. \tag{25}$$

Thus, we can easily obtain

$$\begin{aligned}\left|a_2a_4 - a_3^2\right| = \Bigg|&\frac{-1}{3}e^{-4i\theta}c_1^4(1-\beta)^4\cos^4\theta + \frac{1}{8}e^{-3i\theta}c_1^2(c_2-d_2)(1-\beta)^3\cos^3\theta + \\ &\frac{1}{6}e^{-2i\theta}c_1(c_3-d_3)(1-\beta)^2\cos^2\theta - \frac{1}{16}e^{-2i\theta}(c_2-d_2)^2(1-\beta)^2\cos^2\theta\Bigg|.\end{aligned} \tag{26}$$

According to Lemma (2.2) and Equation (22), we get

$$\left.\begin{array}{l} 2c_2 = c_1^2 + x\left(4 - c_1^2\right) \\ 2d_2 = d_1^2 + x\left(4 - d_1^2\right) \end{array}\right\} \Rightarrow c_2 - d_2 = 0 \tag{27}$$

and

$$c_3 - d_3 = \frac{c_1^3}{2} - c_1\left(4 - c_1^2\right)x - \frac{c_1\left(4 - c_1^2\right)x^2}{2} \tag{28}$$

$$\begin{array}{l} \left|a_2a_4 - a_3^2\right| = \left|-\frac{1}{3}e^{-4i\theta}c_1^4(1-\beta)^4\cos^4\theta + \frac{1}{12}e^{-2i\theta}c_1^4(1-\beta)^2\cos^2\theta - \right. \\ \left. \frac{1}{6}e^{-2i\theta}c_1^2\left(4 - c_1^2\right)x(1-\beta)^2\cos^2\theta - \frac{1}{12}e^{-2i\theta}c_1^2\left(4 - c_1^2\right)x^2(1-\beta)^2\cos^2\theta\right| \end{array}. \tag{29}$$

Since $p \in P$, so $|c_1| \leqslant 2$. Letting $c_1 = c$, we may assume without any restriction that $c \in [0, 2]$. Thus, applying the triangle inequality on the right-hand side of Equation (29), with $\mu = |x| \leqslant 1$, we obtain

$$\begin{aligned} \left|a_2a_4 - a_3^2\right| \leqslant{} & \frac{1}{3}c^4(1-\beta)^4\cos^4\theta + \frac{1}{12}c^4(1-\beta)^2\cos^2\theta + \frac{1}{6}c^2\left(4 - c^2\right)\mu(1-\beta)^2\cos^2\theta \\ & + \frac{1}{12}c^2\left(4 - c^2\right)\mu^2(1-\beta)^2\cos^2\theta = F(\mu). \end{aligned} \tag{30}$$

Differentiating $F(\mu)$, we get

$$F'(\mu) = \frac{c^2\left(4 - c^2\right)(1-\beta)^2\cos^2\theta + c^2\left(4 - c^2\right)\mu(1-\beta)^2\cos^2\theta}{6}. \tag{31}$$

Using elementary calculus, one can show that $F'(\mu) > 0$ for $\mu > 0$. This implies that F is an increasing function, and it therefore cannot have a maximum value at any point in the interior of the closed region $[0, 2] \times [0, 1]$. Further, the upper bound for $F(\mu)$ corresponds to $\mu = 1$, in which case $F(\mu) \leqslant F(1)$

$$\frac{1}{3}c^4(1-\beta)^4\cos^4\theta + \frac{1}{12}c^4(1-\beta)^2\cos^2\theta + \frac{1}{4}c^2\left(4 - c^2\right)(1-\beta)^2\cos^2\theta = G(c).$$

Then

$$G'(c) = \frac{2}{3}c(1-\beta)^2\cos^2\theta\left[\left(2(1-\beta)^2\cos^2\theta - 1\right)c^2 + 1\right]. \tag{32}$$

Setting $G'(c) = 0$, the real critical points are $c_{01} = 0$, $c_{02} = \sqrt{\frac{3}{1 - 2(1-\beta)^2\cos^2\theta}}$.

After some calculations we obtain the following cases:

Case 1: When $\beta \in \left[0, 1 - \frac{1}{2\sqrt{2}\cos\theta}\right]$, we observe that $c_{02} \geqslant 2$, that is c_{02}, is out of the interval $(0, 2)$. Therefore, the maximum value of $G(c)$ occurs at $c_{01} = 0$ or $c = c_{02}$, which contradicts our assumption of having a maximum value at the interior point of $c \in [0, 2]$. Since G is an increasing function, the maximum point of G must be on the boundary of $c \in [0, 2]$, that is $c = 2$. Thus, we have

$$\max_{0 \leqslant c \leqslant 2} G(c) = G(2) = \frac{16(1-\beta)^4\cos^4\theta}{3} + \frac{4}{3}(1-\beta)^2\cos^2\theta.$$

Case 2: When $\beta \in \left(1 - \frac{1}{2\sqrt{2}\cos\theta}, 1\right)$, we observe that $c_{02} < 2$, that is c_{02}, is interior of the interval $[0, 2]$. Since $G''(c_{02}) < 0$, the maximum value of $G(c)$ occurs at $c = c_{02}$. Thus, we have

$$\max_{0\leqslant c\leqslant 2} G(c) = G(c_{02}) = G\left(\sqrt{\frac{1}{\frac{1}{3} - \frac{2}{3}(1-\beta)^2\cos^2\theta}}\right) = \frac{3(1-\beta)^2\cos^2\theta}{2\left[1 - 2(1-\beta)^2\cos^2\theta\right]}.$$

This completes the proof of the theorem.

Corollary 1: Let f given by (1.1) be in the class $S^*_\sigma(\beta)$. Then

$$\left|a_2a_4 - a_3^2\right| \leqslant \begin{cases} \frac{16(1-\beta)^4}{3} + \frac{4}{3}(1-\beta)^2, & \beta \in \left[0, \left(1 - \frac{1}{2\sqrt{2}}\right)\right] \\ \frac{3(1-\beta)^2}{2\left[1 - 2(1-\beta)^2\right]}, & \beta \in \left(1 - \frac{1}{2\sqrt{2}}, 1\right) \end{cases}.$$

Corollary 2: Let f given by (1.1) be in the class S^*_σ. Then

$$\left|a_2a_4 - a_3^2\right| \leqslant \frac{20}{3}.$$

These two corollaries coincide with the results of Deniz *et al.* [28].

Remark 3.1: It is observed that for $\theta = 0$, we get the Hankel determinant $|a_2a_4 - a_3^2|$ for the class $S^*_\sigma(\beta)$ and the Hankel determinant of this class was studied by Deniz *et al.* [28].

4. Hankel Determinants for the Class of Functions $K(\theta, \beta; h)$

We now estimate an upper bound $a_2a_4 - a_3^2$ for the function $f(z)$ in the class $K(\theta, \beta; h)$.

Theorem 4.1. Let the $f(z)$ given by (1.1) be in the class $K(\theta, \beta; h)$. Then

$$\left|a_2a_4 - a_3^2\right| \leqslant \begin{cases} \frac{1}{6}(1-\beta)^4\cos^4\theta + \frac{1}{6}(1-\beta)^2\cos^2\theta, & \beta \in \left[0, 1 - \frac{1}{\sqrt{2}\cos\theta}\right] \\ \frac{3(1-\beta)^2\cos^2\theta}{8\left[2 - (1-\beta)^2\cos^2\theta\right]}, & \beta \in \left(1 - \frac{1}{\sqrt{2}\cos\theta}, 1\right) \end{cases}.$$

Proof: Let $f \in K(\theta, \beta; h)$ and $g = f^{-1}$. From (8) and (9) we have

$$e^{i\theta}\left\{1 + \frac{zf''(z)}{f'(z)}\right\} = [(1-\beta)p(z) + \beta]]\cos\theta + i\sin\theta \tag{33}$$

$$e^{i\theta}\left\{1 + \frac{wg''(w)}{g'(w)}\right\} = [(1-\beta)p(w) + \beta]]\cos\theta + i\sin\theta \tag{34}$$

where $p(z) = 1 + c_1z + c_2z^2 + c_3z^3 + ...$, $(z \in \Delta)$ and $q(w) = 1 + d_1w + d_2w^2 + d_3w^3 + ...$, $(w \in \Delta)$.

Now, equating the coefficients in (33) and (34), we have

$$2e^{i\theta}a_2 = c_1(1-\beta)\cos\theta \tag{35}$$

$$e^{i\theta}\left(6a_3 - 4a_2^2\right) = c_2(1-\beta)\cos\theta \tag{36}$$

$$e^{i\theta}\left(12a_4 - 18a_2a_3 + 8a_2^3\right) = c_3(1-\beta)\cos\theta \tag{37}$$

and

$$-2e^{i\theta}a_2 = d_1(1-\beta)\cos\theta \tag{38}$$

$$e^{i\theta}\left(8a_2^2-6a_3\right)=d_2\left(1-\beta\right)\cos\theta \tag{39}$$

$$e^{i\theta}\left(-32a_2^3+42a_2a_3-12a_4\right)=d_3\left(1-\beta\right)\cos\theta. \tag{40}$$

Now from (35) and (38), we get

$$c_1=-d_1 \tag{41}$$

and

$$a_2=\frac{e^{-i\theta}c_1\left(1-\beta\right)\cos\theta}{2}. \tag{42}$$

Now, from (36) and (39), we get

$$a_3=\frac{e^{-2i\theta}c_1^2\left(1-\rho\right)^2\cos^2\theta}{4}+\frac{e^{-i\theta}\left(c_2-d_2\right)\left(1-\rho\right)\cos\theta}{12}. \tag{43}$$

Additionally, from (37) and (40), we get

$$a_4=\frac{5}{48}e^{-3i\theta}c_1^3\left(1-\beta\right)^3\cos^3\theta+\frac{5}{48}e^{-2i\theta}c_1\left(c_2-d_2\right)\left(1-\beta\right)^2\cos^2\theta+\frac{1}{24}e^{-i\theta}\left(c_3-d_3\right)\left(1-\beta\right)\cos\theta. \tag{44}$$

Thus, we can easily obtain

$$\begin{aligned}\left|a_2a_4-a_3^2\right|=\Bigg|&-\frac{c_1^4}{96}e^{-4i\theta}\left(1-\beta\right)^4\cos^4\theta+\frac{c_1^2}{96}e^{-3i\theta}\left(c_2-d_2\right)\left(1-\beta\right)^3\cos^3\theta+\\&\frac{c_1}{48}e^{-2i\theta}\left(c_3-d_3\right)\left(1-\beta\right)^3\cos^3\theta-\frac{e^{-2i\theta}\left(c_2-d_2\right)^2\left(1-\beta\right)^2\cos^2\theta}{144}\Bigg|\end{aligned}. \tag{45}$$

According to Lemma (2.2), and from Equation (41), we get

$$\left.\begin{aligned}2c_2&=c_1^2+x\left(4-c_1^2\right)\\2d_2&=d_1^2+x\left(4-d_1^2\right)\end{aligned}\right\}\Rightarrow c_2-d_2=0 \tag{46}$$

and

$$c_3-d_3=\frac{c_1^3}{2}-c_1\left(4-c_1^2\right)x-\frac{c_1\left(4-c_1^2\right)x^2}{2} \tag{47}$$

$$\begin{aligned}\left|a_2a_4-a_3^2\right|=\Bigg|&-\frac{c_1^4}{96}e^{-4i\theta}\left(1-\beta\right)^4\cos^4\theta+\frac{c_1^4}{96}e^{-2i\theta}\left(1-\beta\right)^2\cos^2\theta-\\&\frac{e^{-2i\theta}c_1^2\left(4-c_1^2\right)x\left(1-\beta\right)^2\cos^2\theta}{48}-\frac{e^{-2i\theta}c_1^2\left(4-c_1^2\right)x^2\left(1-\beta\right)^2\cos^2\theta}{96}\Bigg|\end{aligned}. \tag{48}$$

Since $p\in P$, $|c_1|\leqslant 2$. Letting $c_1=c$, we may assume without any restriction that $c\in[0,2]$. Thus, applying the triangle inequality on the right-hand side of Equation (4.16), with $\mu=|x|\leqslant 1$, we obtain

$$\begin{aligned}\left|a_2a_4-a_3^2\right|\leqslant&\frac{c^4}{96}e^{-4i\theta}\left(1-\beta\right)^4\cos^4\theta+\frac{c^4}{96}e^{-2i\theta}\left(1-\beta\right)^2\cos^2\theta\\&\frac{e^{-2i\theta}c^2\left(4-c^2\right)\mu\left(1-\beta\right)^2\cos^2\theta}{48}+\frac{e^{-2i\theta}c^2\left(4-c^2\right)\mu^2\left(1-\beta\right)^2\cos^2\theta}{96}\\&=F\left(\mu\right).\end{aligned} \tag{49}$$

Differentiating $F\left(\mu\right)$, we get

$$F'\left(\mu\right)=\frac{e^{-2i\theta}c^2\left(4-c^2\right)\left(1-\beta\right)^2\cos^2\theta}{48}+\frac{e^{-2i\theta}c^2\left(4-c^2\right)\mu\left(1-\beta\right)^2\cos^2\theta}{48}. \tag{50}$$

Using elementary calculus, one can show that $F'(\mu) > 0$ for $\mu > 0$. It implies that F is an increasing function and it hence cannot have a maximum value at any point in the interior of the closed region $[0, 2] \times [0, 1]$. Further, the upper bound for $F(\mu)$ corresponds to $\mu = 1$, in which case

$$F(\mu) \leqslant F(1) \begin{array}{l} \leqslant \frac{c^4}{96}(1-\beta)^4\cos^4\theta + \frac{c^4}{96}(1-\beta)^2\cos^2\theta + \frac{c^2(4-c^2)(1-\beta)^2\cos^2\theta}{48} \\ + \frac{c^2(4-c^2)(1-\beta)^2\cos^2\theta}{96} = G(c)\ \text{(say)} \end{array}.$$

Then

$$G'(c) = \frac{c^3}{24}(1-\beta)^4\cos^4\theta + \frac{c^3}{24}(1-\beta)^2\cos^2\theta + \frac{\left[8c-4c^3\right](1-\beta)^2\cos^2\theta}{32}. \tag{51}$$

Setting $G'(c) = 0$, the real critical points are $c_{01} = 0$, $c_{02} = \sqrt{\frac{6}{\left[2-(1-\beta)^2\cos^2\theta\right]}}$.

After some calculations we obtain the following cases:

Case 1: When $\beta \in \left[0, 1 - \frac{1}{\sqrt{2}\cos\theta}\right]$, we observe that $c_{02} \geqslant 2$, that is c_{02}, is out of the interval $(0, 2)$. Therefore, the maximum value of $G(c)$ occurs at $c_{01} = 0$ or $c = c_{02}$, which contradicts our assumption of having the maximum value at the interior point of $c \in [0, 2]$. Since G is an increasing function, the maximum point of G must be on the boundary of $c \in [0, 2]$, that is $c = 2$. Thus, we have

$$\max_{0\leqslant c\leqslant 2} G(c) = G(2) = \frac{1}{6}(1-\beta)^4\cos^4\theta + \frac{1}{6}(1-\beta)^2\cos^2\theta.$$

Case 2: When $\beta \in \left(1 - \frac{1}{\sqrt{2}\cos\theta}, 1\right)$, we observe that $c_{02} < 2$, that is c_{02}, is interior of the interval $[0, 2]$. Since $G''(c_{02}) < 0$, the maximum value of $G(c)$ occurs at $c = c_{02}$. Thus, we have

$$\begin{aligned}\max_{0\leqslant c\leqslant 2} G(c) = G(c_{02}) &= G\left(\sqrt{\frac{6}{\left[2-(1-\beta)^2\cos^2\theta\right]}}\right) \\ &= \frac{3(1-\beta)^2\cos^2\theta}{8\left[2-(1-\beta)^2\cos^2\theta\right]}\end{aligned}.$$

This completes the proof of the theorem.

Corollary 1: Let f given by (1) be in the class $K_\sigma^*(\beta)$. Then

$$\left|a_2a_4 - a_3^2\right| \leqslant \begin{cases} \frac{(1-\beta)^4}{6} + \frac{(1-\beta)^2}{6}, & \beta \in \left[0, \left(1 - \frac{1}{\sqrt{2}}\right)\right] \\ \frac{3(1-\beta)^2}{8\left[2-(1-\beta)^2\right]}, & \beta \in \left(1 - \frac{1}{\sqrt{2}}, 1\right) \end{cases}.$$

Corollary 2: Let f given by (1) be in the class K_σ^*. Then

$$\left|a_2a_4 - a_3^2\right| \leqslant \frac{1}{3}.$$

These two corollaries coincide with the results of Deniz *et al.* [28].

5. Conclusion

For specific values of α and β, the results obtained in this paper will generalize and unify the results of the earlier researchers in this direction.

Interested researchers can work upon finding an upper bound for $|a_2a_4 - \mu a_3^2|$ and $|a_n|$ for a real or complex μ.

Acknowledgments: The authors are very much thankful to T. Ram Reddy for his valuable guidance in preparing this paper.

Author Contributions: Both authors has read and approved the final paper.

Conflicts of Interest: The authors declare that there are no conflicts of interest regarding the publication of this paper.

References

1. Lewin, M. On a Coefficient problem for bi-univalent functions. *Proc. Am. Math. Soc.* **1967**, *18*, 63–68. [CrossRef]
2. Brannan, D.A., Clunie, J.G., Eds.; Aspects of Contemporary Complex Analysis. In Proceedings of the NATO Advanced Study Institute, University of Durham, Durham, UK, 1–20 July 1979; Academic Press: New York, NY, USA; London, UK, 1980.
3. Netanyahu, E. The minimal distance of the image boundary from the origin and the second coefficient of a univalent function in $|Z| < 1$. *Arch. Ration. Mech. Anal.* **1969**, *32*, 100–112.
4. Kedzierawski, A.W. Some remarks on bi-univalent functions. *Ann. Univ. Mariae Curie Skłodowska Sect. A* **1985**, *39*, 77–81.
5. Tan, D.L. Coefficient estimates for bi-univalent functions. *Chinese Ann. Math. Ser. A* **1984**, *5*, 559–568.
6. Noonan, J.W.; Thomas, D.K. On the second Hankel determinant of a really mean p-valent functions. *Trans. Am. Math. Soc.* **1976**, *223*, 337–346.
7. Dienes, P. *The Taylor Series*; Dover: New York, NY, USA, 1957.
8. Edrei, A. Sur les determinants recurrents et less singularities d'une fonction donee por son develepment de Taylor. *Compos. Math.* **1940**, *7*, 20–88.
9. Pommerenke, C. On the coefficients and Hankel determinants of univalent functions. *J. Lond. Math. Soc.* **1966**, *41*, 111–122. [CrossRef]
10. Pommerenke, C. On the Hankel determinants of univalent functions. *Mathematika* **1967**, *14*, 108–112. [CrossRef]
11. Hayman, W.K. On the second Hankel determinant of mean univalent functions. *Proc. Lond. Math. Soc.* **1968**, *18*, 77–94. [CrossRef]
12. Noonan, J.W.; Thomas, D.K. On the Hankel determinants of a really mean p-valent functions. *Proc. Lond. Math. Soc.* **1972**, *25*, 503–524. [CrossRef]
13. Noonan, J.W. Coefficient differences and Hankel determinants of a really mean p-valent functions. *Proc. Am. Math. Soc.* **1974**, *46*, 29–37.
14. Pommerenke, C. *Univalent Functions*; Vandenhoeck and Rupercht: Gotingen, Germany, 1975.
15. Elhosh, M.M. On the second Hankel determinant of close-to-convex functions. *Bull. Malays. Math. Soc.* **1986**, *9*, 67–68.
16. Elhosh, M.M. On the second Hankel determinant of univalent functions. *Bull. Malays. Math. Soc.* **1986**, *9*, 23–25.
17. Ehrenborg, R. The Hankel determinant of exponential polynomials. *Am. Math. Mon.* **2000**, *107*, 557–560. [CrossRef]
18. Layman, J.W. The Hankel transform and some of its properties. *J. Integer Seq.* **2001**, *4*, 1–11.
19. Cantor, D.G. Power series with integral coefficients. *Bull. Am. Math. Soc.* **1963**, *69*, 362–366. [CrossRef]
20. Zaprawa, P. Estimates of initial coefficients for bi-univalent functions. *Abstr. Appl. Anal.* **2014**, *2014*, 357480. [CrossRef]
21. Ali, R.M. Coefficients of the inverse of strongly starlike functions. *Bull. Malays. Math. Sci. Soc.* **2003**, *26*, 63–71.

22. Lee, S.K.; Ravichandran, V.; Supramaniam, S. Bounds for the second Hankel determinant of certain univalent functions. *J. Inequal. Appl.* **2013**, *2013*, 281. [CrossRef]
23. RamReddy, T.; Vamshee Krishna, D. Hankel determinant for starlike and convex functions with respect to symmetric points. *J. Indian Math. Soc.* **2012**, *79*, 161–171.
24. RamReddy, T.; Vamshee Krishna, D. Hankel determinant for p-valent starlike and convex functions of order α. *Novi Sad J. Math.* **2012**, *42*, 89–96.
25. RamReddy, T.; Vamshee Krishna, D. Certain inequality for certain subclass of p-valent functions. *Palest. J. Math.* **2015**, *4*, 223–228.
26. Janteng, A.; Halim, S.A.; Darus, M. Coefficient inequality for a function whose derivative has a positive real part. *J. Inequal. Pure Appl. Math.* **2006**, *7*, 1–5.
27. Abubaker, A.; Darus, M. Hankel determinant for a class of analytic functions involving a generalized linear differential operator. *Int. J. Pure Appl. Math.* **2011**, *69*, 429–435.
28. Deniz, E.; Çağlar, M.; Orhan, H. Second hankel determinant for bi-starlike and bi-convex functions of order β. *Appl. Math. Comput.* **2015**, *271*, 301–307. [CrossRef]
29. Grenander, U.; Szego, G. *Toeplitz Forms and Their Applications*; University of Californi Press: Berkeley, CA, USA, 1958.

10

A Note on Burg's Modified Entropy in Statistical Mechanics

Amritansu Ray [1,*] and S. K. Majumder [2]

[1] Department of Mathematics, Rajyadharpur Deshbandhu Vidyapith, Serampore, Hooghly 712203, West Bengal, India
[2] Department of Mathematics, Indian Institute of Engineering Science and Technology (IIEST), Shibpur, Howrah 711103, West Bengal, India; majumder_sk@yahoo.co.in
[*] Correspondence: amritansu_ray06@yahoo.co.in

Academic Editor: Reza Abedi

Abstract: Burg's entropy plays an important role in this age of information euphoria, particularly in understanding the emergent behavior of a complex system such as statistical mechanics. For discrete or continuous variable, maximization of Burg's Entropy subject to its only natural and mean constraint always provide us a positive density function though the Entropy is always negative. On the other hand, Burg's modified entropy is a better measure than the standard Burg's entropy measure since this is always positive and there is no computational problem for small probabilistic values. Moreover, the maximum value of Burg's modified entropy increases with the number of possible outcomes. In this paper, a premium has been put on the fact that if Burg's modified entropy is used instead of conventional Burg's entropy in a maximum entropy probability density (MEPD) function, the result yields a better approximation of the probability distribution. An important lemma in basic algebra and a suitable example with tables and graphs in statistical mechanics have been given to illustrate the whole idea appropriately.

Keywords: entropy optimization; probability distribution; Shannon's Entropy; Burg's Entropy

1. Introduction

The concept of entropy [1] figured strongly in the physical sciences during the 19th century, especially in thermodynamics and statistical mechanics [2], as a measure of equilibrium and evolution of thermodynamic systems. Two main views were developed which were the macroscopic view formulated originally by Clausius and Carnot and the microscopic approach associated with Boltzmann and Maxwell. Since then, both the approaches have made introspection in natural thermodynamic and microscopically probabilistic systems possible. Entropy is defined as the measure of a system's thermal energy per unit temperature that is unavailable for doing useful work. Because work is obtained from ordered molecular motion, the amount of entropy is also a measure of molecular disorder or randomness of a system. The concept of entropy provides deep insight into the direction of spontaneous change for many day-to-day phenomena. Now, how entropy was developed by Rudolf Clausius [3] is discussed below.

1.1. Clausius's Entropy

To provide a quantitative measure for the direction of spontaneous change, Clausius introduced the concept of entropy as a precise way of expressing the second law of thermodynamics. The Clausius form of the second law states that spontaneous change for an irreversible process [4] in an isolated system (that is, one that does not exchange heat or work with its surroundings) always proceeds in the direction of increasing entropy. By the Clausius definition, if an amount of heat Q flows into a large heat reservoir at temperature T above absolute zero, then $\Delta s = \frac{Q}{T}$. This equation effectively gives an alternate definition of temperature that agrees with the usual definition. Assume that there are two heat reservoirs R_1 and R_2 at temperatures T_1 and T_2. (such as the stove and the block of ice). If an amount of heat Q flows from R_1 to R_2, then the net entropy change for the two reservoirs is $\Delta s = Q\left(\frac{1}{T_2} - \frac{1}{T_1}\right)$ which is positive, provided that $T_1 > T_2$. Thus, the observation that heat never flows spontaneously from cold to hot is equivalent to requiring the net entropy change to be positive for a spontaneous flow of heat. When the system is in thermodynamic equilibrium, then $dS = 0$, *i.e.*, if $T_1 = T_2$, then the reservoirs are in equilibrium, no heat flows, and $\Delta s = 0$. If the gas absorbs an incremental amount of heat dQ from a heat reservoir at temperature T and expands reversibly against the maximum possible restraining pressure P, then it does the maximum work and $dW = PdV$. The internal energy of the gas might also change by an amount dU as it expands. Then, by conservation of energy, $dQ = dU + PdV$. Because the net entropy change for the system plus reservoir is zero when maximum work is done and the entropy of the reservoir decreases by an amount $dS_{reservoir} = -\frac{dQ}{T}$, this must be counterbalanced by an entropy increase of $dS_{system} = \frac{dU + PdV}{T} = \frac{dQ}{T}$ for the working gas so that $dS_{system} + dS_{reservoir} = 0$. For any real process, less than the maximum work would be done (because of friction, for example), and so the actual amount of heat dQ' absorbed from the heat reservoir would be less than the maximum amount dQ. For example, the gas could be allowed to expand freely into a vacuum and do no work at all. Therefore, it can be stated that $dS_{system} = \frac{dU + PdV}{T} \geqslant \frac{dQ'}{T}$ with, $dQ' = dQ$ in the case of maximum work corresponding to a reversible process. This equation defines S_{system} as a thermodynamic state variable, meaning that its value is completely determined by the current state of the system and not by how the system reached that state. Entropy is a comprehensive property in that its magnitude depends on the amount of material in the system.

In one statistical interpretation of entropy, it is found that for a very large system in thermodynamic equilibrium, entropy S is proportional to the natural logarithm of a quantity Ω corresponding to S and can be realized; that is, $S = K\ln\Omega$, in which K is related to molecular energy. On the other hand, entropy generation analysis [5–11] is used to optimize the thermal engineering devices for higher energy efficiency; it has attracted wide attention to its applications and rates in recent years. In order to access the best thermal design of systems, by minimizing the irreversibility, the second law of thermodynamics could be employed. Entropy generation is a criterion for the destruction of a systematized work.The development of the theory followed two conceptually different lines of thought. Nevertheless, they are symbiotically related, in particular through the work of Boltzmann.

1.2. Boltzmann's Entropy

In addition to thermodynamic (or heat-change) entropy, physicists also study entropy statistically [12,13]. The statistical or probabilistic study of entropy is presented in Boltzmann's law. Boltzmann's equation is somewhat different from the original Clausius (thermodynamic) formulation of entropy. Firstly, the Boltzmann formulation is structured in terms of probabilities, while the thermodynamic formulation does not consist in the calculation of probabilities. The thermodynamic formulation can be characterized as a mathematical formulation, while the Boltzmann formulation is statistical. Secondly, the Boltzmann equation yields a value of entropy S while the thermodynamic formulation yields only a value for the change in entropy (dS). Thirdly, there is a shift in content, as the Boltzmann equation was developed for research on gas molecules rather than thermodynamics. Fourthly, by incorporating probabilities, the Boltzmann equation focuses on microstates, and thus explicitly introduces the question of the relationship between macrostates and microstates. Boltzmann investigated such microstates and defined entropy in a new way such that the macroscopic maximum entropy state corresponded to a thermodynamic configuration which could be formulated by the maximum number of different microstates. He noticed that the entropy of a system can be considered as a measure of the disorder in the system and that in a system having many degrees of freedom, the number measuring the degree of disorder also measured the uncertainty in a probabilistic sense about the particular microstates.

The value W was originally intended to be proportional to the Wahrscheinlichkeit (means probability) of a macrostate for some probability distribution of a possible microstate, in which the thermodynamic state of a system can be realized by assigning different ξ and ρ of different molecules. The Boltzmann formula is the most general formula for thermodynamic entropy; however, his hypothesis was for an ideal gas of N identical particles, of which $N_i (i = 1, 2, \ldots., l)$ are the ith microscopic condition of position and momentum of a given distribution $D_i = (N_1, N_2, \ldots., N_l)$. Here, $D_1 = (N, 0, \ldots., 0)$, $D_2 = (N-1, 1, \ldots., 0)$, ... *etc.* For this state, the probability of each microstate system is equal, so it was equivalent to calculating the number of microstates associated with a macrostate. Then the statistical disorder is given by [14] $W = \frac{N!}{N_1!N_2!\ldots..N_l!}$, $[N_1 + N_2 + \ldots.. + N_l = N]$. Therefore, the entropy given by Boltzmann is: $S_B = K\ln W$ Where $\ln W = \ln\frac{N!}{N_1!N_2!\ldots..N_l!} = \ln N! - \sum_{i=1}^{l} \ln(N_i!)$

Let us now take an approximate value of W for a large $N!$. Using Stirling's approximation $\ln N! \cong N\ln N - N$, we have:

$$\ln W \cong N\ln N - N - \left[\sum_{i=1}^{l} N_i \ln(N_i) - \sum_{i=1}^{l} (N_i)\right]$$
$$= -\sum_{i=1}^{l} N_i \ln N_i = -N\sum_{i=1}^{l} p_i \ln p_i$$

$S_B = -K\sum_{i=1}^{n} p_i \ln p_i$ where $p_i = \frac{N_i}{N}$ is the probability of the occurrence of i th microstates. Boltzmann was the first to emphasize the probabilistic meaning of entropy and the probabilistic nature of thermodynamics.

1.3. Information Theory andShannon's Entropy

Unlike the first two entropy approaches (thermodynamic entropy by Clausius and Boltzmann's entropy), the third major form of entropy did not fall within the field of physics, but was developed instead in a new field known as information theory [15–17] (also known as communication theory). A fundamental step in using entropy in new contexts unrelated to thermodynamics was provided by Shannon [18], who came to conclude that entropy could be used to measure types of disorder other than that of thermodynamic microstates. Shannon was interested in information theory [19,20], particularly in the ways in which information can be conveyed via a message. This led him to examine probability distributions in a very general sense and he worked to find a way of measuring the level of uncertainty in different distributions.

For example, suppose the probability distribution for the outcome of a coin toss experiment is $P(H) = 0.999$ and $P(T) = 0.001$. One is likely to notice that there is much more "certainty" than "uncertainty" about the outcome of this experiment and, consequently, the probability distribution. If, on the other hand, the probability distribution governing that same experiment were $P(H) = 0.5$ and $P(T) = 0.5$, then there is much less "certainty" and much more "uncertainty" when compared to the previous distribution. However, how can these uncertainties can be quantified? Is there some algebraic function which measures the amount of uncertainty in any probabilistic distribution in terms of the individual probabilities? From these types of simple examples and others, Shannon was able to devise a set of criteria which any measure of uncertainty may satisfy. He then tried to find an algebraic form which would satisfy his criteria and discovered that there was only one formula which fit. Let the probabilities of n possible outcomes $E_1, E_2, \ldots\ldots, E_n$ of an experiment be $p_1, p_2, \ldots\ldots, p_n$, giving rise to the probability distribution $P = (p_1, p_2, \ldots, p_n)$; $\sum_{i=1}^{n} p_i = 1; p_i's \geqslant 0$. There is an uncertainty as to the outcome when the experiment is performed. Shannon suggested the measure $-\sum_{i=1}^{n} p_i \ln p_i$, which is identical to the previous entropy relation if the constant of probability is taken as the Boltzmann constant K. Thus, Shannon showed that entropy, which measures the amount of disorder in a thermodynamic system, also measures the amount of uncertainty in any probability distribution. Let us now give the formal definition of Shannon's entropy as follows: Consider a random experiment $P = (p_1, p_2, \ldots, p_n)$ whose possible outcomes have probabilities $p_i, i = 1, 2, \ldots, n$ that are known. Can we guess in advance which outcome we shall obtain? Can we measure the amount of uncertainty? We shall denote such an uncertainty measure by $H(P) = H_n(p_1, p_2, \ldots, p_n)$. The most common as well as the most useful measure of uncertainty is Shannon's informational entropy (which should satisfy some basic requirements), which is defined as follows:

Definition I: Let $(p_1, p_2, \ldots, p_n)$ be the probability of the occurrence of the events $E_1, E_2, \ldots\ldots, E_n$ associated with a random experiment. The Shannon's entropy probability distribution $(p_1, p_2, \ldots, p_n)$ of the random experiment system P is defined by $H(P) = H_n(p_1, p_2, \ldots, p_n) = -\sum_{i=1}^{n} p_i \ln p_i$ where, $0 \ln 0 \cong 0$. The above definition is generalized straightforwardly as the definition of entropy of a random variable.

Definition II: Let $X \in R$ be a discrete random variable which takes the value x_i $(i = 1, 2, 3, \ldots., n)$ with the probability $p_i, i = 1, 2, \ldots, n$; then the entropy $H(X)$ of X is defined by the expression $H(X) = -\sum_{i=1}^{n} p_i \ln p_i$ Examination of H or $H_n(P)$ reveals why Shannon's measure is the most satisfactory measure of entropy because of the following:

(i) $H_n(P)$ is a continuous function of $p_1, p_2, \ldots, p_n$.
(ii) $H_n(P)$ is a symmetric function of its arguments.
(iii) $H_{n+1}(p_1, p_2, \ldots, p_n, 0) = H_n(p_1, p_2, \ldots, p_n)$, *i.e.*, it should not change if there is an impossible outcome to the probability.
(iv) Its minimum is 0 when there is no uncertainty about the outcome. Thus, it should vanish when one of the outcomes is certain to happen so that

$$H_n(p_1, p_2, \ldots, p_n) = 0; p_i = 1, p_j = 0, i \neq j, i = 1, 2, \ldots, n$$

(v) It is the maximum when there is maximum uncertainty, which arises when the outcomes are equally likely so that $H_n(P)$ is the maximum when $p_1 = p_2 = \ldots\ldots = p_n = \frac{1}{n}$.
(vi) The maximum value of $H_n(P)$ increases with n.
(vii) For two independent probability distributions $P(p_1, p_2, \ldots, p_n)$ and $Q(q_1, q_2, \ldots, q_m)$, the uncertainty of the joint scheme $P \cup Q$ should be the sum of their uncertainties: $H_{n+m}(P \cup Q) = H_n(P) + H_m(Q)$

Shannon's entropy has various applications in the field of portfolio analysis, the measurement of economic analysis, transportation, and urban and regional planning as well as in the fields of statistics, thermodynamics, queuing theory, parametric estimation, *etc.* It has been used in the field non-commensurable and conflicting criteria [21] and in the nonlinear complexity of random sequences [22] as well.

2. Discussion

2.1. *Jaynes' Maximum Entropy (MaxEnt) Principle*

Let the random variable of an experiment be X, and assume the probability mass associated with the value x_i is p_i, *i.e.*, $P_X(i) \equiv P(X) = p_i, i = 1,2,\ldots.,n$. The set $(X,P) = \{(x_1,p_2),(x_2,p_2),\ldots.,(x_n,p_n)\}, \sum_{i=1}^{n} p_i = 1$ is called the source ensemble as described by Karmeshu [23]. In general, we may find expected values of the functions $g_1(X), g_2(X), \ldots, g_z(X)$ to get $\sum_{i=1}^{n} p_i g_{ri} = a_r, r = 1,2,\ldots,z, p_i \geqslant 0$ and with natural constraint $\sum_{i=1}^{n} p_i = 1$, given a number of constraints. Thus, we have $z+1$ relations between $(p_1,p_2,\ldots,p_n)$. There may be infinite probability distributions $(p_1,p_2,\ldots,p_n)$ satisfying the above equation. If we know only $g_1(x), g_2(x), \ldots, g_z(x)$, then we get a family of max entropy distributions. If, in addition, we know the values of $a_1,a_2,\ldots,a_z$, we get a specific member of this family and we call it the max entropy probability distribution. According to a great article by Jaynes [24–26], we choose the probability distribution out of all these which maximizes the measure of entropy as shown by Shannon's equation, $S(P) = -\sum_{i=1}^{n} p_i \ln p_i$.

Any distribution of the form $p_i = \exp[-\lambda_0 - \lambda_1 g_1(x_i) - \lambda_2 g^2_{(x_i)} - \ldots - \lambda_z g_z(x_i)]$, $i = 1,2,\ldots,n$ may be regarded as the maximum entropy distribution where $\lambda_0,\lambda_1,\ldots,\lambda_z$ are determined as functions of $a_1,a_2,\ldots,a_z$; then the maximum entropy $S_{\max}$ is given by $S_{\max} = \lambda_0 + \lambda_1 a_1 + \lambda_2 a_2 + \ldots + \lambda_z a_z$.

Kapur [27,28] showed that there is always a concave function of $a_1,a_2,\ldots,a_m$. We also note that all the probabilities given by p_i are always positive. We naturally want to know whether there is another measure of entropy other than Shannon's entropy which, when maximized, subject to, $\sum_{i=1}^{n} p_i = 1, \sum_{i=1}^{n} p_i g_{ri} = a_r, r = 1,2,\ldots,z, p_i \geqslant 0$, gives positive probabilities and for which $S_{\max}$ is possibly a concave function [29] of parameters. Kapur [30] studied that Burg's [31] measure of entropy, which has been very successfully used in spectral analysis, does always give positive probabilities. The maximum entropy principal of Jaynes has been used frequently to derive the distribution of statistical mechanics by maximizing the entropy of the system subject to some given constraints. The Maxwell-Boltzman distribution is obtained when there is only one constraint on a system which prescribes the expected energy per particle of the system by Bose-Einstein (B.E.) distribution, Fermi-Dirac (F.D.) distribution and intermediate statistics (I.S.) distributions; these are obtained by maximizing the entropy subject to two constraints by Kapur and Kesavan, and Kullback [32,33] and also by the present authors [34].

2.2. *Formulation of MEPD in Statistical Mechanics Using Shannon's Measure of Entropy*

Let $p_1,p_2,\ldots,p_n$ be the probabilities of a particle having energy levels $\varepsilon_1,\varepsilon_2,\ldots,\varepsilon_n$, respectively, and let the expected value of energy be prescribed as $\bar{\varepsilon}$; then, to get MEPD, we maximize the Shannon's measure of entropy:

$$S(P) = -\sum_{i=1}^{n} p_i \ln p_i \tag{1}$$

Subject to

$$\sum_{i=1}^{n} p_i = 1, \sum_{i=1}^{n} p_i \varepsilon_i = \bar{\varepsilon} \tag{2}$$

Let the Lagrangian be

$$L = -\sum_{i=1}^{n} p_i \ln p_i - (\lambda - 1)\left(\sum_{i=1}^{n} p_i - 1\right) - \mu\left(\sum_{i=1}^{n} p_i \varepsilon_i - \bar{\varepsilon}\right) \tag{3}$$

Differentiating with respect to p_i's, we get:

$\ln p_i + \lambda + \mu\varepsilon_i = 0 \Rightarrow p_i = \exp(-\lambda - \mu\varepsilon_i)$ where, λ, μ are to be determined by using Equation (2) so that

$$p_i = \frac{\exp(-\mu\varepsilon_i)}{\sum_{i=1}^{n}\exp(-\mu\varepsilon_i)}, i = 1,2,\ldots,n, \tag{4}$$

Where

$$\frac{\sum_{i=1}^{n}\varepsilon_i\exp(-\mu\varepsilon_i)}{\sum_{i=1}^{n}\exp(-\mu\varepsilon_i)} = \overline{\varepsilon} \tag{5}$$

Equation (4) is the well-known Maxwell-Boltzmann distribution from statistical mechanics which is used in many areas [35–37].

2.3. Burg's Entropy Measure and MEPD

When $S(P) = -\sum_{i=1}^{n} p_i \ln p_i$ was replaced by Burg's measure of entropy $B(P) = \sum_i \ln p_i$, it gave interesting results as shown by Kapur. Burg's measure of entropy is always negative, but this does not matter in entropy maximization, where it has been found that a probability distribution with maximum entropy satisfies the same constraint and it does not matter if all the entropies are negative. So, in Equation (1) when we use

$$B(P) = \sum_i \ln p_i, \text{ with the constraints } \sum_{i=1}^{n} p_i = 1, \sum_{i=1}^{n} ip_i = m \tag{6}$$

we get

$$p_i = \frac{1}{\lambda + \mu i}, \; i = 1,2,3,\ldots\ldots,n \tag{7}$$

where $\lambda \& \mu$ are obtained by solving the equations

$$\sum_i \frac{1}{\lambda + \mu i} = 1, \sum_i \frac{i}{\lambda + \mu i} = m \tag{8}$$

Multiplying first and second Equation (8) by $\lambda \& \mu$ respectively then adding

We get

$$\sum_i \frac{\lambda + \mu i}{\lambda + \mu i} = \lambda + \mu m \text{ or, } n = \lambda + \mu m \tag{9}$$

so that from Equation (8),

$$\sum_{i=1}^{n} \frac{1}{n - \mu(m - i)} = 1 \tag{10}$$

Then, $\mu = 0$ is an obvious solution but that will give us $\lambda = n$ and this will satisfy the second equation of (8)

if $m = \frac{n+1}{2}$. Now, Equation (10) is the nth degree polynomial in μ, and one of its roots is zero. Its non-zero solutions will be obtained by solving an equation of $(n-1)$th degree in μ. *Lemma* has been proved by Kapur as the following:

Lemma: All the roots of $\sum_{r=1}^{n} \frac{1}{n - \mu(m - i)} = 1$ are real; in other words, none of the roots can be complex.

Proof: Let $\mu = \alpha + i\beta, i = \sqrt{-1}$ be a pair of complex conjugate roots of Equation (10). Then,

$\sum_{r=1}^{n} \frac{1}{n-(\alpha+i\beta)(m-i)} = 1$ and $\sum_{r=1}^{n} \frac{1}{n-(\alpha-i\beta)(m-i)} = 1$; subtracting the second from the first, we get

$\sum_{r=1}^{n} \frac{2i\beta(m-r)}{\{n-\alpha(m-r)\}^2+\beta^2} = 0$, which gives $\beta = 0$, or $\sum_{r=1}^{n} \frac{(m-r)}{\{n-\alpha(m-r)\}^2+\beta^2} = 0$. The second possibility can easily be ruled out. To find the actual location of the n real roots, let us assume

$f(\mu) \equiv \sum_r \frac{1}{n-\mu(m-r)} - 1$ and this function is discontinuous at the following points:

$-\frac{n}{n-m}, -\frac{n}{n-(m-1)}, \ldots\ldots, -\frac{n}{1-l}, \frac{n}{l}, \ldots\ldots, \frac{n}{m-2}, \frac{n}{m-1}$, where $m = k+l, k = [m]$ and l is a "+" fraction. More precisely, $f(\mu) \to +\infty$ when $\mu \to$ points from one side and $f(\mu) \to -\infty$ when $\mu \to$ points from other side. Again,

$$f(\mu) \equiv \sum_r \frac{1}{n-\mu(m-r)} - 1 = \frac{1}{n}\left[\sum_r \left\{1-\mu\left(\frac{m-r}{n}\right)\right\}^{-1}\right] - 1$$

$$= \frac{1}{n}\left[\sum_r \left\{1+\mu\left(\frac{m-r}{n}\right)\right\} + O\left(\mu^2\right)\right] - 1 = \frac{1}{n}\left[n + \frac{\mu}{n}\left\{mn - \frac{n(n+1)}{2}\right\}\right] - 1 + O\left(\mu^2\right)$$

$$= \frac{\mu}{n}\left\{m - \frac{(n+1)}{2}\right\} + O\left(\mu^2\right)$$

2.4. d Burg's Modifie Entropy (MBE) Measure and MEPD

2.4.1. Monotonic Character of MBE

We propose use of Burg's modified entropy instead of Burg's entropy. Maximizing the Burg's modified measure of entropy:

$$B_{\text{mod}}(P) = \sum_{i=1}^{n} \ln(1+ap_i) - \ln(1+a), a > 0 \tag{11}$$

$$\frac{d}{da}\left(B_{\text{mod}}(P)\right) = \sum_{i=1}^{n} \frac{p_i}{1+ap_i} - \frac{1}{1+a} = \sum_{i=1}^{n} p_i \left[\frac{1}{1+ap_i} - \frac{1}{1+a}\right] = \sum_{i=1}^{n} \frac{ap_i(1-p_i)}{(1+a)(1+ap_i)} > 0 \tag{12}$$

Therefore, $B_{\text{mod}}(P)$ is the monotonic increasing function of a. For the probability distribution

$$(p, 1-p)$$

$$\frac{d}{da}\left(B_{\text{mod}}(P)\right) = \sum_{i=1}^{n} \frac{p_i}{1+ap_i} - \frac{1}{1+a} = \sum_{i=1}^{n} p_i \left[\frac{1}{1+ap_i} - \frac{1}{1+a}\right] = \sum_{i=1}^{n} \frac{ap_i(1-p_i)}{(1+a)(1+ap_i)} > 0 \tag{13}$$

it is showed (see: Table 1.) that $\underset{P}{Max}\left(B_{\text{mod}}(P)\right) = \ln\left(1+\frac{a^2}{4(1+a)}\right) = \ln\left(\frac{(a+2)^2}{4(1+a)}\right)$.

The measure of entropy $\underset{P}{Max}\left(B_{\text{mod}}(P)\right)$ is the Burg's modified entropy. This is a better measure than the standard Burg's measure since it is always positive and there is no computational problem when p_i is very small. In the above case, the maximum value increases with the number of possible outcomes n.

Table 1. The a values & maximum values of Burg's Modified Entropy.

a	$\underset{P}{Max}\,(B_{\mathbf{mod}}\,(P))$
0.5	0.04082
1.0	0.11778
1.5	0.20294
2.0	0.28768
5.0	0.71376
10.0	1.1856
20.0	1.7512
30.0	2.1111
40.0	2.3754
50.0	2.5843
∞	∞

2.4.2. MBE and Its Relation with Burg's Entropy

$$B_{\text{mod}}\,(P) = \sum_{i=1}^{n} \ln(1+ap_i) - \ln(1+a) = \sum_{i=1}^{n}\left[\ln a + \ln\left(p_i + \frac{1}{a}\right)\right] - \ln a - \ln\left(1+\frac{1}{a}\right)$$
$$= (n-1)\ln a + \sum_{i=1}^{n} \ln\left(p_i + \frac{1}{a}\right) - \ln\left(1+\frac{1}{a}\right)$$

So, when $a \to \infty$, maximizing $(B_{\text{mod}}\,(P))$ and $(B\,(P))$ will give the same result in both cases;again, if $(B_{\text{mod}}\,(P))$ is maximized under the constraints

$$\sum_{i=1}^{n} p_i = 1, \sum_{i=1}^{n} p_i g_{ri} = a_r, r = 1,2,\ldots\ldots, m, p_i \geqslant 0 \tag{14}$$

we get
$\frac{1}{1+ap_i} = [\lambda_0 + \lambda_1 g_1(x_i) + \lambda_2 g^2_{(x_i)} + \ldots\ldots + \lambda_m g_m(x_i)], i = 1,2,\ldots\ldots\ldots\ldots, n.$ Letting, we have

$$\frac{1}{p_i} = \lambda'_0 + \lambda'_1 g_1(x_i) + \lambda'_2 g^2_{(x_i)} + \ldots\ldots + \lambda'_m g_m(x_i) \tag{15}$$

The λ'_i's are determined by using Constraints (14) and (15) and this gives the MEPD when Burg's entropy is maximized as subject to Equation (14).

Therefore, when $a \to \infty$, the MEPD of BME $\to$ Burg's MEPD; in fact:

$$(B_{\text{mod}}\,(P)) - (B\,(P)) = (n-1)\ln a + \sum_{i=1}^{n} \ln \frac{\left(p_i + \frac{1}{a}\right)}{p_i} - \ln\left(1+\frac{1}{a}\right) \to \infty \text{ as}$$
$$a \to \infty.$$

2.4.3. MBE and Its Concavity of $S_{\max}$ under Prescribed Mean

$$\text{Maximize } B_{\text{mod}}\,(P) = \sum_{i=1}^{n} \ln(1+ap_i) - \ln(1+a) \tag{16}$$

subjectto $\sum_{i=1}^{n} p_i = 1, \sum_{i=1}^{n} ip_i = m.$

We obtain this using Lagranges multiplier mechanics:

$$\left(p_i + \frac{1}{a}\right) = \frac{1}{\lambda + \mu i} \text{ Where, } \sum_i \frac{1}{\lambda + \mu i} = \sum_i \left(p_i + \frac{1}{a}\right) = 1 + \frac{n}{a} \tag{17}$$

From the above equation:

$$\sum_i \frac{i.1}{\lambda + \mu i} = \sum_i \left(p_i + \frac{1}{a}\right) = \sum_{i=1}^{n} \left(ip_i + \frac{i}{a}\right) \tag{18}$$

$$\sum_i \frac{i}{\lambda + \mu i} = m + \frac{n(n+1)}{2a}, \text{ i.e., } \frac{n(n+1)}{2a} + m = \sum_i \frac{i}{\lambda + \mu i} \text{ and}$$

$$\frac{\mu n(n+1)}{2a} + \mu m = \sum_i \frac{\mu i}{\lambda + \mu i}, \text{ therefore } \sum_i \frac{\lambda}{\lambda + \mu i} = \sum_i \left(p_i + \frac{1}{a}\right) = \lambda\left(1 + \frac{n}{a}\right) \tag{19}$$

So

$$\sum_i \frac{\lambda}{\lambda + \mu i} + \sum_i \frac{\mu i}{\lambda + \mu i} = \lambda\left(1 + \frac{n}{a}\right) + \frac{\mu n(n+1)}{2a} + \mu m = \sum_i \frac{\lambda + \mu i}{\lambda + \mu i} = \sum_i^{n} 1 = n \tag{20}$$

$$\lambda\left(1 + \frac{n}{a}\right) + \frac{\mu n(n+1)}{2a} + \mu m = n \Rightarrow \lambda = \frac{n - \mu\left(m + \frac{n(n+1)}{2a}\right)}{1 + \frac{n}{a}} \tag{21}$$

Therefore,

$$\lambda + \mu i = \frac{n + \mu\left[\left(1 + \frac{n}{a}\right)i - m - \frac{n(n+1)}{2a}\right]}{1 + \frac{n}{a}} \tag{22}$$

and

$$p_i = \frac{a + n}{an + \mu\left[(n+a)i - am - \frac{n(n+1)}{2}\right]} - \frac{1}{a}, i = 1, 2, \ldots, n, \tag{23}$$

where μ is determined as a function of m and that is

$$1 + \frac{n}{a} = \sum_i^{n} \frac{a + n}{an + \mu\left[(n+a)i - am - \frac{n(n+1)}{2}\right]} \;\&$$

$$\frac{1}{a} = \sum_i^{n} \frac{1}{an + \mu\left[(n+a)i - am - \frac{n(n+1)}{2}\right]},$$

Therefore,

$$S_{\max} = \sum_i \ln(1 + ap_i) - \ln(1 + a)$$

$$= \sum_i^{n} \ln \frac{a}{\lambda + \mu i} - \ln(1 + a) = \ln \frac{a^n}{1 + a} - \sum_i \ln(\lambda + \mu i)$$

$$= \ln \frac{a^n}{1 + a} - \sum_i^{n} \ln\left[n + \mu\left(\left(1 + \frac{n}{a}\right)i - m - \frac{n(n+1)}{2}\right)\right] + n\ln\left(1 + \frac{n}{a}\right)$$

$$\frac{dS_{\max}}{dm} = -\sum_{i=1}^{n} \frac{\left[\frac{d\mu}{dm}\left\{\left(1+\frac{n}{a}\right)i - m - \frac{n(n+1)}{2a}\right\} - \mu\right]}{n+\mu\left[\left(1+\frac{n}{a}\right)i - m - \frac{n(n+1)}{2a}\right]}$$

$$= -\sum \frac{\left(p_i + \frac{1}{a}\right)}{a+n} a \left[\frac{d\mu}{dm}\left[\left(1+\frac{n}{a}\right)i - m - \frac{n(n+1)}{2a}\right] - \mu\right]$$

$$= \frac{d\mu}{dm}\left[m + \frac{n(n+1)}{2} - m - \frac{n(n+1)}{2}\right] + \frac{\mu a}{a+n}\left(1+\frac{n}{a}\right)$$

So that

$$\frac{dS_{\max}}{dm} = \frac{d\mu}{dm} \& \frac{d^2 S_{\max}}{dm^2} = \frac{d\mu}{dm} = \frac{\mu \sum_i^n \left(p_i + \frac{1}{a}\right)^2}{\sum_i^n \left(p_i + \frac{1}{a}\right)^2 \left[\left(1+\frac{n}{a}\right)i - m - \frac{n(n+1)}{2a}\right]} \tag{24}$$

$S_{\max}$ will be a concave function of m if $\frac{d\mu}{dm} < 0$, that is, if either $\mu > 0, < 0$ when the denominator $< 0, > 0$, respectively.

In the above case, when $a = 1$ we get $S_{\max}$ and the derivative of $S_{\max}$ as follows:

$$\frac{dS_{\max}}{dm} = -\sum_i \frac{1}{\lambda + \mu i}\left[\frac{d}{dm}(\lambda + \mu i)\right] = \sum_i \left[\frac{n}{2}\frac{d\mu}{dm} + \frac{1}{n+1}\left(m\frac{d\mu}{dm} + \mu\right) - \frac{d\mu}{dm} i\right] \tag{25}$$

$$= \frac{d\mu}{dm}\sum_i (1+p_i)\left(\frac{n}{2} + \frac{m}{n+1} - i\right) + \frac{\mu}{n+1}\sum_i (1+p_i)$$

$$= \frac{d\mu}{dm}\sum_i \left(\frac{n}{2} + \frac{m}{n+1} - i\right) + \frac{d\mu}{dm}\left(\frac{n}{2} + \frac{m}{n+1}\right) - \frac{d\mu}{dm}\sum_i i p_i + \frac{\mu n}{n+1} + \frac{\mu}{n+1}$$

$$= \frac{d\mu}{dm}\sum_i \left(\frac{n}{2}.n + \frac{m}{n+1}.n - \frac{n(n+1)}{2}\right) + \frac{d\mu}{dm}\left(\frac{n}{2} + \frac{m}{n+1}\right) - \frac{d\mu}{dm} m + \frac{\mu n}{n+1} + \frac{\mu}{n+1}$$

$$= \frac{d\mu}{dm}\left(\frac{mn}{n+1} - \frac{m}{n+1} - m\right) + \frac{\mu n}{n+1} + \frac{\mu}{n+1} = \mu\left(\frac{n}{n+1} + \frac{1}{n+1}\right) = \mu \tag{26}$$

So that

$$\frac{d^2 S_{\max}}{dm^2} = \frac{d\mu}{dm} \tag{27}$$

Additionally, we have

$$0 = -\sum_i (1+p_i)^2 \left[\frac{n}{2}\frac{d\mu}{dm} + \frac{1}{n+1}\left(m\frac{d\mu}{dm} + \mu\right) - \frac{d\mu}{dm} i\right] \tag{28}$$

So that

$$\frac{d\mu}{dm}\sum_i (1+p_i)^2 \left(\frac{n}{2} + \frac{m}{n+1} - i\right) + \frac{\mu}{n+1}\sum_i (i+p_i)^2 = 0 \tag{29}$$

Therefore,

$$\frac{d\mu}{dm} = \frac{\frac{\mu}{n+1}\sum_i (1+p_i)^2}{\sum_i (1+p_i)^2 \left(i - \frac{n}{2} - \frac{m}{n+1}\right)} \tag{30}$$

So,

$$\frac{d^2 S_{\max}}{dm^2} < 0 \Rightarrow \frac{d\mu}{dm} < 0 \tag{31}$$

Therefore, from Equations (27) and (30),
$S_{\max}$ will be a concave function of m if either

$$\text{(i) } \mu > 0, \sum_i (1+p_i)^2 \left(i - \frac{n}{2} - \frac{m}{n+1}\right) < 0 \tag{32}$$

Or

$$\text{(ii) } \mu < 0, \sum_i (1+p_i)^2 \left(i - \frac{n}{2} - \frac{m}{n+1}\right) > 0 \tag{33}$$

Additionally, when $\mu = 0$, $\frac{d^2 S_{\max}}{dm^2} = 0$, all the probabilities are equal, and from Equation (24) we have:

We have,

$$m + \frac{n(n+1)}{2} = \frac{1}{\lambda}\frac{n(n+1)}{2} \text{ or, } m = \frac{n(n+1)}{2}\left[\frac{1}{\lambda} - 1\right] \Rightarrow m = \frac{n+1}{2}. \tag{34}$$

Now, if we proceed algebraically as done Section 2.3, we get:

$$1 + p_i = \frac{1}{\lambda + \mu i} = \frac{1}{\frac{1}{n+1}(n - \mu m) - \mu\left(\frac{n}{2} - i\right)}, i = 1, 2, \ldots, n,$$

$\sum\limits_{i=1}^{n} \frac{1}{\frac{1}{n+1}(n - \mu m) - \mu\left(\frac{n}{2} - i\right)} = \sum\limits_{1}^{n}(1 + p_i) = n + 1$. The obvious solution of the above problem is $\mu = 0$ which will give $\lambda = \frac{1}{p_i} = n$, *i.e.*, uniform distribution, and thus we get, $m = \frac{n+1}{2}$.

3. An Illustrative Example in Statistical Mechanics

3.1. Example

Let $p_1, p_2, \ldots, p_{10}$ be the probabilities of a particle having energy levels $\varepsilon_1, \varepsilon_2, \ldots, \varepsilon_{10}$; respectively, and let the expected value of energy be prescribed as m; then, we get the maximum entropy probability distribution (MEPD) with MBE as follows:

$$\text{Maximize} \sum_{i=1}^{10} \ln(1 + p_i) - \ln 2 \text{ such that, } \sum_{i=1}^{10} p_i = 1 \text{ and } \sum_{i=1}^{10} i p_i = m \text{ for } m = 1\,(0.25)\,10$$

Solution: Maximizing the measure of entropy subject to the given constraints, we have $(1 + p_i) = \frac{1}{\lambda + \mu i}$.

When $\sum\limits_i \frac{1}{\lambda + \mu i} = 10 + 1 = 11$, $\frac{10(10+1)}{2} + m = \sum\limits_i \frac{i}{\lambda + \mu i}$ *i.e.*, $\sum\limits_i \frac{i}{\lambda + \mu i} = 55 + m$, we get the probability distribution in the form of a table and also get the values of $S_{\max}$ as described by Kapur and Kesavan.

There may be two cases,

Case (i) when $m < \frac{n+1}{2} = 5.5$.

In this case, when m lies between 1 and 5.5, $\mu > 0$ implies $\frac{dS_{\max}}{dm} > 0$ and then $S_{\max}$ is increasing.

Case (ii) when $m > \frac{n+1}{2} = 5.5$.

In this case, when m lies between 5.5 and 10, $\mu < 0$ implies $\frac{dS_{max}}{dm} < 0$ and then S_{max} is decreasing.

When $m = \frac{n+1}{2} = 5.5$, $\mu = 0$, $\frac{dS_{max}}{dm} = 0$.

it can be shown that S_{max} will be concave if we prescribe $E(g(x))$ instead of $E(x)$, where $g(x)$ is a monotonic increasing function of x; then we will apply the necessary changes. Again, since the concavity of S_{max} has already been proven, this will enable us to handle the inequality constraint of the type $m_1 \leqslant \sum_{i=1}^{n} ip_i \leqslant m_2$.

3.2. Simulated Results

Following Table (Table 2) Found Using LINGO Software 2011 where Different Max-Entropy Values Are Given for Different m Values:

Table 2. Comparative Maximum Entropy Values.

m	Maximum Entropy Value of $B_{mod}(P)$	Maximum Entropy Value of $B(P)$	Maximum Entropy Value of $S(P)$
1.00	0	−158.225	0
1.25	0.0896122	−45.13505	0.6255029
1.50	0.1345336	−38.98478	0.9547543
1.75	0.1633092	−35.43113	1.194875
2.00	0.1836655	−32.94620	1.385892
2.25	0.1989097	−31.05219	1.542705
2.50	0.2107459	−29.53735	1.675885
2.75	0.2201939	−28.28982	1.790029
3.00	0.2279397	−27.24394	1.888477
3.25	0.2344146	−26.35851	1.973503
3.50	0.2399041	−25.60649	2.046725
3.75	0.24461	−24.96958	2.109324
4.00	0.248682	−24.43512	2.162186
4.25	0.252127	−23.99419	2.205980
4.50	0.2549453	−23.64047	2.241209
4.75	0.257137	−23.36942	2.268253
5.00	0.2587024	−23.17789	2.287386
5.25	0.2596416	−23.06376	2.298794
5.50	**0.2599546**	−**23.02585**	**2.302585**
5.75	0.2596416	−23.06376	2.298794
6.00	0.2587024	−23.17789	2.287386
6.25	0.257137	−23.36942	2.268253
6.50	0.2549453	−23.64047	2.241209
6.75	0.252127	−23.99419	2.205980
7.00	0.248682	−24.43512	2.162186
7.25	0.24461	−24.96958	2.109324
7.50	0.2399041	−25.60649	2.046725
7.75	0.2344146	−26.35851	1.973503
8.00	0.2279397	−27.24394	1.888477
8.25	0.2201939	−28.28982	1.790029
8.50	0.2107459	−29.53735	1.675885
8.75	0.1989097	−31.05219	1.542705
9.00	0.1836655	−32.94620	1.385892
9.25	0.1633092	−35.43113	1.194875
9.50	0.1345336	−38.98478	0.9547543
9.75	0.0896122	−45.13505	0.6255029
10.0	0	−141.25285	0

Graphs obtained from the above table are given on the next page (Figures 1–3):

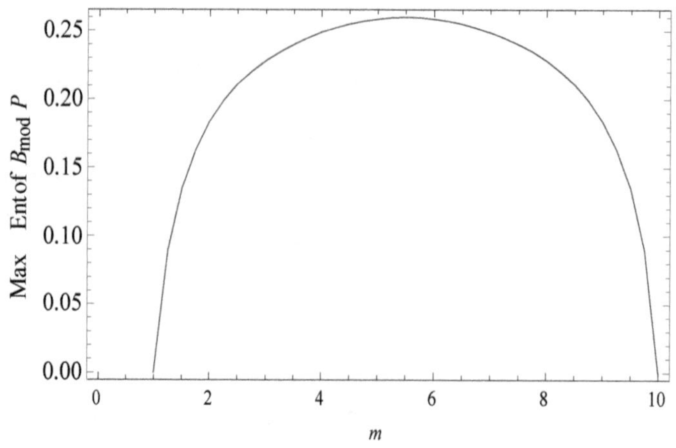

Figure 1. Maximum Burg's modified entropy.

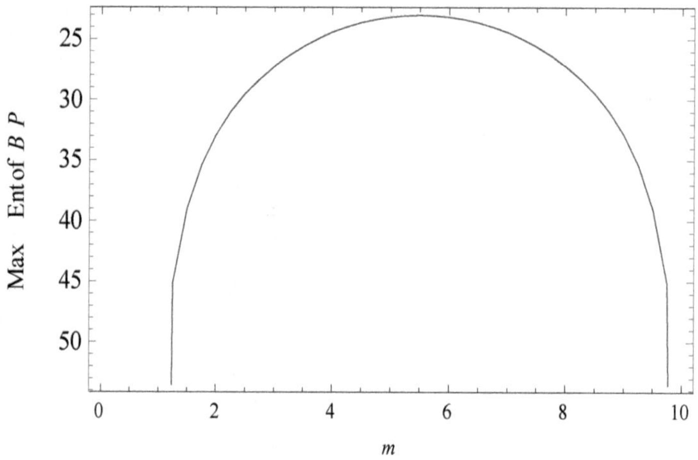

Figure 2. Maximum Burg's entropy.

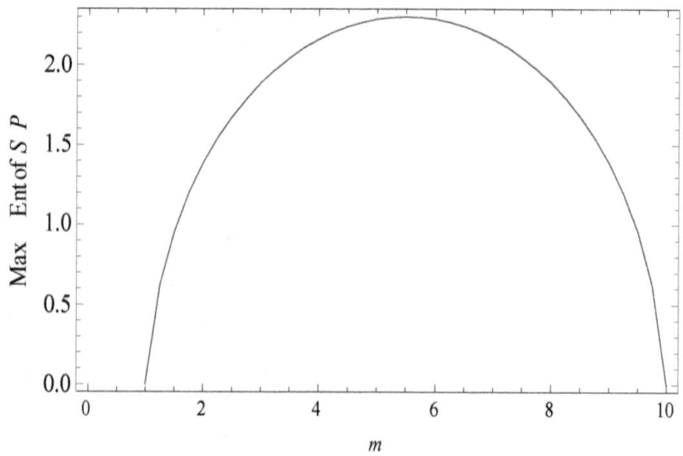

Figure 3. Maximum Shannon's entropy.

If p_i is plotted against i, we get rectangular hyperbolic types of curves (Table 3 and Figure 4).

Table 3. p_i and i graph.

i	p_i
1	1.000000
2	0.2899575
3	0.1747893
4	0.1253783
5	0.1027619
6	0.1027620
7	0.1253782
8	0.1747891
9	0.2899574
10	1.000000

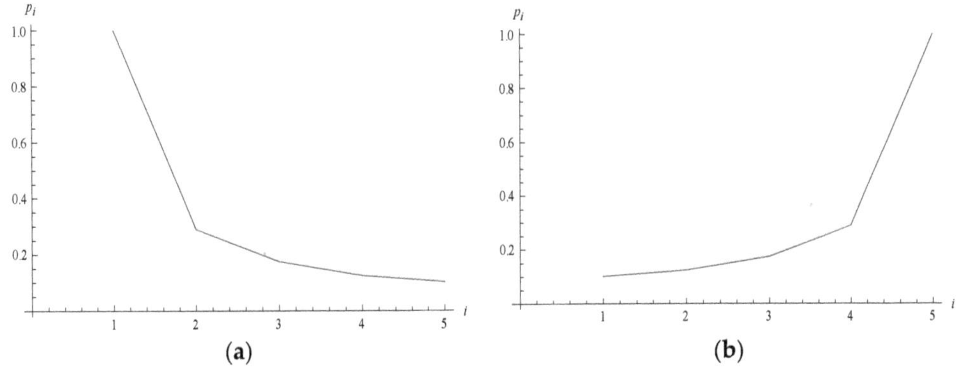

Figure 4. Rectangular hyperbolic types of graphs. (**a**) when $m < \frac{n+1}{2}, \mu > 0$; (**b**) when $m > \frac{n+1}{2}, \mu < 0$.

4. Conclusions

In the present paper we have presented different MEPDs and respective entropy measures with their properties. It has been found that MBE is a better measure than Burg's entropy when the maximized subject to the mean is prescribed, and it also has been shown that unlike Burg's entropy, the maximum value of MBE increases with n. The main problem here will consist of solving $z + 1$ simultaneous transcendental equations for the Lagranges multipliers. An application in statistical mechanics with simulated data has been studied with the help of Lingo11 software and corresponding graphs are provided. Now, one question arises: Will this result continue to hold for other moment constraints also? When we take generalized moment expectation of $g(x)$ instead of expectation of x, then $g(x)$ must be a monotonic increasing function of x, and if p_i becomes negative for some values of the moments, then we have to set those probabilities to zero and reformulate the problem for the remaining probabilities over the remaining range and solve it.

Acknowledgments: Authors would like to express their sincere gratitude and appreciation to the reviewers for their valuable comments and good suggestions to make this paper enriched and impactful, and a special thank you to the Assistant Editor.

Author Contributions: A. Ray and S. K. Majumder conceived and designed the experiments; A. Ray performed the experiments, analyzed the data and both A. Ray and S. K. Majumder contributed in analysis tools; A. Ray wrote the paper.

Conflicts of Interest: The authors declare no conflict of interest.

Nomenclature

S = Entropy

$p_i = \frac{N_i}{N}$, probability of ith event

$H(P) = H_n(p_1, p_2, \ldots, p_n)$ = Information entropy

$S(P)\, or H_n(P) = -\sum_{i=1}^{n} p_i \ln p_i$ = Shannon's entropy

$B(P) = \sum_i \ln p_i$ = Burg's entropy

$B_{mod}(P)$ = Burg's modified entropy

n = Number of energy levels/number of possible outcome

$\mu = \frac{1}{KT}$

K = Boltzmann constant

T = Absolute temperature

N = Identical particle of ideal gas

Δs = Increase of entropy

dS = Change in entropy

$[m]$ = Greatest integer value of m

Where m is the mean value:

$E(x)$ = Expectation

dV= Change in volume

$P \cup Q$ = Union of two sets

$m = 1\,(0.25)\,10$, m ranges from 1 to 10 with step length of 0.25

S_{max} = Maximum entropy

$\underset{P}{Max}\,(B_{mod}(P))$ = Maximum value under the given probability distribution

MBE=Modified Burg's Entropy

Greek Symbols

Ω = The maximum number of microscopic ways in the macroscopic state

ξ = Position of the molecule

ρ = Momentum of the molecule

μ, λ=Lagrangian constant

$\varepsilon_1, \varepsilon_2, \ldots$.=Different energy levels

$\bar{\varepsilon}$ =Mean energy

Subscripts

B=Boltzmann

max =Maximum

mod =modified

Superscript

λ' =New constant different from λ

References

1. Andreas, G.; Keller, G.; Warnecke, G. *Entropy*; Princeton University Press: Princeton, NJ, USA, 2003.
2. Phil, A. *Thermodynamics and Statistical Mechanics: Equilibrium by Entropy Maximisation*; Academic Press: Cambridge, MA, USA, 2002.
3. Rudolf, C. Ueber verschiedene für die Anwendung bequeme Formen der Hauptgleichungen der mechanischen Wärmetheorie. *Annalen der Physik* **1865**. (In German). [CrossRef]
4. Wu, J. Three Factors Causing the Thermal Efficiency of a Heat Engine to Be Less than Unity and Their Relevance to Daily Life. *Eur. J. Phys.* **2015**, *36*, 015008. [CrossRef]
5. Rashidi, M.M.; Shamekhi, L. Entropy Generation Analysis of the Revised Cheng-Minkowycz Problem for Natural Convective Boundary Layer Flow of Nanofluid in a Porous Medium. *J. Thermal Sci.* **2015**, *19*. [CrossRef]
6. Rashidi, M.M.; Mohammadi, F. Entropy Generation Analysis for Stagnation Point Flow in a Porous Medium over a Permeable Stretching Surface. *J. Appl. Fluid Mech.* **2014**, *8*, 753–763.
7. Rashidi, M.M.; Mahmud, S. Analysis of Entropy Generation in an MHD Flow over a Rotating Porous Disk with Variable Physical Properties. *Int. J. Energy* **2014**. [CrossRef]
8. Abolbashari, M.H.; Freidoonimehr, N. Analytical Modeling of Entropy Generation for Casson Nano-Fluid Flow Induced by a Stretching Surface. *Adv. Powder Technol.* **2015**, *231*. [CrossRef]
9. Baag, S.S.R.; Dash, M.G.C.; Acharya, M.R. Entropy Generation Analysis for Viscoelastic MHD Flow over a Stretching Sheet Embedded in a Porous Medium. *Ain Shams Eng. J.* **2016**, *23*. [CrossRef]
10. Shi, Z.; Tao, D. Entropy Generation and Optimization of Laminar Convective Heat Transfer and Fluid Flow in a Microchannel with Staggered Arrays of Pin Fin Structure with Tip Clearance. *Energy Convers. Manag.* **2015**, *94*, 493–504. [CrossRef]
11. Hossein, A.M.; Ahmadi, M.A.; Mellit, A.; Pourfayaz, F.; Feidt, M. Thermodynamic Analysis and Multi Objective Optimization of Performance of Solar Dish Stirling Engine by the Centrality of Entransy and Entropy Generation. *Int. J. Electr. Power Energy Syst.* **2016**, *78*, 88–95. [CrossRef]
12. Giovanni, G. *Statistical Mechanics: A Short Treatise*; Springer Science & Business Media: New York, NY, USA, 2013.
13. Reif, F. *Fundamentals of Statistical and Thermal Physics*; Waveland Press: Long Grove, IL, USA, 2009.
14. Rudolf, C.; Shimony, A. *Two Essays on Entropy*; University of California Press: Berkeley, CA, USA, 1977.
15. Shu-Cherng, F.; Rajasekera, J.R.; Tsao, H.S.J. *Entropy Optimization and Mathematical Programming*; Springer Science & Business Media: New York, NY, USA, 2012.
16. Robert, M.G. *Entropy and Information Theory*; Springer Science & Business Media: New York, NY, USA, 2013.
17. Silviu, G. *Information Theory with New Applications*; MacGraw-Hill Books Company: New York, NY, USA, 1977.
18. Shannon, C.E. A Mathematical Theory of Communication. *SigmobileMob. Comput. Commun. Rev.* **2001**, *5*, 3–55. [CrossRef]
19. Thomas, M.C.; Thomas, J.A. *Elements of Information Theory*; John Wiley & Sons: Hoboken, NJ, USA, 2012.
20. Christodoulos, A.F.; Pardalos, P.M. *Encyclopedia of Optimization*; Springer Science & Business Media: New York, NY, USA, 2008.
21. Arash, M.; Azarnivand, A. Application of Integrated Shannon's Entropy and VIKOR Techniques in Prioritization of Flood Risk in the Shemshak Watershed, Iran. *Water Resour. Manag.* **2015**, *30*, 409–425. [CrossRef]
22. Liu, L.; Miao, S.; Liu, B. On Nonlinear Complexity and Shannon's Entropy of Finite Length Random Sequences. *Entropy* **2015**, *17*, 1936–1945. [CrossRef]
23. Karmeshu. *Entropy Measures, Maximum Entropy Principle and Emerging Applications*; Springer: Berlin, Germany; Heidelberg, Germany, 2012.
24. Jaynes, E.T. Information Theory and Statistical Mechanics. II. *Phys. Rev.* **1957**, *108*, 171–190. [CrossRef]
25. Jaynes, E.T. On the Rationale of Maximum-Entropy Methods. *Proc. IEEE* **1982**, *70*, 939–952. [CrossRef]
26. Jaynes, E.T. Prior Probabilities. *IEEE Trans. Syst. Sci. Cybernet.* **1968**, *4*, 227–241. [CrossRef]
27. Kapur, J.N. *Maximum-Entropy Models in Science and Engineering*; John Wiley & Sons: New York, NY, USA, 1989.
28. Kapur, J.N. Maximum-Entropy Probability Distribution for a Continuous Random Variate over a Finite Interval. *J. Math. Phys. Sci.* **1982**, *16*, 693–714.

29. Ray, A.; Majumder, S.K. Concavity of maximum entropy through modified Burg's entropy subject to its prescribed mean. *Int. J. Math. Oper. Res.* **2016**, *8*, to appear.
30. Kapur, J.N. *Measures of Information and Their Applications*; Wiley: New York, NY, USA, 1994.
31. Burg, J. The Relationship between Maximum Entropy Spectra and Maximum Likelihood Spectra. *Geophysics* **1972**, *37*, 375–376. [CrossRef]
32. Narain, K.J.; Kesavan, H.K. *Entropy Optimization Principles with Applications*; Academic Press: Cambridge, MA, USA, 1992.
33. Solomon, K. *Information Theory and Statistics*; Courier Corporation: North Chelmsford, MA, USA, 2012.
34. Amritansu, R.; Majumder, S.K. Derivation of some new distributions in statistical mechanics using maximum entropy approach. *Yugoslav J. Oper. Res.* **2013**, *24*, 145–155.
35. Ulrych, T.J.; Bishop, T.N. Maximum Entropy Spectral Analysis and Autoregressive Decomposition. *Rev. Geophys.* **1975**, *13*, 183–200. [CrossRef]
36. Michele, P.; Ferrante, A. On the Geometry of Maximum Entropy Problems. **2011**, arXiv:1112.5529.
37. Ke, J.-C.; Lin, C.-H. Maximum Entropy Approach to Machine Repair Problem. *Int. J. Serv. Oper. Inform.* **2010**, *5*, 197–208. [CrossRef]

11

Multiplicative Expression for the Coefficient in Fermionic 3–3 Relation

Igor Korepanov

Academic Editor: Yuli Rudyak

Moscow Technological University, 20 Stromynka Street, Moscow 107076, Russia; korepanov@mirea.ru

Abstract: Recently, a family of fermionic relations were discovered corresponding to Pachner move 3–3 and parameterized by complex-valued 2-cocycles, where the weight of a pentachoron (4-simplex) is a Grassmann–Gaussian exponent. Here, the proportionality coefficient between Berezin integrals in the l.h.s. and r.h.s. of such relations is written in a form multiplicative over simplices.

Keywords: four-dimensional TQFT; piecewise-linear TQFT; algebraic realizations of Pachner moves; Grassmann–Gaussian exponents

MSC: 57R56 (Primary); 57Q99; 13P25 (Secondary)

---

1. Introduction

This paper continues a series of papers [1–3]. The reader is referred especially to [3] for definitions and facts that are only briefly mentioned here. In addition, the reader is referred to [4] for a concise exposition of Grassmann–Berezin calculus of anticommuting variables (or to [5] for a more modern and detailed exposition), and to [6] for a pedagogical introduction to Pachner moves.

In [2], a large family was discovered of Grassmann–Gaussian relations corresponding to Pachner move 3–3, with pentachoron (4-simplex) weights depending on a single Grassmann variable attached to each 3-face. In [3], a full parameterization was given for (a Zariski open set of) such relations, in terms of a 2-cocycle given on both l.h.s. and r.h.s. of the Pachner move. Many questions still remain, however, to be solved before we arrive at a full-fledged four-dimensional topological quantum field theory (TQFT) on piecewise-linear manifolds.

In the present paper, we solve one such question and show that the answer is remarkably nontrivial. It consists in finding the coefficient called 'const' in ([3] formula (53)) (as well as ([2], formula (6)) in a form that would make possible further construction of a manifold invariant. Namely, the coefficient should be represented as a ratio, const $= c_r/c_l$ (compare relation Equation (2) below), of two expressions belonging to the two sides of the move, and each of these must be *multiplicative*—have the form of a product over simplices belonging to the corresponding side. This was the case in an earlier paper [1], see formula (1) and Theorem 1 there, also reproduced in ([2] Section 6), although the 3–3 relations in these papers must be regarded as degenerate from the viewpoint of the present paper. This was also the case in ([7], formula (38)) and ([8], formula (12)), where different but similar relations were considered.

1.1. PL Manifold Invariants and Pachner Moves

In order to construct invariants of piecewise linear (PL) manifolds, it makes sense to construct algebraic relations corresponding to *Pachner moves*, see, for instance, ([6], Section 1). Pachner's theorem states that a triangulation of a PL manifold can be transformed into any other triangulation using a finite sequence of these moves [9], so there is a hope to pass then from such relations to some quantities characterizing the whole manifold.

More specifically, a popular idea consists in constructing *state sum*-like invariants. This means the following: a set X of "states", also often called "colors", is introduced for every simplex of some fixed dimension d (for instance, for edges—simplices of dimension $d = 1$. Of course, one can also imagine more complicated cases, where simplices of different dimensions are involved). A *coloring* is then any mapping from the set of all d-simplices in triangulation into X. In addition, there is a "Boltzmann weight" W_u, assigned most often to every simplex u of the maximal dimension and depending on the colors of d-simplices contained in u. The values of W_u are supposed to lie in some ring, because the invariant is composed from them using multiplication and addition, as in formula (1) below.

If there is a manifold M with *colored boundary* ∂M (that is, colors are assigned to all d-simplices lying in ∂M), then its weight is defined, typically, as the following sum of products, each corresponding to a coloring of *inner* d-simplices:

$$W_M = \sum_{\substack{\text{over all colorings} \\ \text{of inner } d\text{-simplices}}} \prod_{\text{over all } u} W_u \tag{1}$$

If M is now a result of gluing together two manifolds, M_1 and M_2, by means of identifying some (identically triangulated) parts $N_1 \subset \partial M_1$ and $N_2 \subset \partial M_2$ of their boundaries, the weight:

$$W_M = \sum_{\substack{\text{over all colorings} \\ \text{of inner } d\text{-simplices} \\ \text{in } N_1 = N_2}} W_{M_1} W_{M_2}$$

has the following obvious property: if we change in any way the triangulation *within* M_1 or/and M_2, but not changing the triangulation of its boundary and not changing its weight, then W_M remains the same.

Pachner moves are elementary re-buildings of small clusters of simplices, not changing the topology. In a usual notation, move m–n replaces m simplices of maximal dimension with n simplices. If we have invented the simplex weight such that the cluster weight does not change under all relevant Pachner moves, then Equation (1) gives a manifold invariant.

Such straightforward scheme may work in four dimensions, see, for instance, [10], where all pentachora are assigned the same "constant" weight satisfying the fundamental equation ([10] formula (22)). On the other hand, that "constant" equation may be too restrictive; it may make sense to consider simplex weights depending on parameters having, e.g., some (co)homological meaning.

In the four-dimensional case, the Pachner moves are 3–3, 2–4 and 1–5. The first of them is usually regarded as 'central', and we will be dealing with it in this paper. Here, we describe this move and fix notations for the involved vertices and simplices.

Let there be a cluster of three pentachora (4-simplices) 12345, 12346 and 12356 situated around the 2-face 123. Move 3–3 transforms it into the cluster of three other pentachora, 12456, 13456 and 23456, situated around the 2-face 456. The inner 3-faces (tetrahedra) are 1234, 1235 and 1236 in the l.h.s., and 1456, 2456 and 3456 in the r.h.s. The boundary of both sides consists of nine tetrahedra.

Note that we have listed in the previous paragraph exactly *all* simplices in which the l.h.s. of move 3–3 differs from its r.h.s. The boundary of both sides is, of course, the same, and consists of nine tetrahedra.

1.2. Discrete Field Theory

Our relation corresponding to Pachner move 3–3, or its *algebraic realization*, that will hopefully lead to a very interesting and new TQFT on PL manifolds, is

$$c_l \iiint \mathcal{W}_{12345}\mathcal{W}_{12346}\mathcal{W}_{12356}\, \mathrm{d}x_{1234}\, \mathrm{d}x_{1235}\, \mathrm{d}x_{1236} = c_r \iiint \mathcal{W}_{12456}\mathcal{W}_{13456}\mathcal{W}_{23456}\, \mathrm{d}x_{1456}\, \mathrm{d}x_{2456}\, \mathrm{d}x_{3456} \quad (2)$$

This relation has already appeared in (practically) this general form in [1–3]. The integrals in Equation (2) are Berezin integrals [4,5] in Grassmann (anticommuting) variables, and $\mathcal{W}_{ijklm}$ are Grassmann–Gaussian pentachoron weights explained below; x_t are Grassmann variables living on tetrahedra t. Thus, the summation over all colorings in Equation (1) has been replaced in Equation (2) by integration in anticommuting variables. The basic reasons behind this are as follows:

- Although it is generally believed that anything "fermionic", like a theory with anticommuting variables, has its "bosonic" parallel—a theory with usual commuting variables, there seems to be no "bosonic" theory as yet with nonconstant relations similar to our relation (2), where, let us recall, the weight parameters are determined by a 2-cocycle. In addition, it was exactly anticommuting variables that appeared naturally in the author's work, although it took quite a while to understand that they are actually hidden behind such formulas as in the short notes [11, 12],
- Grassmann integration is quite a natural operation, it actually leads to finite summation; the basic difference from Equation (1) can be interpreted as "grading": minus signs are inserted in a proper way. Note that, already in 1988, Atiyah in his fundamental paper "Topological quantum field theory" [13] mentioned such a possibility. Here is the exact quotation: "the vector spaces Z(Σ) may be mod 2 graded with appropriate signs then inserted", see ([13] p. 181).

As formula (2) contains factors c_l and c_r, it is important, in order to obtain a manifold invariant, to have them expressed in a form multiplicative over simplices. Consider, heuristically, a simplified problem, where a manifold invariant is written in the form analogous to Equation (1), although now it *must* include some factor c:

$$c \int \cdots \int \prod_u \mathcal{W}_u \prod_{\substack{\text{inner}\\ \text{tetrahedra } t}} \mathrm{d}x_t \quad (3)$$

and it is very desirable to have an explicit expression for c. Now let c_l and c_r have a multiplicative form, for instance, such as we actually find in this paper, see formulas (60) and (61):

$$\frac{\prod(\text{some values belonging to inner tetrahedra}) \prod(\text{some values belonging to pentachora})}{\prod(\text{some values belonging to inner 2-faces})} \quad (4)$$

Then, one can readily see that invariant Equation (3) works *for moves 3–3* with c having the same form (4).

Remark 1. The actual invariant of *all* Pachner moves will include some more factors in the integrand compared with its "light" version Equation (3), as can be seen, for example, from paper [7], where similar but simpler invariants are considered. The multiplicative form retains, of course, its full importance (compare, for instance, ([7], formula (43))). The work on the extension of our present formulas to all Pachner moves is now in progress.

1.3. The Results of This Paper and How They Are Explained

The results are explicit formulas for everything in Equation (2)—that is, Grassmann weights $\mathcal{W}_{ijklm}$ and coefficients c_l and c_r—in terms of a 2-cocycle ω introduced in [3] (where not everything was calculated explicitly. In particular, only the existence of the mentioned coefficients was shown, see ([3] Theorem 9)). As all formulas are algebraic, the author might have presented just these formulas, saying: and now the validity of Equation (2) can be checked using computer algebra. The formulas look, however, rather intricate, so we follow another way, focusing on the actual author's reasonings.

2. Explicit Formulas for Matrix Elements

In this Section, as well as in the next Sections 3 and 4, we work within a single pentachoron $u = 12345$. The changes to be made for other u are quite simple and will be explained later.

Convention 1. We denote triangles (2-simplices) by the letter s, tetrahedra (3-simplices) by t, and pentachora (4-simplices)—by u. As for edges (1-simplices), we tend to use the letter b for them, while vertices (0-simplices) are $i, j, k, \dots$.

Convention 2. We also write the simplices by their vertices, e.g., $s = ijk$ or, as we have written above, $u = 12345$. The vertices are thus given by their numbers, and in writing so, we assume by default that the vertices are *ordered*: $i < j < k$, *etc*. If, however, we need a triangle whose order of vertices in unknown or unessential, we write s as $\{ijk\}$, as in Lemma 1 below.

2.1. Edge Operators

Our Grassmann–Gaussian pentachoron weight is

$$\mathcal{W}_u = \mathcal{W}_{12345} = \exp\left(-\frac{1}{2}\mathrm{x}^{\mathrm{T}}F\mathrm{x}\right)$$

where

$$\mathrm{x} = \begin{pmatrix} x_{2345} & x_{1345} & x_{1245} & x_{1235} & x_{1234} \end{pmatrix}^{\mathrm{T}} \tag{5}$$

is the column of Grassmann variables corresponding to the 3-faces $t \subset u$, and F—a 5×5 antisymmetric matrix.

We are going to recall the construction of matrix F from [3]. Moreover, we will write down some specific explicit expressions for the entries of F that do not appear in [3]. On the other hand, we will skip some details for which the reader is referred to [3].

Our starting point is a *2-cocycle* ω: it takes complex values $\omega_s = \omega_{ijk}$ on triangles $s = ijk \subset u$ such that:

$$\omega_{jkl} - \omega_{ikl} + \omega_{ijl} - \omega_{ijk} = 0 \tag{6}$$

Then, there are *edge operators* d_b for the ten edges $b = ij \subset u$ that make the bridge between ω and matrix F. Edge operators have the following properties:

- They belong to the 10-dimensional space of operators:

$$d = \sum_{t \subset u} (\beta_t \partial_t + \gamma_t x_t) \tag{7}$$

  where x_t means left multiplication by this Grassmann variable; $\partial_t = \partial/\partial x_t$; and β_t and γ_t are arbitrary complex coefficients,
- More specifically, the sum (7) for a given d_b runs *only over such three tetrahedra* t *that* $t \supset b$,
- Each of the edge operators annihilates the pentachoron weight:

$$d_b \mathcal{W}_u = 0$$

- They are antisymmetric with respect to changing the edge orientation:

$$d_{ij} = -d_{ji} \tag{8}$$

- They obey the following linear relations for each vertex $i \in u$:

$$\sum_{\substack{j \in u \\ j \neq i}} d_{ij} = 0 \tag{9}$$

- In addition, there is one more linear relation:

$$\sum_{b \subset u} \nu_b d_b = 0 \tag{10}$$

where ν is any 1-cocycle such that ω makes its coboundary:

$$\omega = \delta \nu, \quad \text{i.e.,} \quad \omega_{ijk} = \nu_{jk} - \nu_{ik} + \nu_{ij} \tag{11}$$

- They form a maximal (5-dimensional) *isotropic* subspace in the (10-dimensional) space of all operators of the form (7), where the scalar product is, by definition, the anticommutator:

$$\langle d', d'' \rangle = [d', d'']_+ = d'd'' + d''d'$$

2.2. Partial Scalar Products of Edge Operators

Due to the form (7), we have *t-components*

$$d_b|_t = \beta_t \partial_t + \gamma_t x_t$$

of edge operators, and the (vanishing) scalar product of two edge operators is a sum over tetrahedra:

$$0 = \langle d_{b_1}, d_{b_2} \rangle = \sum_{t \subset u} \langle d_{b_1}, d_{b_2} \rangle_t$$

where $\langle d_{b_1}, d_{b_2} \rangle_t$—we call it the *partial scalar product* of d_{b_1} and d_{b_2} with respect to tetrahedron t—is by definition the same as $\langle d_{b_1}|_t, d_{b_2}|_t \rangle$.

Lemma 1. *Choose a tetrahedron $t \subset u$ and a triangle $\{ijk\} \subset t$ (see Convention 2 for this notation). Then the partial scalar product $\langle d_{ij}, d_{ik} \rangle_t$ remains the same under any permutation of i, j, k.*

Proof. Let us prove, for instance, that

$$\langle d_{12}, d_{13} \rangle_{1234} = \langle d_{21}, d_{23} \rangle_{1234} \tag{12}$$

Setting $i = 3$ in Equation (9) and taking its t-component, we have (keeping in mind also Equation (8)):

$$-d_{13}|_{1234} - d_{23}|_{1234} + d_{34}|_{1234} = 0 \tag{13}$$

We want to take the scalar product of Equation (13) with d_{12}. As 1234 is the only tetrahedron common for the edges 12 and 34, and all edge operators are orthogonal to each other, we get:

$$\langle d_{12}, d_{34} \rangle_{1234} = \langle d_{12}, d_{34} \rangle = 0 \tag{14}$$

So, the mentioned scalar product, together with Equation (14), gives Equation (12) at once. □

Lemma 2. *For a tetrahedron $t \subset u$, construct the expression*

$$\omega_s \langle d_{b_1}, d_{b_2} \rangle_t \tag{15}$$

Here tetrahedron t is considered as oriented, s is any of its 2-faces with the induced orientation, and $b_1, b_2 \subset s$ are two edges sharing the same initial vertex (thus also oriented). Then, the expression (15) *does not depend on a specific choice of s, b_1 and b_2, and thus pertains solely to t.*

Proof. Let us prove, for instance, that

$$-\omega_{123} \langle d_{12}, d_{13} \rangle_{1234} = \omega_{124} \langle d_{12}, d_{14} \rangle_{1234} \tag{16}$$

(the minus sign accounts for opposite orientations of 123 and 124). A small exercise shows that the following linear relation is a consequence of Equation (10):

$$-\omega_{123} d_{13}|_{1234} - \omega_{124} d_{14}|_{1234} + \omega_{234} d_{34}|_{1234} = 0$$

Multiplying this scalarly by d_{12} and using once again orthogonality Equation (14), we get Equation (16). □

Lemma 3. *Expression* (15) *also remains the same* for all tetrahedra *t forming the boundary of pentachoron u, if these tetrahedra are oriented consistently (as parts of the boundary ∂u).*

Proof. It is enough to consider the situation where s is the common 2-face of two tetrahedra $t, t' \subset u$, and show that

$$\langle d_{b_1}, d_{b_2} \rangle_t = -\langle d_{b_1}, d_{b_2} \rangle_{t'} \tag{17}$$

Indeed, as the orientation of s as part of ∂t is different from its orientation as part of $\partial t'$, there are two values ω_s differing in sign, and Equation (17) will yield at once that Equation (15) is the same for t and t'.

To prove (17), we note that t and t' are the only tetrahedra containing both b_1 and b_2, so

$$0 = \langle d_{b_1}, d_{b_2} \rangle = \langle d_{b_1}, d_{b_2} \rangle_t + \langle d_{b_1}, d_{b_2} \rangle_{t'}$$

□

Convention 3. We normalize edge operators in such way that quantity (15) becomes unity.

Here is the matrix of scalar products $\langle d_a, d_b \rangle_{1234}$ calculated according to Convention 3. The rows (resp. columns) correspond to edge a (resp. b) taking values in lexicographic order: 12, 13, 14, 23, 24, 34:

$$\begin{pmatrix} \omega_{124}^{-1}-\omega_{123}^{-1} & \omega_{123}^{-1} & -\omega_{124}^{-1} & -\omega_{123}^{-1} & \omega_{124}^{-1} & 0 \\ \omega_{123}^{-1} & -\omega_{134}^{-1}-\omega_{123}^{-1} & \omega_{134}^{-1} & \omega_{123}^{-1} & 0 & -\omega_{134}^{-1} \\ -\omega_{124}^{-1} & \omega_{134}^{-1} & \omega_{124}^{-1}-\omega_{134}^{-1} & 0 & -\omega_{124}^{-1} & \omega_{134}^{-1} \\ -\omega_{123}^{-1} & \omega_{123}^{-1} & 0 & \omega_{234}^{-1}-\omega_{123}^{-1} & -\omega_{234}^{-1} & \omega_{234}^{-1} \\ \omega_{124}^{-1} & 0 & -\omega_{124}^{-1} & -\omega_{234}^{-1} & \omega_{124}^{-1}+\omega_{234}^{-1} & -\omega_{234}^{-1} \\ 0 & -\omega_{134}^{-1} & \omega_{134}^{-1} & \omega_{234}^{-1} & -\omega_{234}^{-1} & \omega_{234}^{-1}-\omega_{134}^{-1}. \end{pmatrix} \tag{18}$$

Remark 2. To calculate *diagonal* elements in matrix (18) is an easy exercise using linear relations similar to Equation (13).

Remark 3. As for tetrahedron 1235, we must not only replace '4' by '5' in matrix (18), but also change all signs—due to its different orientation! Similarly, analogues of matrix (18) for other tetrahedra can be calculated.

2.3. Superisotropic Operators and Matrix F

Superisotropic operators are such operators of the form (7) that annihilate the weight $\mathcal{W}_u$ and whose *each t-component* is isotropic, *i.e.*, either $\gamma_t = 0$ or $\beta_t = 0$. The rows of matrix F correspond to superisotropic operators in the following sense: every component of the column

$$\mathsf{p} + F\mathsf{x} \tag{19}$$

where x is given by Equation (5) and p, similarly, by

$$\mathsf{p} = \begin{pmatrix} \partial_{2345} & \partial_{1345} & \partial_{1245} & \partial_{1235} & \partial_{1234} \end{pmatrix}^{\mathrm{T}}$$

is superisotropic.

We recall ([3] Subsection 4.2) how superisotropic operators *proportional* to entries of the column (19) are constructed in terms of edge operators. They all are linear combinations written as

$$g = \sum_{1 \le i < j \le 5} \alpha_{ij} d_{ij}, \qquad \alpha_{ij} \in \mathbb{C} \tag{20}$$

First, we choose and fix one of two square roots of each ω_s:

$$q_s \overset{\text{def}}{=} \sqrt{\omega_s}$$

Second, we define "initial" α_{ij} as

$$\alpha_b = \prod_{\substack{s \supset b \\ \text{or } s \cap b = \varnothing}} q_s \tag{21}$$

Example 1. As the 2-faces $s \subset 12345$ containing edge 12 are 123, 124 and 125, and the only 2-face not intersecting with 12 is 345, such "initial" α_{12} is

$$\alpha_{12} = q_{123} q_{124} q_{125} q_{345}$$

Finally, the operator proportional to the i-th entry in the column (19), and thus corresponding to the tetrahedron t *not containing* the vertex i, is obtained by the following change of signs:

$$\alpha_b \text{ remains the same if } b \subset t, \text{ else } \alpha_b \mapsto -\alpha_b \tag{22}$$

We want to identify the entries in the column (19) with the operators given by Equations (20)–(22). Such identifications are determined to within a renormalization $x_t \mapsto x_t' = \lambda_t x_t$ of Grassmann variables, implying also $\partial_i \mapsto \partial_i' = (1/\lambda_i)\partial_i$.

Remark 4. This renormalization leads to multiplying matrix F from *both sides* by the diagonal matrix $\operatorname{diag}(\lambda_{2345}^{-1}, \ldots, \lambda_{1234}^{-1})$.

To fix the mentioned arbitrariness, we choose a distinguished edge a in every tetrahedron t and assume that the restriction of d_a onto t has a unit coefficient before ∂_t:

$$d_a|_t = \partial_t + \gamma x_t$$

As $\langle \partial_t, x_t \rangle = 1$, this implies

$$\gamma = \frac{1}{2} \langle d_a, d_a \rangle_t$$

Convention 4. In this paper, the distinguished edge a in any tetrahedron t will always be the lexicographically first one, for example, $a = 12$ in tetrahedron $t = 1234$.

We now denote $g^{(t)}$ the superisotropic operator defined according to Equations (20)–(22). If such operator contains a summand $\gamma x_{t'}$, then $\gamma = \langle g^{(t)}, d_a \rangle_{t'}$, and if it contains $\beta \partial_t$, then $\beta = 2 \frac{\langle g^{(t)}, d_a \rangle_t}{\langle d_a, d_a \rangle_t}$. Hence, the matrix element $F_{tt'} = \gamma / \beta$ (because the coefficient at ∂_t must be set to unity, according to Equation (19)), *i.e.*,

$$F_{tt'} = \frac{\langle g^{(t)}, d_a \rangle_{t'} \langle d_a, d_a \rangle_t}{2 \langle g^{(t)}, d_a \rangle_t} \tag{23}$$

The scalar products are calculated according to Equation (18) and Remark 3.

Example 2. Here is a typical matrix element:

$$F_{12} = F_{2345,1345} = -\frac{(q_{235}^2 - q_{234}^2)}{2 q_{134} q_{135} q_{234} q_{235}} \cdot \frac{f_{12}^{(\mathrm{n})}}{f_{12}^{(\mathrm{d})}} \tag{24}$$

where

$$\begin{aligned} f_{12}^{(\mathrm{n})} = {} & q_{124} q_{134} q_{235} q_{345} - q_{125} q_{135} q_{234} q_{345} + q_{123} q_{135}^2 q_{245} \\ & - q_{123} q_{134}^2 q_{245} - q_{124} q_{135} q_{145} q_{235} + q_{125} q_{134} q_{145} q_{234} \end{aligned} \tag{25}$$

and

$$\begin{aligned} f_{12}^{(\mathrm{d})} = {} & q_{125} q_{134} q_{235} q_{345} - q_{124} q_{135} q_{234} q_{345} - q_{124} q_{135} q_{235} q_{245} \\ & + q_{125} q_{134} q_{234} q_{245} + q_{123} q_{145} q_{235}^2 - q_{123} q_{145} q_{234}^2 \end{aligned} \tag{26}$$

3. Divisors of Matrix Elements

The central part of the present work consisted in finding a nice description for poles and zeros of matrix elements $F_{tt'}$ of the typical form (24). The point is, of course, that the quantities $\omega_{ijk} = q_{ijk}^2$ make a cocycle, so there are *dependencies*

$$q_{jkl}^2 - q_{ikl}^2 + q_{ijl}^2 - q_{ijk}^2 = 0 \tag{27}$$

for all tetrahedra $ijkl$.

3.1. Variables a_{ij} and Their Relation to "Initial" α_{ij}

Recall that we are working within *one pentachoron* 12345. It has ten 2-faces, as well as ten edges. This fact, together with the accumulated experience (compare [3] formula (50)), suggests the idea to introduce a *1-chain* a_{ik} such that ω_{ijk} is written as a product of its three values, namely:

$$\omega_{ijk} = a_{ij} a_{ik} a_{jk} \tag{28}$$

Given all ω_{ijk}, the a_{ij} are found from the system of equations which become linear after taking logarithms and are easily solved. Interestingly, the result is, up to an overall factor, our old alphas from formula (21):

$$a_{ij} = p \cdot \alpha_{ij} \tag{29}$$

where

$$p = \left(\prod_{\substack{\text{over all 2-faces } ijk \\ \text{of pentachoron } 12345}} \omega_{ijk} \right)^{-1/6}.$$

The cocycle relations are now written (instead of Equation (27)) as

$$a_{kl}a_{jl}a_{jk} - a_{kl}a_{il}a_{ik} + a_{jl}a_{il}a_{ij} - a_{jk}a_{ik}a_{ij} = 0 \tag{30}$$

Remark 5. We do not permute the indices of a_{ij} in this paper, but if needed, the natural idea is to assume that

$$a_{ij} = -a_{ji}$$

3.2. Matrix Elements in Terms of a_{ij}

Matrix elements $F_{tt'}$ can now be calculated in terms of a_{ij}. To be exact, here is what we do: set $\alpha_{ij} = a_{ij}/p$ according to Equation (29); the value of p is not of great importance because it will soon cancel out. Then, apply formula (23) with $g^{(t)}$ expressed using Equations (20) and (22); the scalar products are, of course, calculated according to Equation (18), Remark 3, and Equation (28). The following two examples show what we get.

Example 3.

$$F_{12} = \frac{a_{25}a_{35} - a_{24}a_{34}}{2a_{13}a_{14}a_{15}a_{34}a_{35}} \cdot \frac{a_{15}a_{34}a_{35} - a_{14}a_{34}a_{35} + a_{14}a_{15}a_{35} - a_{13}a_{15}a_{35} - a_{14}a_{15}a_{34} + a_{13}a_{14}a_{34}}{a_{25}a_{34}a_{35} - a_{24}a_{34}a_{35} - a_{24}a_{25}a_{35} + a_{23}a_{25}a_{35} + a_{24}a_{25}a_{34} - a_{23}a_{24}a_{34}} \tag{31}$$

Example 4.

$$F_{21} = -\frac{a_{15}a_{35} - a_{14}a_{34}}{2a_{23}a_{24}a_{25}a_{34}a_{35}} \cdot \frac{a_{25}a_{34}a_{35} - a_{24}a_{34}a_{35} + a_{24}a_{25}a_{35} - a_{23}a_{25}a_{35} - a_{24}a_{25}a_{34} + a_{23}a_{24}a_{34}}{a_{15}a_{34}a_{35} - a_{14}a_{34}a_{35} - a_{14}a_{15}a_{35} + a_{13}a_{15}a_{35} + a_{14}a_{15}a_{34} - a_{13}a_{14}a_{34}} \tag{32}$$

Of course,

$$F_{12} = -F_{21} \tag{33}$$

even if it is not immediately obvious from Equations (31) and (32). We will shed some light on this by studying the poles and zeros of these expressions.

3.3. The Variety of Zeros of the Main Factor in the Denominator of a Matrix Element as Function of Six Variables

The main factor in the denominator of Equation (31) is

$$a_{25}a_{34}a_{35} - a_{24}a_{34}a_{35} - a_{24}a_{25}a_{35} + a_{23}a_{25}a_{35} + a_{24}a_{25}a_{34} - a_{23}a_{24}a_{34} \tag{34}$$

and its pleasing feature is that it depends on the six variables a_{ij} belonging to just one tetrahedron 2345. There is just one dependence between these a_{ij}:

$$a_{34}a_{35}a_{45} - a_{24}a_{25}a_{45} + a_{23}a_{25}a_{35} - a_{23}a_{24}a_{34} \tag{35}$$

Lemma 4. *The primary decomposition of the affine algebraic variety determined by Equations* (31) *and* (35), *and lying in the affine space of six variables* $a_{23}, \dots, a_{45}$, *consists of the four irreducible components given by the following primary ideals, which are also already prime:*

$$(a_{25} + a_{34},\ a_{24} + a_{35}) \tag{36}$$

$$(a_{35},\ a_{24}) \tag{37}$$

$$(a_{34},\ a_{25}) \tag{38}$$

and

$$\begin{aligned}(&a_{24}a_{25}a_{34} - a_{24}a_{25}a_{35} - a_{24}a_{34}a_{35} + a_{25}a_{34}a_{35} + a_{24}a_{25}a_{45} - a_{34}a_{35}a_{45},\\ &a_{23}a_{25}a_{34} - a_{23}a_{25}a_{35} + a_{23}a_{25}a_{45} - a_{23}a_{34}a_{45} + a_{25}a_{34}a_{45} - a_{34}a_{35}a_{45},\\ &a_{23}a_{24}a_{35} - a_{23}a_{25}a_{35} - a_{23}a_{24}a_{45} + a_{24}a_{25}a_{45} + a_{23}a_{35}a_{45} - a_{24}a_{35}a_{45},\\ &a_{23}a_{24}a_{34} - a_{23}a_{25}a_{35} + a_{24}a_{25}a_{45} - a_{34}a_{35}a_{45})\end{aligned} \tag{39}$$

Proof. Direct calculation using *Singular* computer algebra system. □

Remark 6. The reader may notice that some more computer calculations of primary decompositions might have been helpful in the process of doing this work. They are, however, more difficult, and the calculation in Lemma 4 is typical of what the available computer capabilities allowed us to do—and this proved to be enough for achieving the goal of this work.

While there is no problem understanding the structure of components (36)–(38), the component (39) deserves the following lemma.

Lemma 5. *The affine algebraic variety determined by the ideal* (39), *and lying in the space of six variables* $a_{23}, \dots, a_{45}$, *admits the following trigonometric parameterization:*

$$a_{ij} = c \cdot \tan(x_i - x_j) \tag{40}$$

It is thus rational, because parameterization Equation (40) *becomes rational if re-written in terms of* c *and tangents of three independent differences of* x_i.

Proof. It is enough to substitute Equation (40) into each of the four generators of the prime ideal (39), and check that all of them vanish. □

3.4. *Divisor of a Matrix Element: Almost Full Description, Excluding Only Subvarieties* $a_{ij} = 0$

We consider the affine algebraic variety $\mathcal{M}$ in the space of *ten* variables $a_{12}, \dots, a_{45}$, defined by the relations (30) for all tetrahedra $ijkl \subset u = 12345$. Then, we consider its Zariski open subspace $\mathcal{M}'$ defined as follows:

$$\mathcal{M}' = (\mathcal{M} \text{ minus all subvarieties where some } a_{ij} = 0) \tag{41}$$

Remark 7. The goal of this paper consists in finding the expressions (60) and (61) below, for example, by guess. It looks hardly possible to guess these expressions if based on nothing, while studying divisors on $\mathcal{M}'$ has proved to provide a good basis for the correct guess, as we will see. So, we content ourself with $\mathcal{M}'$. Nevertheless, studying divisors on the whole $\mathcal{M}$ might be also of interest, because, for instance, divisors (37) and (38) lie exactly in $\mathcal{M} \setminus \mathcal{M}'$.

In $\mathcal{M}'$, we introduce the following subvarieties of *codimension 1*, denoted as D with indices because we think of them as Weil divisors:

- D_u: this is the subvariety given by the old formulas (40), but now ten of them: $1 \le i \le j \le 5$,
- $(D_u)_K$: choose now subset $K \subset \{1,2,3,4,5\}$, and define $(D_u)_K$ by the same formulas (40) except that we change the signs of those a_{ij} whose exactly one subscript i or j is in K. We write also $(D_u)_1$, $(D_u)_{12}$, *etc.* instead of $(D_u)_{\{1\}}$, $(D_u)_{\{1,2\}}$, *etc.*,
- D_t^-: for a tetrahedron $t = ijkl \subset u$, let $b = ij$ be the distinguished edge. Then D_t^- is given by the following equations (compare to (36)):

$$D_t^-: \quad \begin{cases} a_{ik} = -a_{jl} \\ a_{jk} = -a_{il} \end{cases} \tag{42}$$

- D_t^+, similarly:

$$D_t^+: \quad \begin{cases} a_{ik} = a_{jl} \\ a_{jk} = a_{il} \end{cases} \tag{43}$$

Lemma 6. *For a tetrahedron $t = ijkl$ and its distinguished edge ij, the sum $D_t^- + D_t^+$ gives, on $\mathcal{M}'$, exactly the zero divisor of $\dfrac{\sigma_t}{a_{ij}} = a_{jk}a_{ik} - a_{jl}a_{il}$ (compare with the first factor in the numerator of either Equation* (31) *or* (32)*!), where we denoted*

$$\sigma_t = \omega_{ijk} - \omega_{ijl} \tag{44}$$

Proof. Due to the cocycle relation (6),

$$(\sigma_t = 0 \ \text{ on } \ \mathcal{M}') \Leftrightarrow (a_{ik}a_{jk} - a_{il}a_{jl} = 0 \ \text{ and } \ a_{ik}a_{il} - a_{jk}a_{jl} = 0)$$

and the r.h.s. clearly gives Equation (42) or Equation (43). □

Lemma 7. *For every triangle $s = ijk$, introduce the quantity*

$$\psi_s = a_{jk} - a_{ik} + a_{ij}$$

Trigonometric parameterization (40) *specifies, on the subset where all $a_{ij} \neq 0$, the variety that can be given by the system of equations of the following form:*

$$\frac{\psi_s}{\omega_s} \ \text{ is the same for all } \ s$$

This applies to the case where the indices in Equation (40) *take four (like in Lemma 5) as well as five values (or, in fact, any number of them).*

Proof. Direct calculation. □

Theorem 8. (i) *The pole divisor of matrix element* (31)*, restricted to $\mathcal{M}'$, is D_u.*
(ii) *The zero divisor of* (31)*, restricted to $\mathcal{M}'$, is $(D_u)_{12} + D_{2345}^+ + D_{1345}^+$ (the last two are defined in* (43)*).*

Proof. First, note that the component (36) of the divisor of function (34) cancels out with the first factor in the numerator of (31), that is,

$$a_{25}a_{35} - a_{24}a_{34} \tag{45}$$

and what remains of the zero divisor of (45) after this canceling is D_{2345}^+, according to Lemma 6.

For item (i), this means that, on the pole variety of (31), all the expressions ψ_s/ω_s are the same for $s \subset 2345$. In addition, analyzing (32) similarly (and taking into account (33)), we arrive at the conclusion that the same are also ψ_s/ω_s for $s \subset 1345$. It is not hard to deduce now (through a small calculation) that ψ_s/ω_s are the same *for the whole pentachoron* 12345, *including* $s = 123$, 124 *and* 125. So, item (i) is proved.

For item (ii), we first notice that the main factor in the numerator of F_{12} (resp. F_{21}) is the same (up to an overall sign) as the main factor in the denominator of F_{21} (resp. F_{12}) except that the sign is changed of all a_{ij} with $i = 1$ (resp. $i = 2$). For F_{12}, this means that D^+_{1345} appears as a component of zero divisor, in analogy with (36), while the first paragraph of this proof means that D^+_{2345} is also there. The rest, namely $(D_u)_{12}$, appears in full analogy with D_u in the previous paragraph. So, item (ii) is also proved. □

Remark 8. Theorem 8 speaks about a specific matrix element and divisors. It applies, however, to all similar objects, with obvious changes.

4. Function φ_{12345}

On our subvariety $\mathcal{M}' \subset \mathcal{M}$ (41), we can express all a_{ij} in terms of q_{ijk} according to (29), where the factor p never vanishes and can be ignored as long as we are considering the zero or pole varieties of expressions *homogeneous* in variables a_{ij}.

Remark 9. All functions of a_{ij} or q_{ijk} in this paper *are* homogeneous.

Remark 10. In addition, the fact that p is multivalued makes no obstacle in our way.

Convention 5. We will denote, taking some liberty, that part of the variety of variables q_{ijk}, $1 \le i < j < k \le 5$, where all $q_{ijk} \ne 0$, by the same letter $\mathcal{M}'$ as the similar variety in variables a_{ij}. It is implied of course that the q_{ijk} obey the cocycle relations (27). In addition, we will use the old notations like $(D_u)_K$ and $D_t^\pm$ for codimension one subvarieties in $\mathcal{M}'$ that are like in Section 3.4 except that we made the substitution (29) in their defining equations.

For every 3-face t of pentachoron 12345, we define expression $f^{(t)}$ as the biggest factor in the denominator of type (24), namely:

$$\begin{aligned} f^{(2345)} &= q_{125}q_{134}q_{235}q_{345} - q_{124}q_{135}q_{234}q_{345} - q_{124}q_{135}q_{235}q_{245} \\ &\quad + q_{125}q_{134}q_{234}q_{245} + q_{123}q_{145}q_{235}^2 - q_{123}q_{145}q_{234}^2 \end{aligned} \quad (46)$$

$$\begin{aligned} f^{(1345)} &= q_{124}q_{134}q_{235}q_{345} - q_{125}q_{135}q_{234}q_{345} - q_{123}q_{135}^2q_{245} \\ &\quad + q_{123}q_{134}^2q_{245} + q_{124}q_{135}q_{145}q_{235} - q_{125}q_{134}q_{145}q_{234} \end{aligned} \quad (47)$$

$$\begin{aligned} f^{(1245)} &= q_{123}q_{125}^2q_{345} - q_{123}q_{124}^2q_{345} - q_{124}q_{134}q_{235}q_{245} \\ &\quad + q_{125}q_{135}q_{234}q_{245} - q_{125}q_{134}q_{145}q_{235} + q_{124}q_{135}q_{145}q_{234} \end{aligned} \quad (48)$$

$$\begin{aligned} f^{(1235)} &= q_{124}q_{125}^2q_{345} - q_{123}^2q_{124}q_{345} - q_{123}q_{134}q_{235}q_{245} \\ &\quad - q_{125}q_{134}q_{135}q_{245} + q_{125}q_{145}q_{234}q_{235} + q_{123}q_{135}q_{145}q_{234} \end{aligned} \quad (49)$$

$$\begin{aligned} f^{(1234)} &= q_{124}^2q_{125}q_{345} - q_{123}^2q_{125}q_{345} - q_{123}q_{135}q_{234}q_{245} \\ &\quad - q_{124}q_{134}q_{135}q_{245} + q_{124}q_{145}q_{234}q_{235} + q_{123}q_{134}q_{145}q_{235} \end{aligned} \quad (50)$$

Remark 11. The overall sign of any of expressions (46)–(50) is not now of big importance.

In addition, for every subset $K \subset \{i, j, k, l\}$ consider function $f_K^{(t)}$ made from (46)–(50) as follows: change the signs at those q_{ijk} having an odd number of subscripts (one or all three of i, j and k) is in K. In the next Lemma 9 we go through the 3-faces of 12345 in their natural order, and write down the zero divisors of some interesting functions on $\mathcal{M}'$.

Lemma 9.
- $f^{(2345)}$ *has zero divisor* $D_u + (D_u)_1 + D^-_{2345}$,
- $f_1^{(1345)}$ *has zero divisor* $(D_u)_1 + (D_u)_{12} + D^+_{1345}$,
- $f_{12}^{(1245)}$ *has zero divisor* $(D_u)_{12} + (D_u)_{123} + D^-_{1245}$,
- $f_{123}^{(1235)}$ *has zero divisor* $(D_u)_{123} + (D_u)_{1234} + D^+_{1235}$,
- $f_{1234}^{(1234)} - f^{(1234)}$ *has zero divisor* $(D_u)_{1234} + (D_u)_{12345} + D^-_{1234} = (D_u)_5 + D_u + D^-_{1234}$.

Thus, on $\mathcal{M}'$, *the function*

$$\frac{f^{(2345)} f_{12}^{(1245)} f^{(1234)}}{f_1^{(1345)} f_{123}^{(1235)}} \tag{51}$$

has the divisor (zeros with sign plus, poles with sign minus)

$$2D_u + D^-_{2345} - D^+_{1345} + D^-_{1245} - D^+_{1235} + D^-_{1234} \tag{52}$$

Proof. The formulas for divisors of the first five functions make a simple variations on the theme of Lemma 4, where, of course, Convention 5 must be also taken into account. Then, (52) follows by adding/subtracting relevant divisors. □

Motivated by Lemma 6, we divide the expression (51) by

$$\sigma_{2345}\sigma_{1245}\sigma_{1234} = (\omega_{234} - \omega_{235})(\omega_{124} - \omega_{125})(\omega_{123} - \omega_{124})$$

Theorem 10. *The divisor of the so obtained expression*

$$\varphi_{12345} = \frac{f^{(2345)} f_{12}^{(1245)} f^{(1234)}}{\sigma_{2345}\sigma_{1245}\sigma_{1234} f_1^{(1345)} f_{123}^{(1235)}} \tag{53}$$

considered as a function on $\mathcal{M}'$, *is*

$$2D_u - \sum_{t \subset u} D_t^+ \tag{54}$$

Proof. This follows from (52) and Lemma 6. □

Divisor (54) is thus symmetric under all permutations of 3-faces t of pentachoron u. This suggests the idea that function φ_{12345} may also remain the same, to within a possible sign change, under any permutation of vertices $1, \dots, 5$. Basically, this idea turns out to be right, but we do not go into details here; these details include, in particular, choosing the right signs of $q_{ijk} = \sqrt{\omega_{ijk}}$ *after we have changed the sign of* ω_{ijk} *itself* due to permuting its indices.

5. The Poles and Zeros of the Coefficient in 3–3 Relation, and Its Explicit Form

We now pass from the single pentachoron 12345 to Pachner move 3–3, where six pentachora are involved.

The l.h.s. and r.h.s. of move 3–3 are triangulated manifolds with boundary. We can orient the pentachora in both these manifolds consistently, and also so that these orientations induce the same orientation on the boundary $\partial(\text{l.h.s.}) = \partial(\text{r.h.s.})$. For one such orientation (of two), the signs in the following table show whether this consistent orientation of a pentachoron coincides with the orientation given by the natural order of its vertices:

left-hand side			right-hand side		
12345	12346	12356	12456	13456	23456
+	−	+	+	−	+

(55)

We do now all calculations in terms of variables q_{ijk} and not a_{ij}. This is due to the following important remark.

Remark 12. Variables a_{ij} depend on a pentachoron (*i.e.*, two a_{ij} for the same edge ij, but calculated within two different pentachora containing this edge, are different), while q_{ijk} do not.

5.1. Matrix Elements for All Six Pentachora Involved in Move 3–3

In Section 2, we explained how to calculate matrix F elements for pentachoron 12345. For *any* pentachoron $ijklm$ (recall that $i < \cdots < m$, according to Convention 2), the obvious substitution $1 \mapsto i, \ldots, 5 \mapsto m$ must be made. Besides this, the sign of matrix element must be changed for the pentachora marked with minus sign in table (55), as we are going to explain in (the proof of) Lemma 11, where we study the way how our normalization of edge operators, given by Convention 3, propagates from one pentachoron to another.

Lemma 11. *Expression* (15) *can be normalized to unity for a whole oriented triangulated manifold.*

Proof. Let tetrahedron t be the common 3 face of two pentachora, $t = u_1 \cap u_2$. Let $a \subset t$ be its edge, and $d_a^{(u_1)}$ and $d_a^{(u_1)}$—the corresponding edge operators in our two pentachora. Then,

$$\begin{array}{ll} \text{if} & d_a^{(u_1)}|_t = \beta_t \partial_t + \gamma_t x_t \\ \text{then} & d_a^{(u_2)}|_t = \beta_t \partial_t - \gamma_t x_t \end{array} \quad (56)$$

see ([3] formulas (58)).

We see now that, on passing to a neighboring pentachoron, first, the orientation of t changes (and this affects the orientation of s in(15)!), and second—partial scalar products of edge operators also change their signs because of (56). Hence, the quantity (15) remains the same, as before in Lemmas 1, 2 and 3. This means that it pertains to the whole triangulated manifold, if it is orientable and connected. Hence, we can normalize all edge operators globally so that quantity (15) stays always equal to unity. □

In addition, it is clear from (23) that, indeed, changing the sign of partial scalar products implies changing the sign of matrix elements.

5.2. Components in the l.h.s. and r.h.s., Their Poles and Zeros

Triple integrals in (2) are polynomials in Grassmann variables, and their coefficients are proportional. A typical coefficient, namely one at x_{1245} (the Grassmann variable corresponding to tetrahedron 1245), is

$$L = F_{1234,1236} F_{1235,1245} - F_{1235,1236} F_{1234,1245} \quad (57)$$

in the l.h.s., and

$$R = F_{1456,3456} F_{2456,1245} - F_{2456,3456} F_{1456,1245} \quad (58)$$

in the r.h.s.

Remark 13. Two tetrahedra in the subscripts of a matrix element in (57) or (58) clearly determine the relevant pentachoron.

Our goal is now to guess the form of c_l and c_r. As we can then check the correctness of our guess with a direct computer calculation, informal reasoning will be quite enough for us at this moment.

So, we analyze poles and zeros of L, R, and other similar Grassmann polynomial coefficients, in order to invent such c_l and c_r that will compensate these poles and zeros. First, we do so assuming that no one of values q_{ijk} vanish, that is, within the 'global analogue' of set $\mathcal{M}'$ (41). The poles of at least one component in the l.h.s. are relevant, while the zeros must be *common* for all components; similarly for r.h.s. We see this way that the poles are situated on divisors D_u (see Section 3.4) for *all* pentachora in the relevant side of Pachner move, while the zeros are situated on divisors D_t^+ of all *inner* tetrahedra, again in the relevant side of Pachner move.

5.3. Fitting the Divisors, and the Formulas for c_l and c_r

The above analysis of poles and zeros of triple Berezin integrals in (2), when confronted with the divisor (54) of function φ_{12345}, suggests that *square roots* of such functions may be the key ingredient of our c_l and c_r. So, we introduce, in analogy with φ_{12345}, quantities φ_u for each pentachoron u (simply making relevant subscript substitutions).

Now, we look at what may happen where some q_{ijk} do vanish. Motivated by the products of q_{ijk} factored out in the denominators of expressions like (24), we introduce the quantities

$$\varrho_t = q_{ijk} q_{ijl} \tag{59}$$

for tetrahedra $t = ijkl$. These ϱ_t are expected to compensate the poles appearing where the mentioned denominators vanish.

Remark 14. Note that 2-faces ijk and ijl in (59) both contain the distinguished edge $ij \subset t$, see Section 2.3.

What remains is a bit more guessing, trying, scrutinizing formulas, and we arrive at the following theorem:

Theorem 12. *For the 3–3 relation* (2) *to hold, it is enough to set its left-hand-side coefficient to*

$$c_l = \varrho_{1234}\, \varrho_{1235}\, \varrho_{1236} \sqrt{\varphi_{12345}} \sqrt{\varphi_{12346}} \sqrt{\varphi_{12356}} \,/\, q_{123} \tag{60}$$

and its right-hand-side coefficient to

$$c_r = \varrho_{1456}\, \varrho_{2456}\, \varrho_{3456} \sqrt{\varphi_{12456}} \sqrt{\varphi_{13456}} \sqrt{\varphi_{23456}} \,/\, q_{456} \tag{61}$$

Proof. As the proportionality of Grassmann polynomials in the two sides in (2) has been already established in [3], it is enough to compare the coefficients in the l.h.s. and r.h.s. at *any one* specific monomial. For instance, it is enough to show that

$$c_l L = c_r R \tag{62}$$

where L and R are given by (57) and (58). Now, we note that both sides in (62) are functions of the values ω_{ijk} of our cocycle ω, obtained using arithmetic operations (addition, subtraction, multiplication and division) and also taking square roots—but nothing more involved. The values ω_{ijk} are not independent, but we can pass to independent variables ν_{ij}, see (11). Of the fifteen ν_{ij}, $1 \le i < j \le 6$, five can be set to zero without loss of generality, *e.g.*, all those with $i = 1$.

Expressing everything in (62) in terms of the ten remaining ν_{ij} and using some computer algebra, we see that (62) indeed holds. □

A small miracle in formulas (60) and (61) is the denominators q_{123} and q_{456}, appearing because exactly such a value factors out in a non-obvious way in the numerator of L or R during the reduction to common denominator.

Acknowledgments: I thank Evgeniy Martyushev for his interest in this work. He made also some interesting calculations; although they are not used directly in this paper, they showed me some beautiful things and added thus to my inspiration. I am also grateful to creators and maintainers of *Maxima* and *Singular* computer algebra systems for their great work.

Conflicts of Interest: The author declares no conflict of interest.

References

1. Korepanov, I.G. Special 2-cocycles and 3–3 Pachner move relations in Grassmann algebra. **2013**, arXiv:1301.5581.
2. Korepanov, I.G.; Sadykov, N.M. Parameterizing the simplest Grassmann–Gaussian relations for Pachner move 3–3. *SIGMA* **2013**, *9*, 53, arXiv:1305.3246.
3. Korepanov, I.G. Two-cocycles give a full nonlinear parameterization of the simplest 3–3 relation. *Lett. Math. Phys.* **2014**, *104:10*, 1235–1261.
4. Berezin, F.A. *The Method of Second Quantization*; Academic Press: New York, NY, USA, 1966.
5. Berezin, F.A. Introduction to Superanalysis. In *Mathematical Physics and Applied Mathematics*; Kirillov, A.A., Ed.; D. Reidel: Dordrecht, The Netherlands, 1987; p. 424.
6. Lickorish, W.B.R. Simplicial moves on complexes and manifolds. *Geom. Topol. Monogr.* **1999**, *2*, 299–320.
7. Korepanov, I.G. Relations in Grassmann algebra corresponding to three- and four-dimensional Pachner moves. *SIGMA* **2011**, *7*, 117, arXiv:1105.0782.
8. Korepanov, I.G. Deformation of a $3 \to 3$ Pachner move relation capturing exotic second homologies. **2012**, arXiv:1201.4762.
9. Pachner, U. PL homeomorphic manifolds are equivalent by elementary shellings. *Eur. J. Combinatorics* **1991**, *12*, 129–145.
10. Kashaev, R.M. On realizations of Pachner moves in 4D. **2015**, arXiv:1504.01979.
11. Korepanov, I.G. A formula with volumes of five tetrahedra and discrete curvature. **2000**, arXiv:nlin/0003001.
12. Korepanov, I.G. A formula with hypervolumes of six 4-simplices and two discrete curvatures. **2000**, arXiv:nlin/0003024.
13. Atiyah, M.F. Topological quantum field theory. *Publ. Mathématiques de l'I.H.É.S.* **1988**, *68*, 175–186.

12

Qualitative Properties of Difference Equation of Order Six

Abdul Khaliq [1,*] and E.M. Elsayed [1,2]

1 Department of Mathematics, Faculty of Science, King Abdulaziz University, P.O. Box 80203, Jeddah 21589, Saudi Arabia; emmelsayed@yahoo.com
2 Department of Mathematics, Faculty of Science, Mansoura University, Mansoura 35516, Egypt
* Correspondence: khaliqsyed@gmail.com

Academic Editor: Johnny Henderson

Abstract: In this paper we study the qualitative properties and the periodic nature of the solutions of the difference equation

$$x_{n+1} = \alpha x_{n-2} + \frac{\beta x_{n-2}^2}{\gamma x_{n-2} + \delta x_{n-5}}, \quad n = 0, 1, ...,$$

where the initial conditions $x_{-5}, x_{-4}, x_{-3}, x_{-2}, x_{-1}, x_0$ are arbitrary positive real numbers and $\alpha, \beta, \gamma, \delta$ are positive constants. In addition, we derive the form of the solutions of some special cases of this equation.

Keywords: periodicity; global attractor; boundedness; rational difference equations

Mathematics Subject Classification: 39A10

---

1. Introduction

This paper deals with behavior of the solutions of the difference equation

$$x_{n+1} = \alpha x_{n-2} + \frac{\beta x_{n-2}^2}{\gamma x_{n-2} + \delta x_{n-5}}, \quad n = 0, 1, ..., \tag{1.1}$$

where the initial conditions $x_{-5}, x_{-4}, x_{-3}, x_{-2}, x_{-1}, x_0$ are arbitrary positive real numbers and $\alpha, \beta, \gamma, \delta$ are constants. In addition, we obtain the form of solution of some special cases.
Recently, there has been great interest in studying difference equation systems. One of the reasons for this is a necessity for some techniques which can be used in investigating equations arising in mathematical models describing real life situations in population biology, economics, probability theory, genetics, psychology, and so forth.

Difference equations appear naturally as discrete analogues and as numerical solutions of differential and delay differential equations having applications in biology, ecology, economy, physics, and so on. Although difference equations are very simple in form, it is extremely difficult to understand thoroughly the behaviors of their solution (see [1–9] and the references cited therein). Recently, a great effort has been made in studying the qualitative analysis of rational difference equations and rational difference system (see [10–23]).

Elabbasy *et al.* [8] studied the boundedness, global stability, periodicity character and gave the solution of some special cases of the difference equation.

$$x_{n+1} = \frac{\alpha x_{n-l} + \beta x_{n-k}}{A x_{n-l} + B x_{n-k}}$$

Elabbasy and Elsayed [9] investigated the local and global stability, boundedness, and gave the solution of some special cases of the difference equation

$$x_{n+1} = \frac{ax_{n-l}x_{n-k}}{bx_{n-p} + cx_{n-q}}.$$

In [13], Elsayed investigated the solution of the following non-linear difference equation

$$x_{n+1} = ax_n + \frac{bx_n^2}{cx_n + dx_{n-1}^2}.$$

Keratas *et al.* [24] obtained the solution of the following difference equation

$$x_{n+1} = \frac{x_{n-5}}{1 + x_{n-2}x_{n-5}}.$$

Saleh *et al.* [25] investigated the dynamics of the solution of difference equation

$$y_{n+1} = A + \frac{y_n}{y_{n-k}}.$$

Yalçınkaya [26] has studied the following difference equation

$$x_{n+1} = \alpha + \frac{x_{n-m}}{x_n^k}.$$

For other related work on rational difference equations, see [24–49].

Below, we outline some basic definitions and some theorems that we will need to establish our results.

Let I be some interval of real numbers and let

$$F : I^{k+1} \to I,$$

be a continuously differentiable function. Then for every set of initial conditions $x_{-k}, x_{-k+1}, ..., x_0 \in I$, the difference equation

$$x_{n+1} = F(x_n, x_{n-1}, ..., x_{n-k}), \ \ n = 0, 1, ..., \tag{1.2}$$

has a unique solution $\{x_n\}_{n=-k}^{\infty}$.

A point $\overline{x} \in I$ is called an equilibrium point of Equation (1.2) if

$$\overline{x} = f(\overline{x}, \overline{x}, ..., \overline{x}).$$

That is, $x_n = \overline{x}$ for $n \geq 0$, is a solution of Equation (1.2), or equivalently, $\overline{x}$ is a fixed point of f.

DEFINITION 1.1. *(Equilibrium Point)*

A point $\overline{x} \in I$ is called an equilibrium point of Equation (1.2) if

$$\overline{x} = f(\overline{x}, \overline{x}, ..., \overline{x}).$$

That is, $x_n = \overline{x}$ for $n \geq 0$, is a solution of Equation (1.2), or equivalently, $\overline{x}$ is a fixed point of f.

DEFINITION 1.2. *(Periodicity)*

A Sequence $\{x_n\}_{n=-k}^{\infty}$ is said to be periodic with period p if $x_{n+p} = x_n$ for all $n \geq -k$.

DEFINITION 1.3. *(Fibonacci Sequence)*

The sequence

$$\{F_m\}_{m=1}^{\infty} = \{1,2,3,5,8,13,....\} \quad i.e., \ F_m = F_{m-1} + F_{m-2} \geq 0,$$
$$F_{-2} = 0, \ F_{-1} = 1,$$

is called Fibonacci Sequence.

DEFINITION 1.4. *(Stability)*

(i) The equilibrium point $\overline{x}$ of Equation (1.2) is locally stable if for every $\epsilon > 0$, there exists $\delta > 0$ such that for all $x_{-k}, x_{-k+1}, ..., x_{-1}, x_0 \in I$ with

$$|x_{-k} - \overline{x}| + |x_{-k+1} - \overline{x}| + ... + |x_0 - \overline{x}| < \delta,$$

we have

$$|x_n - \overline{x}| < \epsilon \quad \text{for all} \quad n \geq -k.$$

(ii) The equilibrium point $\overline{x}$ of Equation (1.2) is locally asymptotically stable if $\overline{x}$ is locally stable solution of Equation (1.2) and there exists $\gamma > 0$, such that for all $x_{-k}, x_{-k+1}, ..., x_{-1}, x_0 \in I$ with

$$|x_{-k} - \overline{x}| + |x_{-k+1} - \overline{x}| + ... + |x_0 - \overline{x}| < \gamma,$$

we have

$$\lim_{n \to \infty} x_n = \overline{x}.$$

(iii) The equilibrium point $\overline{x}$ of Equation (1.2) is global attractor if for all $x_{-k}, x_{-k+1}, ..., x_{-1}, x_0 \in I$, we have

$$\lim_{n \to \infty} x_n = \overline{x}.$$

(iv) The equilibrium point $\overline{x}$ of Equation (1.2) is globally asymptotically stable if $\overline{x}$ is locally stable, and $\overline{x}$ is also a global attractor of Equation (1.2).

(v) The equilibrium point $\overline{x}$ of Equation (1.2) is unstable if $\overline{x}$ is not locally stable.

(vi) The linearized equation of Equation (1.2) about the equilibrium $\overline{x}$ is the linear difference equation

$$y_{n+1} = \sum_{i=0}^{k} \frac{\partial F(\overline{x}, \overline{x}, ..., \overline{x})}{\partial x_{n-i}} y_{n-i}. \tag{1.3}$$

Theorem A. [38] Assume that $p, q \in R$ and $k \in \{0, 1, 2, ...\}$. Then

$$|p| + |q| < 1,$$

is a sufficient condition for the asymptotic stability of the difference equation

$$x_{n+1} + px_n + qx_{n-k} = 0, \quad n = 0, 1, ...$$

The following theorem will be useful for establishing the results in this paper.

Theorem B. [39] Let $[\alpha, \beta]$ be an interval of real numbers assume that $g : [\alpha, \beta]^2 \to [\alpha, \beta]$, is a continuous function and consider the following equation

$$x_{n+1} = g(x_n, x_{n-1}), \qquad n = 0, 1, ..., \tag{1.4}$$

satisfying the following conditions :

(a) $g(x, y)$ is non-decreasing in $x \in [\alpha, \beta]$ for each fixed $y \in [\alpha, \beta]$ and $g(x, y)$ is non-increasing in $y \in [\alpha, \beta]$ for each fixed $x \in [\alpha, \beta]$.

(b) For any $(m, M) \in [\alpha, \beta] \times [\alpha, \beta]$ that is a solution of the system

$$M = g(M, m) \qquad \text{and} \qquad m = g(m, M),$$

we have that

$$m = M.$$

Then Equation (1.4) has a unique equilibrium $\overline{x} \in [\alpha, \beta]$ and every solution of Equation (1.4) converges to $\overline{x}$.

2. Local Stability of the Equilibrium Point of Equation (1.1)

In this section we study the local stability properties of the equilibrium point of Equation (1.1). The equilibrium points of Equation (1.1) are given by the relation

$$\overline{x} = \alpha\overline{x} + \frac{\beta\overline{x}^2}{\gamma\overline{x} + \delta\overline{x}},$$

or

$$\overline{x}^2(1-\alpha)(\gamma+\delta) = \beta\overline{x}^2.$$

If $(1-\alpha)(\gamma+\delta) \neq \beta$, then the unique equilibrium point is $\overline{x} = 0$.
Let $f : (0, \infty)^2 \longrightarrow (0, \infty)$ be a continuously differentiable function defined by

$$f(u, v) = \alpha u + \frac{\beta u^2}{\gamma u + \delta v}. \tag{2.1}$$

Therefore

$$\left(\frac{\partial f}{\partial u}\right)_{\overline{x}} = \alpha + \frac{\beta\gamma + 2\beta\delta}{(\gamma+\delta)^2}, \quad \left(\frac{\partial f}{\partial v}\right)_{\overline{x}} = \frac{-\beta\delta}{(\gamma+\delta)^2}.$$

Then the linearized equation of Equation (1.1) about $\overline{x}$ is

$$y_{n+1} - \left(\alpha + \frac{\beta\gamma + 2\beta\delta}{(\gamma+\delta)^2}\right) y_n + \left(\frac{\beta\delta}{(\gamma+\delta)^2}\right) y_{n-1} = 0. \tag{2.2}$$

Theorem 1. *Assume that* $\beta(\gamma + 3\delta) < (\gamma+\delta)^2(1-\alpha)$, $\alpha < 1$. *Then the equilibrium point* $\overline{x} = 0$ *of Equation (1.1) is locally asymptotically stable.*

Proof: From Theorem A, it follows that Equation (2.2) is asymptotically stable if

$$\left|\alpha + \frac{\beta\gamma + 2\beta\delta}{(\gamma+\delta)^2}\right| + \left|\frac{\beta\delta}{(\gamma+\delta)^2}\right| < 1,$$

or

$$\alpha + \frac{\beta\gamma + 3\beta\delta}{(\gamma+\delta)^2},$$

and so

$$\beta(\gamma + 3\delta) < (\gamma+\delta)^2(1-\alpha),$$

which completes the proof.

3. Global Attractivity of the Equilibrium Point of Equation (1.1)

In this section we investigate the global attractivity character of solutions of Equation (1.1).

Theorem 2. *The equilibrium point $\bar{x}$ of Equation (1.1) is a global attractor if $\gamma(1-\alpha) \neq \beta$.*

Proof: Let $\alpha,\ \beta$ are real numbers and assume that $g\ :\ [\alpha, \beta]^2 \to [\alpha, \beta]$, be a function defined by Equation (2.1).

Suppose that (m, M) is a solution of the system

$$M = g(M, m) \quad and \quad m = g(m, M).$$

Then from Equation (1.1), we see that

$$M = \alpha M + \frac{\beta M^2}{\gamma M + \delta m}, \qquad m = \alpha m + \frac{\beta m^2}{\gamma m + \delta M}.$$

Therefore

$$M(1-\alpha) = \frac{\beta M^2}{\gamma M + \delta m}, \quad m(1-\alpha) = \frac{\beta m^2}{\gamma m + \delta M},$$

or,

$$\gamma(1-\alpha)(M^2 - m^2) = b(M^2 - m^2), \qquad c(1-\alpha) \neq \beta,$$

thus

$$M = m.$$

From Theorem B, it follows that $\bar{x}$ is a global attractor of Equation (1.1) and then the proof is complete.

4. Boundedness of Solutions of Equation (1.1)

In this section we study the boundedness of solution of Equation (1.1)

Theorem 3. *Every solution of Equation(1.1) is bounded if* $\left(\alpha + \frac{\beta}{\gamma}\right) < 1$.

Proof: Let $\{x_n\}_{n=-5}^{\infty}$ be a solution of Equation (1.1). It follows from Equation (1.1) that

$$x_{n+1} = \alpha x_{n-2} + \frac{\beta x_{n-2}^2}{\gamma x_{n-2} + \delta x_{n-5}} \leq \alpha x_{n-2} + \frac{\beta x_{n-2}^2}{\gamma x_{n-2}} = \left(\alpha + \frac{\beta}{\gamma}\right) x_{n-2}.$$

Then

$$x_{n+1} \leq x_{n-2}, \text{ for all } n \geq 0.$$

Then the sub-sequences $\{x_{3n-1}\}_{n=-5}^{\infty}$, $\{x_{3n-2}\}_{n=-5}^{\infty}$ and $\{x_{3n}\}_{n=-5}^{\infty}$ are decreasing and so are bounded from above by $M = \max\{x_{-5}, x_{-4}, x_{-3}, x_{-2}, x_{-1}, x_0\}$.

In order to confirm the result in this section we consider some numerical examples for $x_{-5} = 10,\ x_{-4} = 5,\ x_{-3} = 8,\ x_{-2} = 2,\ x_{-1} = 9,\ x_0 = 7,\ \alpha = 0.5,\ \beta = 6,\ \gamma = 9,\ \delta = 10$. (See Figure 1) and $x_{-5} = 10,\ x_{-4} = 5,\ x_{-3} = 8,\ x_{-2} = 2,\ x_{-1} = 9,\ x_0 = 7,\ \alpha = 0.6,\ \beta = 6,\ \gamma = 7,\ \delta = 12$. (See Figure 2).

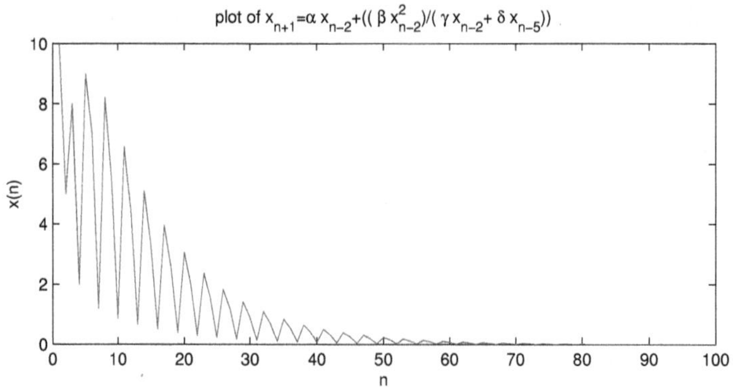

Figure 1. Expresses the solution of $x_{n+1} = \alpha x_{n-2} + \dfrac{\beta x_{n-2}^2}{\gamma x_{n-2} + \delta x_{n-5}}$, when we put initials and constants $x_{-5} = 10$, $x_{-4} = 5$, $x_{-3} = 8$, $x_{-2} = 2$, $x_{-1} = 9$, $x_0 = 7$, $\alpha = 0.5$, $\beta = 6$, $\gamma = 9$, $\delta = 10$.

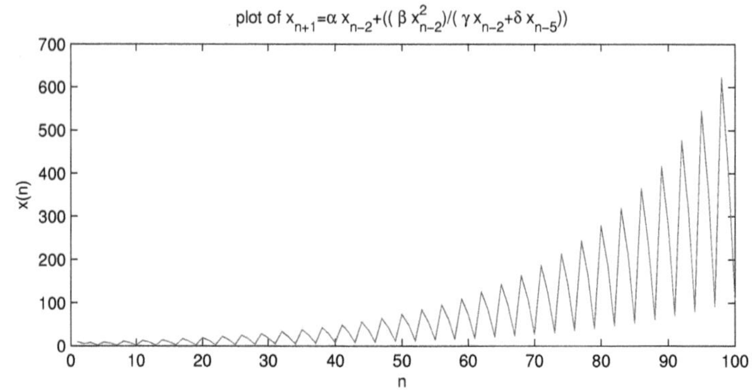

Figure 2. Represents behavior of Equation (1.1) when $x_{-5} = 10$, $x_{-4} = 5$, $x_{-3} = 8$, $x_{-2} = 2$, $x_{-1} = 9$, $x_0 = 7$, $\alpha = 0.6$, $\beta = 6$, $\gamma = 7$, $\delta = 12$.

5. Special Cases of Equation (1.1)

5.1. First Equation

In this section we study the following special case of Equation (1.1)

$$x_{n+1} = x_{n-2} + \frac{x_{n-2}^2}{x_{n-2} + x_{n-5}} \tag{5.1}$$

where the initial conditions x_{-5}, x_{-4}, x_{-3}, x_{-2}, x_{-1}, x_0 are arbitrary real numbers.

Theorem 4. *Let $\{x_n\}_{n=-5}^{\infty}$ be a solution of Equation (5.1) then for $n = 0, 1, 2, \ldots$*

$$x_{3n-1} = k\prod_{i=1}^{n}\left(\frac{f_{2i+1}k + f_{2i}q}{f_{2i}k + f_{2i-1}q}\right), \quad x_{3n-2} = h\prod_{i=1}^{n}\left(\frac{f_{2i+1}h + f_{2i}p}{f_{2i}h + f_{2i-1}p}\right),$$

$$x_{3n} = r\prod_{i=1}^{n}\left(\frac{f_{2i+1}r + f_{2i}t}{f_{2i}r + f_{2i-1}t}\right),$$

where $x_{-5} = t$, $x_{-4} = q$, $x_{-3} = p$, $x_{-2} = r$, $x_{-1} = k$, $x_0 = h$, $\{f_m\}_{m=1}^{\infty} = \{1, 1, 2, 3, 5, 8, 13, \ldots\}$.

Proof: We prove that the forms given are solutions of Equation (5.1) by using mathematical induction. First, we let $n = 0$, then the result holds. Second, we assume that the expressions are satisfied for $n-1$, $n-2$. Our objective is to show that the expressions are satisfied for n. That is;

$$x_{3n-8} = r\prod_{i=1}^{n-2}\left(\frac{f_{2i+1}r+f_{2i}t}{f_{2i}r+f_{2i-1}t}\right), \qquad x_{3n-7} = k\prod_{i=1}^{n-2}\left(\frac{f_{2i+1}k+f_{2i}q}{f_{2i}k+f_{2i-1}q}\right),$$

$$x_{3n-6} = h\prod_{i=1}^{n-2}\left(\frac{f_{2i+1}h+f_{2i}p}{f_{2i}h+f_{2i-1}p}\right), \qquad x_{3n-5} = r\prod_{i=1}^{n-1}\left(\frac{f_{2i+1}r+f_{2i}t}{f_{2i}r+f_{2i-1}t}\right),$$

$$x_{3n-4} = k\prod_{i=1}^{n-1}\left(\frac{f_{2i+1}k+f_{2i}q}{f_{2i}k+f_{2i-1}q}\right), \qquad x_{3n-3} = h\prod_{i=1}^{n-1}\left(\frac{f_{2i+1}h+f_{2i}p}{f_{2i}h+f_{2i-1}p}\right).$$

Now, it follows from Equation (5.1) that,

$$\begin{aligned}
x_{3n-1} &= x_{3n-4} + \frac{x_{3n-4}^2}{x_{3n-4}+x_{3n-7}} \\
&= k\prod_{i=1}^{n-1}\left(\frac{f_{2i+1}k+f_{2i}q}{f_{2i}k+f_{2i-1}q}\right) + \frac{k\prod_{i=1}^{n-1}\left(\frac{f_{2i+1}k+f_{2i}q}{f_{2i}k+f_{2i-1}q}\right) k\prod_{i=1}^{n-1}\left(\frac{f_{2i+1}k+f_{2i}q}{f_{2i}k+f_{2i-1}q}\right)}{k\prod_{i=1}^{n-1}\left(\frac{f_{2i+1}k+f_{2i}q}{f_{2i}k+f_{2i-1}q}\right) + k\prod_{i=1}^{n-2}\left(\frac{f_{2i+1}k+f_{2i}q}{f_{2i}k+f_{2i-1}q}\right)} \\
&= k\prod_{i=1}^{n-1}\left(\frac{f_{2i+1}k+f_{2i}q}{f_{2i}k+f_{2i-1}q}\right) + \frac{\prod_{i=1}^{n-1}\left(\frac{f_{2i+1}k+f_{2i}q}{f_{2i}k+f_{2i-1}q}\right)\left(\frac{f_{2n-1}k+f_{2n-2}q}{f_{2n-2}k+f_{2n-3}q}\right)}{\left(\frac{f_{2n-1}k+f_{2n-2}q}{f_{2n-2}k+f_{2n-3}q}\right)+1} \\
&= k\prod_{i=1}^{n-1}\left(\frac{f_{2i+1}k+f_{2i}q}{f_{2i}k+f_{2i-1}q}\right)\left(1+\frac{f_{2n-1}k+f_{2n-2}q}{f_{2n-1}k+f_{2n-2}q+f_{2n-2}k+f_{2n-3}q}\right) \\
&= k\prod_{i=1}^{n-1}\left(\frac{f_{2i+1}k+f_{2i}q}{f_{2i}k+f_{2i-1}q}\right)\left(1+\frac{f_{2n-1}k+f_{2n-2}q}{f_{2n}k+f_{2n-1}q}\right) \\
&= k\prod_{i=1}^{n-1}\left(\frac{f_{2i+1}k+f_{2i}q}{f_{2i}k+f_{2i-1}q}\right)\left(\frac{f_{2n+1}k+f_{2n}q}{f_{2n}k+f_{2n-1}q}\right).
\end{aligned}$$

Therefore

$$x_{3n-1} = k\prod_{i=1}^{n}\left(\frac{f_{2i+1}k+f_{2i}q}{f_{2i}k+f_{2i-1}q}\right).$$

Also, we see from Equation (5.1) that,

$$\begin{aligned}
x_{3n-2} &= x_{3n-5} + \frac{x_{3n-5}^2}{x_{3n-5}+x_{3n-8}} \\
&= r\prod_{i=1}^{n-1}\left(\frac{f_{2i+1}r+f_{2i}t}{f_{2i}r+f_{2i-1}t}\right) + \frac{r\prod_{i=1}^{n-1}\left(\frac{f_{2i+1}r+f_{2i}t}{f_{2i}r+f_{2i-1}t}\right)\left(\frac{f_{2n-1}r+f_{2n-2}t}{f_{2n-2}r+f_{2n-3}t}\right)}{\left(\frac{f_{2n-1}r+f_{2n-2}t}{f_{2n-2}r+f_{2n-3}t}\right)+1} \\
&= r\prod_{i=1}^{n-1}\left(\frac{f_{2i+1}r+f_{2i}t}{f_{2i}r+f_{2i-1}t}\right)\left(1+\frac{f_{2n-1}r+f_{2n-2}t}{f_{2n-1}r+f_{2n-2}t+f_{2n-2}r+f_{2n-3}t}\right) \\
&= r\prod_{i=1}^{n-1}\left(\frac{f_{2i+1}r+f_{2i}t}{f_{2i}r+f_{2i-1}t}\right)\left(\frac{f_{2n+1}r+f_{2n}t}{f_{2n}r+f_{2n-1}t}\right).
\end{aligned}$$

Then

$$x_{3n-2} = h\prod_{i=1}^{n}\left(\frac{f_{2i+1}h + f_{2i}p}{f_{2i}h + f_{2i-1}p}\right).$$

Also, wee see from Equation (5.1) that,

$$\begin{aligned}
x_{3n} &= x_{3n-3} + \frac{x_{3n-3}^2}{x_{3n-3} + x_{3n-6}} \\
&= h\prod_{i=1}^{n-1}\left(\frac{f_{2i+1}h + f_{2i}p}{f_{2i}r + f_{2i-1}t}\right) + \frac{h\prod_{i=1}^{n-1}\left(\frac{f_{2i+1}h + f_{2i}p}{f_{2i}h + f_{2i-1}p}\right)\left(\frac{f_{2n-1}h + f_{2n-2}p}{f_{2n-2}h + f_{2n-3}p}\right)}{\left(\frac{f_{2n-1}h + f_{2n-2}p}{f_{2n-2}h + f_{2n-3}p}\right) + 1} \\
&= h\prod_{i=1}^{n-1}\left(\frac{f_{2i+1}h + f_{2i}p}{f_{2i}h + f_{2i-1}p}\right)\left(1 + \frac{f_{2n-1}h + f_{2n-2}p}{f_{2n-1}h + f_{2n-2}p + f_{2n-2}h + f_{2n-3}p}\right) \\
&= h\prod_{i=1}^{n-1}\left(\frac{f_{2i+1}h + f_{2i}p}{f_{2i}h + f_{2i-1}p}\right)\left(\frac{f_{2n+1}h + f_{2n}p}{f_{2n}h + f_{2n-1}p}\right)
\end{aligned}$$

Thus

$$x_{3n} = h\prod_{i=1}^{n}\left(\frac{f_{2i+1}h + f_{2i}p}{f_{2i}h + f_{2i-1}p}\right).$$

Hence, the proof is complete.

We will confirm our result by considering some numerical examples assume $x_{-5} = 3$, $x_{-4} = 5$, $x_{-3} = 9$, $x_{-2} = 2$, $x_{-1} = 9$, $x_0 = 4$. (See Figure 3).

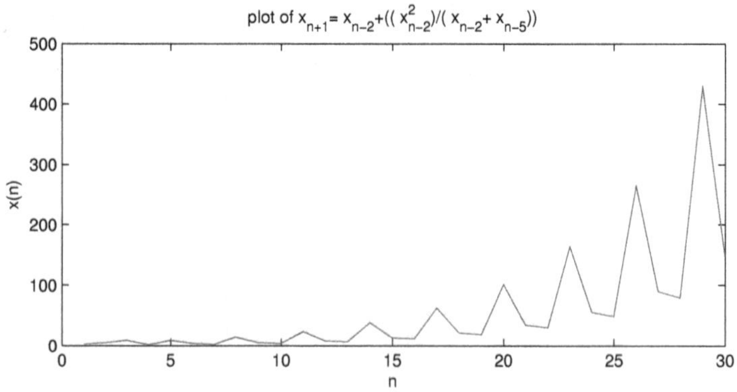

Figure 3. Shows the behavior for Equation (5.1) with $x_{-5} = 3$, $x_{-4} = 5$, $x_{-3} = 9$, $x_{-2} = 2$, $x_{-1} = 9$, $x_0 = 4$.

5.2. *Second Equation*

In this section we solve a more specific form of Equation (1.1)

$$x_{n+1} = x_{n-2} + \frac{x_{n-2}^2}{x_{n-2} - x_{n-5}}, \tag{5.2}$$

where the initial conditions $x_{-5}, x_{-4}, x_{-3}, x_{-2}, x_{-1}, x_0$ are arbitrary real numbers.

Theorem 5. *Let $\{x_n\}_{n=-5}^{\infty}$ be a solution of Equation (5.3). Then for $n = 0, 1, 2, \ldots$*

$$x_{3n-1} = k\prod_{i=1}^{n}\left(\frac{f_{i+2}k - f_i q}{f_i k - f_{i-2}q}\right), \quad x_{3n-2} = h\prod_{i=1}^{n}\left(\frac{f_{i+2}h - f_i p}{f_i h - f_{i-2}p}\right),$$

$$x_{3n} = r\prod_{i=1}^{n}\left(\frac{f_{i+2}r - f_i t}{f_i r - f_{i-2}t}\right)$$

where $x_{-5} = t,\ x_{-4} = q,\ x_{-3} = p,\ x_{-2} = r,\ x_{-1} = k,\ x_0 = h,\ \{f_m\}_{m=-1}^{\infty} = \{1, 0, 1, 1, 2, 3, 5, 8, 13, \ldots\}$.

Proof: The proof is the same as for Theorem 4 and is therefore omitted.

To confirm our result assume $x_{-5} = 3,\ x_{-4} = 5,\ x_{-3} = 9,\ x_{-2} = 2,\ x_{-1} = 9,\ x_0 = 4$. (See Figure 4).

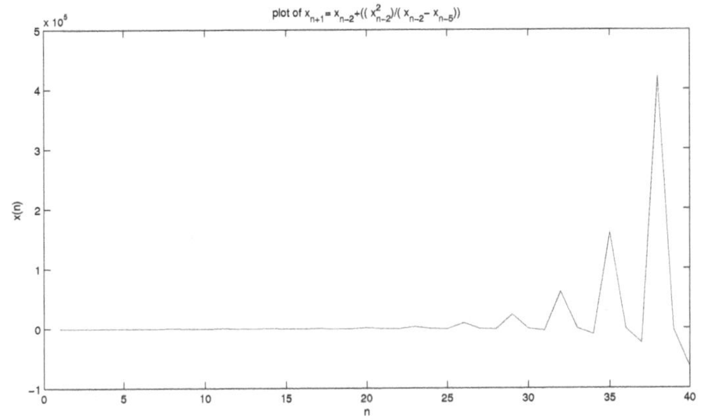

Figure 4. Shows solution of Equation (5.2) with $x_{-5} = 3,\ x_{-4} = 5,\ x_{-3} = 9,\ x_{-2} = 2,\ x_{-1} = 9, x_0 = 4$.

5.3. Third Equation

In this section we deal with the following special case of Equation (1.1)

$$x_{n+1} = x_{n-2} - \frac{x_{n-2}^2}{x_{n-2} + x_{n-5}}, \tag{5.3}$$

where the initial conditions $x_{-5}, x_{-4}, x_{-3}, x_{-2}, x_{-1}, x_0$ are arbitrary real numbers.

Theorem 6. *Let $\{x_n\}_{n=-5}^{\infty}$ be a solution of Equation(5.5) then for $n = 0, 1, 2, \ldots.$*

$$x_{3n-1} = \frac{kq}{f_n k + f_{n+1}q}, \quad x_{3n-2} = \frac{rt}{f_n r + f_{n+1}t}, \quad x_{3n} = \frac{hp}{f_n h + f_{n+1}p},$$

where $x_{-5} = t,\ x_{-4} = q,\ x_{-3} = p,\ x_{-2} = r,\ x_{-1} = k,\ x_0 = h,\ \{f_m\}_{m=-1}^{\infty} = \{1, 0, 1, 1, 2, 3, 5, 8, 13, \ldots\}$.

Proof: For $n = 0$, the result holds. Now suppose that $n > 0$ and that our assumption holds for $n-2,\ n-3$. That is,

$$\begin{aligned} x_{3n-8} &= \frac{rt}{f_{n-2}r + f_{n-1}t}, & x_{3n-7} &= \frac{kq}{f_{n-2}k + f_{n-1}q}, \\ x_{3n-6} &= \frac{hp}{f_{n-2}h + f_{n-1}p}, & x_{3n-5} &= \frac{rt}{f_{n-1}r + f_n t}, \\ x_{3n-4} &= \frac{kq}{f_{n-1}k + f_n q}, & x_{3n-3} &= \frac{hp}{f_{n-1}k + f_n q}. \end{aligned}$$

Now, it follows from Equation (5.3) that,

$$\begin{aligned} x_{3n-1} &= x_{3n-4} - \frac{x_{3n-4}^2}{x_{3n-4} + x_{3n-7}} \\ &= \frac{kq}{f_{n-1}k + f_n q} - \left(\frac{\frac{kq}{f_{n-2}k + f_{n-1}q} \frac{kq}{f_{n-2}k + f_{n-1}q}}{\frac{kq}{f_{n-2}k + f_{n-1}q} + \frac{kq}{f_{n-2}k + f_{n-1}q}} \right) \\ &= \frac{kq}{f_{n-1}k + f_n q} - \left(\frac{\frac{kq}{f_{n-2}k + f_{n-1}q} (f_{n-2}k + f_{n-1}q)}{f_{n-2}k + f_{n-1}q + f_{n-1}k + f_n q} \right) \\ &= \frac{kq}{f_{n-1}k + f_n q} \left(1 - \frac{f_{n-2}k + f_{n-1}q}{f_{n-2}k + f_{n-1}q + f_{n-1}k + f_n q} \right) \\ &= \frac{kq}{f_{n-1}k + f_n q} \left(\frac{f_{n-2}k + f_{n-1}q + f_{n-1}k + f_n q + f_{n-2}k + f_{n-1}q}{f_{n-2}k + f_{n-1}q + f_{n-1}k + f_n q} \right) \\ &= \frac{kq}{f_{n-1}k + f_n q} \left(\frac{f_{n-1}k + f_n q}{f_n k + f_{n+1}q} \right) = \frac{kq}{f_n k + f_{n+1}q}. \end{aligned}$$

Also, from Equation (5.3), we see that,

$$\begin{aligned} x_{3n} &= x_{3n-3} - \frac{x_{3n-3}^2}{x_{3n-3} + x_{3n-6}} \\ &= \frac{hp}{f_{n-1}k + f_n q} - \left(\frac{\frac{hp}{f_{n-1}k + f_n q} \frac{hp}{f_{n-1}k + f_n q}}{\frac{hp}{f_{n-1}k + f_n q} + \frac{hp}{f_{n-2}h + f_{n-1}p}} \right) \\ &= \frac{hp}{f_{n-1}k + f_n q} - \left(\frac{\frac{hp}{f_{n-2}h + f_{n-1}p} (f_{n-2}h + f_{n-1}p)}{f_{n-2}h + f_{n-1}p + f_{n-1}h + f_n p} \right) \\ &= \frac{hp}{f_{n-1}h + f_n p} \left(1 - \frac{f_{n-2}h + f_{n-1}p}{f_{n-2}h + f_{n-1}p + f_{n-1}h + f_n p} \right) \\ &= \frac{hp}{f_{n-1}h + f_n p} \left(\frac{f_{n-1}h + f_n p}{f_n h + f_{n+1}p} \right) = \frac{hp}{f_n h + f_{n+1}p}. \end{aligned}$$

Also, from Equation (5.3), we get,

$$\begin{aligned} x_{3n-2} &= x_{3n-5} - \frac{x_{3n-5}^2}{x_{3n-5} + x_{3n-8}} \\ &= \frac{rt}{f_{n-1}r + f_n t} - \left(\frac{\frac{rt}{f_{n-1}r + f_n t}\frac{rt}{f_{n-1}r + f_n t}}{\frac{rt}{f_{n-1}r + f_n t} + \frac{rt}{f_{n-2}r + f_{n-1}t}} \right) \\ &= \frac{rt}{f_{n-1}r + f_n t} - \left(\frac{\frac{rt}{f_{n-2}r + f_{n-1}t}(f_{n-2}r + f_{n-1}t)}{f_{n-2}r + f_{n-1}t + f_{n-1}r + f_n t} \right) \\ &= \frac{rt}{f_{n-1}r + f_n t}\left(\frac{f_{n-1}r + f_n t}{f_n r + f_{n+1}t} \right) = \frac{rt}{f_n r + f_{n+1}t}. \end{aligned}$$

Hence, the proof is complete.

We consider a numerical example of this special case, assume $x_{-5} = 2, x_{-4} = 5, x_{-3} = 6$, $x_{-2} = 12, x_{-1} = 9, x_0 = 18$. (See Figure 5).

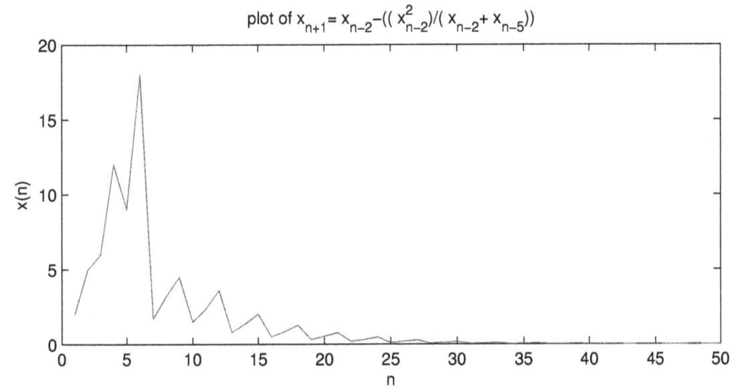

Figure 5. Shows the dynamics of Equaton (5.3) when $x_{-5} = 2$, $x_{-4} = 5$, $x_{-3} = 6, x_{-2} = 12$, $x_{-1} = 9$, $x_0 = 18$.

5.4. Fourth Equation

In this section we deal with the form of solution of the following equation

$$x_{n+1} = x_{n-2} - \frac{x_{n-2}^2}{x_{n-2} - x_{n-5}}, \tag{5.4}$$

where the initial conditions x_{-5}, x_{-4}, x_{-3}, x_{-2}, x_{-1}, x_0 are arbitrary real numbers.

Theorem 7. *Let $\{x_n\}_{n=-5}^{\infty}$be a solution of Equation (5.4) Then every solution of Equation (5.4) is periodic with period 18. Moreover, $\{x_n\}_{n=-5}^{\infty}$ takes the form*

$$\left\{ \begin{array}{c} t, q, p, r, k, h, \frac{-rt}{r-t}, \frac{-kq}{k-q}, \frac{-hp}{h-p}, -t, -q, -p, \\ -r, -k, -h, \frac{rt}{r-t}, \frac{kq}{k-q}, \frac{hp}{h-p}, t, q, p, r, k, h, \ldots \end{array} \right\},$$

or,

$$\begin{aligned}
x_{18n-5} &= t, & x_{18n-4} &= q, & x_{18n-3} &= p, & x_{18n-2} &= r, & x_{18n-1} &= k,\\
x_{18n} &= h, & x_{18n+1} &= \frac{-rt}{r-t}, & x_{18n+2} &= \frac{-kq}{k-q}, & x_{18n+3} &= \frac{-hp}{h-p},\\
x_{18n+4} &= -t, & x_{18n+5} &= -q, & x_{18n+6} &= -p, & x_{18n+7} &= -r, & x_{18n+8} &= -k,\\
x_{18n+9} &= -h, & x_{18n+10} &= \frac{rt}{r-t}, & x_{18n+11} &= \frac{kq}{k-q}, & x_{18n+12} &= \frac{hp}{h-p},
\end{aligned}$$

where $x_{-5} = t,\ x_{-4} = q,\ x_{-3} = p,\ x_{-2} = r,\ x_{-1} = k,\ x_0 = h,\ x_{-5} \neq x_{-2},\ x_{-4} \neq x_0,\ x_{-3} \neq x_{-1}$.
Proof: The proof is the same as the proof of Theorem 6 and thus will be omitted.

Figure 6 shows the solution when $x_{-5} = 4,\ x_{-4} = 7,\ x_{-3} = 5,\ x_{-2} = 14,\ x_{-1} = 19,\ x_0 = 11$.

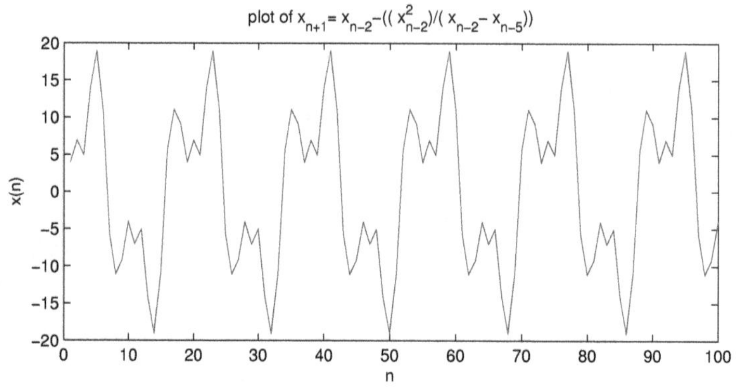

Figure 6. Shows the periodic behavior of solution of $x_{n+1} = \alpha x_{n-2} - \frac{\beta x_{n-2}^2}{\gamma x_{n-2} - \delta x_{n-5}}$, with $x_{-5} = 4,\ x_{-4} = 7,\ x_{-3} = 5,\ x_{-2} = 14,\ x_{-1} = 19,\ x_0 = 11$.

6. Conclusions

In this paper we investigated the global attractivity, boundedness and the solutions of some special cases of Equation (1.1). In Section 2 we proved when $\beta(\gamma + 3\delta) < (\gamma + \delta)^2(1 - \alpha)$, Equation (1.1) has local stability. In Section 3 we showed that the unique equilibrium of Equation (1.1) is globally asymptotically stable if $\gamma(1 - \alpha) \neq \beta$. In Section 4 we investigated that the solution of Equation (1.1) is bounded if $\left(\alpha + \frac{\beta}{\gamma}\right) < 1$. In Section 5 we obtained the form of the solution of four special cases of Equation (1.1) and gave numerical examples of each of the case with different initial values.

Acknowledgments: The authors thank the main editor and anonymous referees for their valuable suggestions and comments leading to improvement of this paper.

Author Contributions: All authors carried out the proofs of the main results and approved the final manuscript.

Conflicts of Interest: The authors declare no conflict of interest.

References

1. Ahmed, A.M.; Youssef, A.M. A solution form of a class of higher-order rational difference equations. *J. Egyp. Math. Soc.* **2013**, *21*, 248–253.
2. Alghamdi, M.; Elsayed, E.M.; El-Dessoky, M.M. On the Solutions of Some Systems of Second Order Rational Difference Equations. *Life Sci. J.* **2013**, *10*, 344–351.
3. Asiri, A.; Elsayed, E.M.; El-Dessoky, M.M. On the Solutions and Periodic Nature of Some Systems of Difference Equations. *J. Comput. Theor. Nanosci.* **2015**, *12*, 3697–3704.
4. Das, S.E.; Bayram, M. On a System of Rational Difference Equations. *World Appl. Sci. J.* **2010**, *10*, 1306–1312.

5. Din, Q. Qualitative nature of a discrete predator-prey system. *Contemp. Methods Math. Phys. Gravit.* **2015**, *1*, 27–42.
6. Din, Q. Stability analysis of a biological network. *Netw. Biol.* **2014**, *4*, 123–129.
7. Din, Q.; Elsayed, E.M. Stability analysis of a discrete ecological model. *Comput. Ecol. Softw.* **2014**, *4*, 89–103.
8. Elabbasy, E.M.; El-Metwally, H.; Elsayed, E.M. On the Difference Equation $x_{n+1} = \frac{\alpha x_{n-l} + \beta x_{n-k}}{A x_{n-l} + B x_{n-k}}$. *Acta Math. Vietnam.* **2008**, *33*, 85–94.
9. Elabbasy, E.M.; Elsayed, E.M. Dynamics of a rational difference equation. *Chin. Ann. Math. Ser. B* **2009**, *30*, 187–198.
10. El-Dessoky, M.M.; Elsayed, E.M. On the solutions and periodic nature of some systems of rational difference equations. *J. Comput. Anal. Appl.* **2015**, *18*, 206–218.
11. El-Metwally, H.; Elsayed, E.M. Qualitative Behavior of some Rational Difference Equations. *J. Comput. Anal. Appl.* **2016**, *20*, 226–236.
12. El-Moneam, M.A. On the dynamics of the solutions of the rational recursive sequences. *Br. J. Math. Comput. Sci.* **2015**, *5*, 654–665.
13. Elsayed, E.M. Qualitative behaviour of difference equation of order two. *Math. Comput. Model.* **2009**, *50*, 1130–1141.
14. Elsayed, E.M. Solution and attractivity for a rational recursive sequence. *Discret. Dyn. Nat. Soc.* **2011**, *2011*, 982309.
15. Elsayed, E.M. Solutions of rational difference system of order two. *Math. Comput. Mod.* **2012**, *55*, 378–384.
16. Elsayed, E.M. Behavior and expression of the solutions of some rational difference equations. *J. Comput. Anal. Appl.* **2013**, *15*, 73–81.
17. Elsayed, E.M. Solution for systems of difference equations of rational form of order two. *Comput. Appl. Math.* **2014**, *33*, 751–765.
18. Elsayed, E.M. On the solutions and periodic nature of some systems of difference equations. *Int. J. Biomath.* **2014**, *7*, 1450067.
19. Elsayed, E.M. New method to obtain periodic solutions of period two and three of a rational difference equation. *Nonlinear Dyn.* **2015**, *79*, 241–250.
20. Elsayed, E.M. Dynamics and Behavior of a Higher Order Rational Difference Equation. *J. Nonlinear Sci. Appl.* **2016**, *9*, 1463–1474.
21. Elsayed, E.M.; Ahmed, A.M. Dynamics of a three-dimensional systems of rational difference equations. *Math. Methods Appl. Sci.* **2016**, *39*, 1026–1038.
22. Elsayed, E.M.; Alghamdi, A. Dynamics and Global Stability of Higher Order Nonlinear Difference Equation. *J. Comput. Anal. Appl.* **2016**, *21*, 493–503.
23. Elsayed, E.M.; El-Dessoky, M.M. Dynamics and global behavior for a fourth-order rational difference equation. *Hacet. J. Math. Stat.* **2013**, *42*, 479–494.
24. Karatas, R.; Cinar, C.; Simsek, D. On positive solutions of the difference equation $x_{n+1} = \frac{x_{n-5}}{1 + x_{n-2} x_{n-5}}$. *Int. J. Contemp. Math. Sci.* **2006**, *1*, 495–500.
25. Saleh, M.; Aloqeili, M. On the difference equation $y_{n+1} = A + \frac{y_n}{y_{n-k}}$. *Appl. Math. Comput.* **2006**, *176*, 359–363.
26. Yalçınkaya, I. On the difference equation $x_{n+1} = \alpha + \frac{x_{n-m}}{x_n^k}$. *Discret. Dyn. Nat. Soc.* **2008**, *2008*, 805460, doi:10.1155/2008/805460.
27. Elsayed, E.M.; El-Dessoky, M.M.; Alzahrani, E.O. The Form of The Solution and Dynamics of a Rational Recursive Sequence. *J. Comput. Anal. Appl.* **2014**, *17*, 172–186.
28. Elsayed, E.M.; El-Dessoky, M.M.; Asiri, A. Dynamics and Behavior of a Second Order Rational Difference equation. *J. Comput. Anal. Appl.* **2014**, *16*, 794–807.
29. Elsayed, E.M.; El-Metwally, H. Stability and solutions for rational recursive sequence of order three. *J. Comput. Anal. Appl.* **2014**, *17*, 305–315.
30. Elsayed, E.M.; El-Metwally, H. Global behavior and periodicity of some difference equations. *J. Comput. Anal. Appl.* **2015**, *19*, 298–309.
31. Elsayed, E.M.; Ibrahim, T. F. Solutions and periodicity of a rational recursive sequences of order five. *Bull. Malays. Math. Sci. Soc.* **2015**, *38*, 95–112.

32. Elsayed, E.M.; Ibrahim, T.F. Periodicity and solutions for some systems of nonlinear rational difference equations. *Hacet. J. Math. Stat.* **2015**, *44*, 1361–1390.
33. Halim, Y.; Bayram, M. On the solutions of a higher-order difference equation in terms of generalized Fibonacci sequences. *Math. Methods Appl. Sci.* **2015**, doi:10.1002/mma.3745.
34. Ibrahim, T.F.; Touafek, N. Max-type system of difference equations with positive two-periodic sequences. *Math. Methods Appl. Sci.* **2014**, *37*, 2562–2569.
35. Ibrahim, T.F.; Touafek, N. On a third order rational difference equation with variable coefficients. *Dyn. Contol Disc Imput Syst. Appl. Algo.* **2013**, *20*, 251–264.
36. Jana, D.; Elsayed, E.M. Interplay between strong Allee effect, harvesting and hydra effect of a single population discrete-time system. *Int. J. Biomath.* **2016**, *9*, 1650004.
37. Khaliq, A.; Elsayed, E.M. The Dynamics and Solution of some Difference Equations. *J. Nonlinear Sci. Appl.* **2016**, *9*, 1052–1063.
38. Khan, A.Q.; Qureshi, M.N. Global dynamics of a competitive system of rational difference equations. *Math. Methods Appl. Sci.* **2015**, *38*, 4786–4796.
39. Khan, A.Q.; Qureshi, M.N.; Din, Q. Asymptotic behavior of an anti-competitive system of rational difference equations. *Life Sci. J.* **2014**, *11*, 16–20.
40. Kocic, V.L.; Ladas, G. *Global Behavior of Nonlinear Difference Equations of Higher Order with Applications*; Kluwer Academic Publishers: Dordrecht, The Netherlands, 1993.
41. Kulenovic, M.R.S.; Ladas, G. *Dynamics of Second Order Rational Difference Equations with Open Problems and Conjectures*; Chapman & Hall / CRC Press: London, UK, 2001.
42. Kurbanli, A.S. On the Behavior of Solutions of the System of Rational Difference Equations. *World Appl. Sci. J.* **2010**, *10*, 1344–1350.
43. Simsek, D.; Cinar, C.; Yalcinkaya, I. On the recursive sequence $x_{n+1} = \frac{x_{n-3}}{1+x_{n-1}}$. *Int. J. Contemp. Math. Sci.* **2006**, *1*, 475–480.
44. Touafek, N. On a second order rational difference equation. *Hacet. J. Math. Stat.* **2012**, *41*, 867–874.
45. Yalcinkaya, I.; Cinar, C. On the dynamics of difference equation $\frac{ax_{n-k}}{b+cx_n^p}$. *Fasc. Math.* **2009**, *42*, 141–148.
46. Yazlik, Y.; Elsayed, E.M.; Taskara, N. On the Behaviour of the Solutions of Difference Equation Systems. *J. Comput. Anal. Appl.* **2014**, *16*, 932–941.
47. Yazlik, Y.; Tollu, D.T.; Taskara, N. On the solutions of a max-type difference equation system. *Math. Methods Appl. Sci.* **2015**, *38*, 4388–4410.
48. Zayed, E.M.E.; El-Moneam, M.A. On the rational recursive sequence $x_{n+1} = \frac{\alpha+\beta x_n+\gamma x_{n-1}}{A+Bx_n+Cx_{n-1}}$. *Commun. Appl. Nonlinear Anal.* **2005**, *12*, 15–28.
49. Zayed, E.M.E.; El-Moneam, M.A. On the rational recursive sequence $x_{n+1} = ax_n - \frac{bx_n}{cx_n - dx_{n-k}}$. *Commun. Appl. Nonlinear Anal.* **2008**, *15*, 47–57.

13

Skew Continuous Morphisms of Ordered Lattice Ringoids

Sergey Victor Ludkowski

Department of Applied Mathematics, Moscow State Technical University MIREA, avenue Vernadsky 78, Moscow 119454, Russia; Ludkowski@mirea.ru

Academic Editor: Hvedri Inassaridze

Abstract: Skew continuous morphisms of ordered lattice semirings and ringoids are studied. Different associative semirings and non-associative ringoids are considered. Theorems about properties of skew morphisms are proved. Examples are given. One of the main similarities between them is related to cones in algebras of non locally compact groups.

Keywords: non-associative; algebra; morphism; idempotent; skew; semiring; ringoid

Mathematics Subject Classification 2010: 08A99; 16Y60; 17A01; 17D99; 22A26; 22A30

---

1. Introduction

Semirings, ringoids, algebroids and non-associative algebras play important role in algebra and among them ordered semirings and lattices as well [1–8]. This is also motivated by idempotent mathematical physics naturally appearing in quantum mechanics and quantum field theory (see, for example, [9] and references therein). They also arise from the consideration of algebroids and ringoids associated with non locally compact groups. Namely, this appears while the studies of representations of non locally compact groups, quasi-invariant measures on them and convolution algebras of functions and measures on them [10–13]. The background for this is A. Weil's theorem (see [14]) asserting that if a topological group has a quasi-invariant σ-additive non trivial measure relative to the entire group, then it is locally compact. Therefore, it appears natural to study inverse mapping systems of non locally compact groups and their dense subgroups. Such spectra lead to structures of algebroids and ringoids. Investigations of such objects are also important for making advances in representation theory of non locally compact groups.

In this paper methods of categorial topology are used (see [15–18] and references therein).

This article is devoted to ordered ringoids and semirings with an additional lattice structure. Their continuous morphisms are investigated in Section 3. Preliminaries are given in Section 2. Necessary definitions 2.1–2.4 are recalled. For a topological ringoid K and a completely regular topological space X new ringoids $C(X,K)$ are studied, where $C(X,K)$ consists of all continuous mappings $f : X \to K$ with point-wise algebraic operations. Their ideals, topological directed structures and idempotent operations are considered in Lemmas 2.6, 2.8, 2.9, 2.12 and Corollary 2.7. There are also given several examples 2.13–2.18 of objects. One of the main examples between them is related to cones in algebras of non locally compact groups. Another example is based on ordinals. Construction of ringoids with the help of inductive limits is also considered.

Structure and properties of these objects are described in Section 3. Definitions of morphisms of ordered semirings and some their preliminaries are described in Subsection 3.1. An existence of idempotent K-homogeneous morphisms under definite conditions is proved in Lemma 3.4. A relation between order preserving weakly additive morphisms and non-expanding morphisms is given in

Lemma 3.7. An extension of an order preserving weakly additive morphism is considered in Lemma 3.9.

Then a weak* topology on a family $\mathcal{O}(X,K)$ of all order preserving weakly additive morphisms on a Hausdorff topological space X with values in K is taken. The weak* compactness of $\mathcal{O}(X,K)$ under definite conditions is proved in Theorem 3.10. Further in Proposition 3.11 there is proved that $I(X,K)$ and $I_h(X,K)$ are closed in $\mathcal{O}(X,K)$, where $I(X,K)$ denotes the set of all idempotent K-valued morphisms, also $I_h(X,K)$ denotes its subset of idempotent homogeneous morphisms.

Categories related to morphisms and ringoids are presented in Subsection 3.2. An existence of covariant functors, their ranges and continuity of morphisms are studied in Lemmas 3.14, 3.16, 3.21, 3.34 and Propositions 3.15, 3.22. In Propositions 3.24, 3.26 and 3.29 such properties of functors as being monomorphic and epimorphic are investigated. Supports of functors are studied in Proposition 3.31. Moreover, in Proposition 3.32 it is proved that definite functors preserve intersections of closed subsets. Then functors for inverse systems are described in Proposition 3.33. Bi-functors preserving pre-images are considered in Proposition 3.35. Monads in certain categories are investigated in Theorem 3.38. Exact sequences in categories are considered in Proposition 3.39.

Lattices associated with actions of groupoids on topological spaces are investigated in subsection 3.3. Supports of (T,G)-invariant semi-idempotent continuous morphisms are estimated in Proposition 3.42, where G is a topological groupoid and T is its representation described in Lemma 3.40. Structures of families of all semi-idempotent continuous morphisms associated with a groupoid G and a ringoid K are investigated in Proposition 3.43 and Theorems 3.44, 3.45.

The main results are Propositions 3.22, 3.24, 3.29, 3.32, 3.33, 3.35, 3.39, 3.43, Theorems 3.38, 3.44 and 3.45. All main results of this paper are obtained for the first time. The obtained results can be used for further studies of such objects, their classes and classification. They can be applied to investigations of non locally compact group algebras also.

2. Ringoids and Lattice Structure

2.1. Preliminaries

To avoid misunderstandings we first present our definitions.

1. Definitions. Let K be a set and let two operations $+ : K^2 \to K$ the addition and $\times : K^2 \to K$ the multiplication be given so that K is a semigroup (with associative binary operations) or a quasigroup (with may be non-associative binary operations) relative to $+$ and $\times$ with neutral elements $e_+ =: 0$ and $e_\times =: 1$ so that $a \times 0 = 0 \times a = 0$ for each $a \in K$ and either the right distributivity $a(b+c) = ab + ac$ for every $a,b,c \in K$ or the left distributivity $(b+c)a = ba + ca$ for every $a,b,c \in K$ is accomplished, then K is called a semiring or a ringoid respectively with either right or left distributivity correspondingly. If it is simultaneously right and left distributive, then it is called simply a semiring or a ringoid respectively.

A semiring K (or a ringoid, or a ring, or a non-associative ring) having also a structure of a linear space over a field $\mathbf{F}$ and such that $\alpha(a+b) = \alpha a + \alpha b$, $1a = a$, $\alpha(ab) = (\alpha a)b = a(\alpha b)$ and $(\alpha\beta)a = \alpha(\beta a)$ for each $\alpha, \beta \in \mathbf{F}$ and $a, b \in K$ is called a semialgebra (or an algebroid, or an algebra or a non-associative algebra correspondingly).

A semiring K (or a semialgebra and so on) supplied with a topology on K (or on K and $\mathbf{F}$ correspondingly) relative to which algebraic operations are continuous is called a topological semiring (or a topological semialgebra and so forth correspondingly).

A set K with binary operations $\mu_1, ..., \mu_n$ will also be called an algebraic object. An algebraic object is commutative relative to an operation μ_p if $\mu_p(a,b) = \mu_p(b,a)$ for each $a, b \in K$.

An algebraic object K with binary operations $\mu_1, ..., \mu_n$ is called either directed or linearly ordered or well-ordered if it is such as a set correspondingly and its binary operations preserve an ordering: $\mu_p(a,b) \leq \mu_p(c,d)$ for each $p = 1, ..., n$ and for every $a,b,c,d \in K$ so that $a \leq c$ and $b \leq d$ when a,b,c,d belong to the same linearly ordered set Z in K.

Henceforward, we suppose that the minimal element in an ordered K is zero.

Henceforth, for semialgebras, non-associative algebras or algebroids A speaking about ordering on them we mean that only their non-negative cones $K = \{y : y \in A, 0 \leq y\}$ are considered. For non-negative cones K in semialgebras, non-associative algebras or algebroids only the case over the real field will be considered.

2. Definition. A (non-associative) topological algebra or a topological ringoid, etc., we call topologically simple if it does not contain closed ideals different from $\{0\}$ and K, where $K \neq \{0\}$.

3. Definition. We consider a directed set K which satisfies the condition:

(DW) for each linearly ordered subset A in K there exists a well-ordered subset B in K such that $A \subset B$.

4. Definitions. Let K be a well-ordered (or directed satisfying condition $3(DW)$) either semiring or ringoid (or a non-negative cone in a algebroid over the real field $\mathbf{R}$) such that

(1) $\sup E \in K$ for each $E \in T$, where T is a family of subsets of K.

If K is a directed topological either semiring or ringoid, we shall suppose that it is supplied with a topology

(2) $\tau = \tau_K$ so that every set

$L_b := \{y : y \in K, y < b \text{ or } y \text{ is not comparable with } b\}$ and
$G_b := \{y : y \in K, b < y \text{ or } y \text{ is not comparable with } b\}$
is open relative to it.

That is, if a set Z is linearly ordered in K this topology τ_K provides the hereditary topology on Z which is not weaker than the interval topology on Z generated by the base $\{(a,b)_Z : a < b \in Z\}$, where $(a,b)_Z := \{c : c \in Z, a < c < b\}$.

For a completely regular topological space X and a topological semiring (or ringoid) K let $C(X,K)$ denote a semiring (or a ringoid respectively) of all continuous mappings $f : X \to K$ with the element-wise addition $(f+g)(x) = f(x) + g(x)$ and the element-wise multiplication $(fg)(x) = f(x)g(x)$ operations for every $f, g \in C(X,K)$ and $x \in X$.

If K is a directed semiring (or a directed ringoid) and X is a linearly ordered set, $C_+(X,K)$ (or $C_-(X,K)$) will denote the set of all monotone non-decreasing (or non-increasing correspondingly) maps $f \in C(X,K)$.

For the space $C(X,K)$ (or $C_+(X,K)$ or $C_-(X,K)$) we suppose that

(3) a family T of subsets of K contains the family $\{f(X) : f \in C(X,K)\}$ (or $\{f(X) : f \in C_+(X,K)\}$ or $\{f(X) : f \in C_-(X,K)\}$ correspondingly) and K satisfies Condition (1).

5. Remark. For example, the class On of all ordinals has the addition $\mu_1 = +_o$ and the multiplication $\mu_2 = \times_o$ operations which are generally non-commutative, associative, with unit elements 0 and 1 respectively, on On the right distributivity is satisfied (see Propositions 4.29–4.31 and Examples 1–3 in [19]). Relative to the interval topology generated by the base $\{(a,b) : a < b \in On\}$ the class On is the topological well-ordered semiring, where $(a,b) = \{c : c \in On, a < c < b\}$. For each non-void set A in On there exists $\sup A \in On$ (see [20]).

If K is a linearly ordered non-commutative relative to the addition semiring (or a ringoid), then the new operation $(a,b) \mapsto \max(a,b) =: a +_2 b$ defines the commutative addition. Then $c(a +_2 b) = \max(ca, cb) = ca +_2 cb$ and $(a +_2 b)c = \max(ac, bc) = ac +_2 bc$ for every $a, b, c \in K$, that is $(T, +_2, \times)$ is left and right distributive.

As an example of a semiring (or a ringoid) K in Definitions 4 one can take $K = On$ or $K = \{A : A \in On, |A| \leq b\}$, where b is a cardinal number such that $\aleph_0 \leq b$. Each segment $[a,b] := \{c : c \in On, a \leq c \leq b\}$ is compact in On, where $a < b \in On$. Evidently, $K = On$ satisfies Condition 4(1), since $\sup E$ exists for each set E in On (see [20]).

Particularly, if a topological space X is compact and $C(X,K)$ is a semiring (or a ringoid) of all continuous mappings $f : X \to K$, then a family T contains the family of compact subsets

$\{f(X) : f \in C(X,K)\}$, since a continuous image of a compact space is compact (see Theorem 3.1.10 [21]).

It is possible to modify Definition 4 in the following manner. For a well-ordered K without Condition 4(1) one can take the family of all continuous bounded functions $f : X \to K$ and denote this family of functions by $C(X,K)$ for the uniformity of the notation.

For a directed K satisfying Condition 3(DW) without Condition 4(1) it is possible to take the family of all monotone non-decreasing (or non-increasing) bounded functions $f : X \to K$ for a linearly ordered set X and denote this family by $C_+(X,K)$ ($C_-(X,K)$ correspondingly) also.

Naturally, $C(X,K)$ has also the structure of the left and right module over the semiring (or the ringoid correspondingly) K, *i.e.*, af and fa belong to $C(X,K)$ for each $a \in K$ and $f \in C(X,K)$. To any element $a \in K$ the constant mapping $g^a \in C(X,K)$ corresponds such that $g^a(x) = a$ for each $x \in X$. If K is right (or left) distributive, then $q(f+h) = qf + qh$ (or $(f+h)q = fq + hq$ correspondingly) for every $q, f, h \in C(X,K)$.

The semiring (or the ringoid) $C(X,K)$ will be considered directed:

(1) $f \le g$ if and only if $f(x) \le g(x)$ for each $x \in X$.

Indeed, if $f,h \in C(X,K)$, then $a = \sup(f(X)) \in K$ and $b = \sup(h(X)) \in K$ according to Condition 4(1). Then there exists $c \in K$ so that $a \le c$ and $b \le c$, consequently, $f \le g^c$ and $h \le g^c$. Thus for each $f,h \in C(X,K)$ there exists $q \in C(X,K)$ so that $f \le q$ and $h \le q$. From $a+b \le c+d$ and $ab \le cd$ for each $a \le c$ and $b \le d$ in K it follows that $f+q \le g+h$ and $fq \le gh$ for each $f \le g$ and $q \le h$ in $C(X,K)$.

If $f \le g$ and $f \ne g$ (i.e. $\exists x \in X\ f(x) \ne g(x)$), then we put $f < g$.

For a mapping $f \in C(X,K)$ its support $supp(f)$ is defined as usually

(2) $supp(f) := cl_X\{x : x \in X, f(x) \ne 0\}$, where $cl_X A$ denotes the closure of A in X when $A \subset X$.

Henceforth, we consider cases, when

(3) a topology on X is sufficiently fine so that functions separate points in X, *i.e.*, for each $x \ne z$ in X there exists f in $C(X,K)$(or $C_-(X,K)$ or $C_+(X,K)$ correspondingly) such that $f(x) \ne f(z)$.

The latter is always accomplished in the purely algebraic discrete case.

2.2. Directed Ringoids $C(X,K)$ of Mappings

6. Lemma. *If E is a closed subspace in a topological space X, then $C(X,K|E) := \{f : f \in C(X,K), supp(f) \subset E\}$ is an ideal in $C(X,K)$.*

Proof. If $f \in C(X,K|E)$ and $g \in C(X,K)$, then $f(x)g(x) = 0$ and $g(x)f(x) = 0$ when $f(x) = 0$, consequently, $supp(fg)$ and $supp(gf)$ are contained in E. Moreover, if $f,h \in C(X,K|E)$, then $supp(f+h)$ and $supp(h+f)$ are contained in E, since $f(x)+h(x) = 0$ and $h(x)+f(x) = 0$ for each $x \in X \setminus E$, while $X \setminus E$ is open in X. Thus $C(X,K|E)$ is a semiring (or a ringoid respectively) and $C(X,K|E)C(X,K) \subseteq C(X,K|E)$ and $C(X,K)C(X,K|E) \subseteq C(X,K|E)$.

7. Corollary. *If E is clopen (i.e. closed and open simultaneously) in X, then $C(E,K)$ is an ideal in $C(X,K)$.*

Proof. For a clopen topological subspace E in X one gets $C(E,K)$ isomorphic with $C(X,K|E)$, since each $f \in C(E,K)$ has the zero extension on $X \setminus E$.

8. Lemma. *For a linearly ordered set X and a directed semiring (ringoid) K there are directed semirings (or ringoids correspondingly) $C_+(X,K)$ and $C_-(X,K)$.*

Proof. The sets $C_+(X,K)$ and $C_-(X,K)$ are directed according to Condition 5(1) with a partial ordering inherited from $C(X,K)$. Since $a+b \le c+d$ and $ab \le cd$ for each $a \le c$ and $b \le d$ in K, then $f+q \le g+h$ and $fq \le gh$ for each $f \le g$ and $q \le h$ all either in $C_+(X,K)$ or $C_-(X,K)$. On the other hand, for each $f,h \in C(X,K)$ there exists $g^c \in C(X,K)$ so that $f \le g^c$ and $h \le g^c$

(see § 5). If $f(x) \le f(y)$ and $h(x) \le h(y)$ for $f,h \in C_+(X,K)$ and each $x \le y$ in K, then $f(x)+h(x) \le f(y)+h(y)$ and $f(x)h(x) \le f(y)h(y)$, consequently, $f+h$ and fh are in $C_+(X,K)$. Analogously, if $f,h \in C_-(X,K)$, then $f+h$ and fh are in $C_-(X,K)$. But a constant mapping g^c belongs to $C_+(X,K)$ and $C_-(X,K)$. Thus $C_+(X,K)$ and $C_-(X,K)$ are directed semirings (or ringoids correspondingly).

9. Lemma. *If $H = H_X$ is a covering of X and τ_K is a topology on K satisfying Conditions 3(DW) and 4(1 − 3), then a semiring (or ringoid or a non-negative cone in a algebroid over* **R***) $C(X,K)$ can be supplied with a topology relative to which it is a topological directed (TD) semiring (or a TD ringoid or a TD algebroid respectively).*

Proof. Take a topology τ_C on $C(X,K)$ with the base β_C formed by the following sets and their finite intersections:

$$
\begin{aligned}
&(1)\ B_1(g,A,V) := \{f : f \in C(X,K), f(A) \subset g+V\},\\
&B_2(g,A,V) := \{f : f \in C(X,K), f(A) \subset V+g\},\\
&B_3(g,A,V) := \{f : f \in C(X,K), f(A) \subset gV\},\\
&B_4(g,A,V) := \{f : f \in C(X,K), f(A) \subset Vg\},
\end{aligned}
$$

where $g \in C(X,K)$, $A \in H$, $V \in \tau_K$. Evidently, the addition $+ = \mu_1$ and the multiplication $\times = \mu_2$ are continuous relative to this topology, since each $U \in \tau_C$ is the union of base sets $P \in \beta_C$. In view of Section 5 $C(X,K)$ is directed: $\mu_p(f,h) \le \mu_p(g,u)$ for each $p = 1,2$ and for every $f,g,h,u \in C(X,K)$ so that $f \le g$ and $h \le u$ when f,g,h,u belong to the same linearly ordered set in $C(X,K)$, since element-wise these inequalities are satisfied in K, i.e. for $f(x), g(x), h(x), u(x)$ with $x \in X$ (see §1).

10. Note. Henceforward, it will be supposed that $C(X,K)$ is supplied with the topology τ_C of Lemma 9, while $C_+(X,K)$ and $C_-(X,K)$ are considered relative to the topology inherited from $C(X,K)$. Particularly, if $X \in H_X$, then it provides the topology of the uniform convergence on $C(X,K)$.

11. Corollary. *If the conditions of Lemma 9 are satisfied and $H = 2^X$ is the family of all subsets in X and a topology τ_K on K is discrete, then τ_C is the discrete topology on $C(X,K)$.*

12. Lemma. *Suppose that the conditions of Lemma 9 are satisfied. Then the functions*

(1) $f \vee g(x) := \max(f(x), g(x))$ and
(2) $f \wedge g(x) := \min(f(x), g(x))$

are in $C(X,K)$ (or in $C_+(X,K)$ or in $C_-(X,K)$) for every pair of functions $f,g \in C(X,K)$ (or in $C_+(X,K)$ or in $C_-(X,K)$ correspondingly) satisfying the condition:

(3) for each $x \in X$ either $f(x) < g(x)$ or $g(x) < f(x)$ or $f(x) = g(x)$.

Proof. Let $f,g \in C(X,K)$ satisfy Condition (3). Then the sets $\{x : x \in X, f(x) \le g(x)\}$ and $\{x : x \in X, f(x) \le g(x)\}$ are closed in X, since f and g are continuous functions on X and the topology τ_K on K satisfies Condition 4(2). For each closed set E in K the sets

$(f \vee g)^{-1}(E) = [f^{-1}(E) \cap \{x : x \in X, g(x) \le f(x)\}] \cup [g^{-1}(E) \cap \{x : x \in X, f(x) \le g(x)\}]$ and
$(f \wedge g)^{-1}(E) = [f^{-1}(E) \cap \{x : x \in X, f(x) \le g(x)\}] \cup [g^{-1}(E) \cap \{x : x \in X, g(x) \le f(x)\}]$

are closed in X, consequently, the mappings $f \vee g$ and $f \wedge g$ are continuous. If $f,g \in C_+(X,K)$ and $x < y \in X$, then $f(x) \le f(y)$ and $g(x) \le g(y)$. If $f(x) \le g(x)$ and $g(y) \le f(y)$, then $(f \vee g)(x) = g(x) \le g(y) \le f(y) = (f \vee g)(y)$ and $(f \wedge g)(x) = f(x) \le g(x) \le g(y) = (f \wedge g)(y)$. If $f(x) \le g(x)$ and $f(y) \le g(y)$, then $(f \vee g)(x) = g(x) \le g(y) = (f \vee g)(y)$ and $(f \wedge g)(x) = f(x) \le f(y) = (f \wedge g)(y)$. Therefore, $(f \vee g)(x) \le (f \vee g)(y)$ and $(f \wedge g)(x) \le (f \wedge g)(y)$ for each $x < y \in X$. Thus $f \vee g$ and $f \wedge g \in C_+(X,K)$. Analogously if $f,g \in C_-(X,K)$, then $f \vee g$ and $f \wedge g \in C_-(X,K)$.

Relative to the topology of §9 on $C(X,K)$ operations $\vee$ and $\wedge$ are continuous on $C(X,K)$, $C_+(X,K)$ and $C_-(X,K)$.

2.3. Examples of Directed Ringoids

13. Example. Ringoids and ordinals. The class On of all ordinals has the addition $\mu_1 = +_o$ and the multiplication $\mu_2 = \times_o$ operations which are generally non-commutative, associative, with unit elements 0 and 1 respectively, on On the right distributivity is satisfied (see Propositions 4.29–4.31

and Examples 1–3 in [19,22]). Relative to the interval topology generated by the base $\{(a,b) : a < b \in On\}$ the class On is the topological well-ordered semiring, where $(a,b) = \{c : c \in On, a < c < b\}$. For each non-void set A in On there exists $\sup A \in On$ (see [20]).

14. Example. Construction of ringoids with the help of inductive limits. Let J be a directed set of the cardinality $card(J) \geq \aleph_0$ such that for each $l, k \in J$ there exists $j \in J$ with $l \leq j$ and $k \leq j$ (see also §I.3 [21]), and let $\phi : J \to J$ be a monotone decreasing map, $G_j \subseteq [0,\infty)$, let also $p_j^k : G_k \to G_j$ be an embedding for each $k \leq j \in J$. There is considered G_j as a ringoid with the addition, the multiplication, with neutral elements $0_j = 0$ by addition and $1_j = 1$ by multiplication and the linear ordering $x_j < y_j$ inherited from $[0,\infty) = \{t : t \in \mathbf{R}, 0 \leq t < \infty\}$ for each $j \in J$. Put $G_0 = \lim\{G_j, p_j^k, J\}$ to be the inductive limit of the direct mapping system so that G is the quotient $(\bigoplus_j G_j)/\Xi$ of the direct sum $\bigoplus_j G_j$ by the equivalence relation Ξ caused by mappings p_j^k. Then consider $G := \{x : x \in G_0, \sup_{j\in J} x_j < \infty\}$, where $x_j = p_j(x)$, $p_j : G \to G_j$ notates the projection.

Then we define $g +_1 h := \{v_j : v_j = g_j + h_j \forall j \in J\}$ and $g \times_1 h := \{w_j : w_j = g_j p_j^k(h_k) \forall j \in J$ with $k = \phi(j)\}$ for all $g, h \in G$, where $g_j = p_j(g)$ for each $j \in J$. Let also $x <_1 y$ in G if and only if $x_j < y_j$ for each $j \in J$. Certainly for each $x, y \in G$ there exists $z \in G$ so that $x \leq_1 z$ and $y \leq_1 z$, for example, $z_j = \max(x_j, y_j)$ for each $j \in J$. Therefore we get that if $x <_1 y$ and $u <_1 z$ in G, then $x +_1 u <_1 y +_1 z$ and $x \times_1 u <_1 y \times_1 z$. We supply G with a topology τ_b inherited from the inductive limit topology on G_0, where $[0,\infty)$ is supplied with the standard metric of $\mathbf{R}$ and G_j has the topology inherited from $[0,\infty)$. Then we deduce that $U(x_j, b, j) + U(z_j, b, j) \subset U(x_j + z_j, 2b, j)$ and $U(x_j, b, j)U(z_k, b, k) \subset U(x_j p_j^k(z_k), b(1 + x_j + z_k), j)$ for every $x, z \in G$ and $b > 0$ and $j \in J$ with $k = \phi(j)$, where $U(x_j, b, j) := \{y_j : y_j \in G_j, x_j - b < y_j < x_j + b\}$. Since $\sup_{j\in J} x_j < \infty$ for each $x \in G$, then the addition and the multiplication in G are continuous. Thus $(G, +_1, \times_1, <_1, \tau_b)$ is the topological directed ringoid with the left and the right distributivity in which the multiplication $\times_1$ is non-associative, since $\phi(j) < j$ for each $j \in J$. It is worth to note that each set of the form $S(x) := \{y : y \in G$, either $x < y$ or x is incomparable with $y\}$ is open in (G, τ_b), where $x \in G$.

15. Example. The case of $G_j \subseteq [0,\infty)^\omega$ for each $j \in J$, where ω is a directed set, can be considered analogously to Example 14, taking the lexicographic ordering on the Cartesian product $M := \omega \times J$ and considering M instead of J.

16. Example. On G from Example 14 one can take also $x +_2 y := \{v_j : v_j = \max(x_j, y_j) \forall j \in J\}$ and $x \times_2 y := \{w_j : w_j = \min(x_j, y_k) \forall j \in J$ with $k = \phi(j)\}$. Then $(G, +_2, \times_2, <_1, \tau_b)$ is a topological non-associative ringoid with the left and right distributivity.

17. Example. Ringoids associated with families of measures. Let G_j be a Boolean algebra on a set H_j and let $p_j^k : G_k \to G_j$ be an embedding for each $j \in J$ with $k = \phi(j)$ so that $m_j(p_j^k(C_k)) \leq m_k(C_k)$ for each $C_k \in G_k$, where J and ϕ are as in subsection 14. Suppose that on each Boolean algebra G_j there is a probability (finitely additive) measure $m_j : G_j \to [0,1]$ so that G_j is metrizable by the metric $d_j(A_j, B_j) := m_j(A_j \triangle B_j)$, where $A_j \triangle B_j := (A_j \setminus B_j) \cup (B_j \setminus A_j)$. Otherwise it is possible to consider the quotient algebra G_j/Ξ_j, where $A_j \Xi_j B_j$ if and only if $d_j(A_j, B_j) = 0$. Put $A + B := \{C : C_j = A_j \cup B_j \forall j \in J\}$ and $A \times B := \{C : C_j = A_j \cap p_j^k(B_k) \forall j \in J$ with $k = \phi(j)\}$, where $A, B \in G$, $A_j, B_j \in G_j$, $G = \lim\{G_j, p_j^k, J\}$ is the inductive limit of Boolean algebras, $A_j = p_j(A)$, $p_j : G \to G_j$ denotes the projection.

Consider on G the inductive limit topology τ_b, where G_j is supplied with the metric d_j for each $j \in J$. Naturally it is possible to put $A \leq B$ in G if and only if $A_j \subseteq B_j$ for each $j \in J$. Then the inequalities

$m_j((A_j \cup C_j) \triangle (B_j \cup D_j)) \leq m_j((A_j \triangle B_j) \cup (C_j \triangle D_j)) \leq m_j(A_j \triangle B_j) + m_j(C_j \triangle D_j)$ and

$m_j((A_j \cap p_j^k(C_k)) \triangle (B_j \cap p_j^k(D_k))) \leq m_j(A_j \triangle B_j) + m_j(p_j^k(C_k) \triangle p_j^k(D_k)) \leq m_j(A_j \triangle B_j) + m_k(C_k \triangle D_k)$

are fulfilled for each $A_j, B_j \in G_j$ and $C_k, D_k \in G_k$. Therefore $(G, +, \times, <, \tau_b)$ is the topological ringoid with the left and right distributivity and the non-associative multiplication.

Instead of measures it is possible more generally to consider submeasures m_j, that is possessing the subadditivity property: $m_j(C_j) \le m_j(A_j) + m_j(B_j)$ for each $A_j, B_j, C_j \in G_j$ satisfying the inclusion $C_j \subset A_j \cup B_j$.

18. Example. Ringoids induced by spectra of non locally compact groups. Let $\{G_j, p^k_j, J\}$ be a family of topological non locally compact groups G_j, where J is a directed set, $p^k_j : G_k \to G_j$ is a continuous injective homomorphism for each $j < k$ in J. Let also $\phi : J \to J$ be an increasing map and let $m_j : \mathcal{B}_j \to [0,1]$ be a Radon probability σ-additive measure on the Borel σ-algebra $\mathcal{B}_j$ of G_j such that m_j is left quasi-invariant relative to $p^k_j(G_k)$ for each $j \in J$ with $k = \phi(j)$. That is there exists the Radon-Nikodym derivative (i.e., the left quasi-invariance factor) $d_m(v,g) := m^v(dg)/m(dg)$ for each $m = m_j$, where $v \in G_k$, $g \in G_j$, $m^v(A) := m((p^k_j(v))^{-1}A)$ for each $A \in Af(G_j, m_j)$, where $Af(G_j, m_j)$ denotes a σ-algebra which is the completion of $\mathcal{B}_j$ by m_j-null sets.

It is assumed that a uniformity τ_{G_j} on G_j is such that $\tau_{G_j}|G_k \subset \tau_{G_k}$ and (G_j, τ_{G_j}) is complete for each $j \in J$ with $k = \phi(j)$. Suppose also that there exists an open base of neighborhoods of $e_k \in G_k$ such that their closures in G_j are compact.

It is known that such systems exist for loop groups and groups of diffeomorphisms and Banach-Lie groups.

Then $L^p_{G_k}(G_j, m_j, \mathbf{R})$ for $1 \le p \le \infty$ denotes the Banach space of all m_j-measurable functions $f : G_j \to \mathbf{R}$ such that $f^h(g) \in L^p(G_j, m_j, \mathbf{R})$ for each $h \in G_k$ and

$$\|f\|_{L^p_{G_k}(G_j,m_j,\mathbf{R})} := \sup_{h \in G_k} \|f^h\|_{L^p(G_j,m_j,\mathbf{R})} < \infty,$$

where $f^h(g) := f((p^k_j(h))^{-1}g)$ for each $g \in G_j$ and $h \in G_k$, $j \in J$ with $k = \phi(j)$. Next we consider the space

$L^\infty(L^1_{G_k}(G_j, m_j, \mathbf{R}) : j \in J, k = \phi(j)) := \{f = (f_j : j \in J);\ f_j \in L^1_{G_k}(G_j, m_j, \mathbf{R})$ for each $j \in J$; $\|f\|_\infty := \sup_{j \in J} \|f_j\|_{L^1_{G_k}(G_j)} < \infty$, where $k = \phi(j)\}$.

There exists the non-associative normed algebra $\mathcal{E} := L^\infty(L^1_{G_k}(G_j, m_j, \mathbf{R}) : j \in J, k = \phi(j))$ supplied with the multiplication

$f \tilde{*} u = w$ such that

$$w_j(g_j) = (f_k \tilde{*} u_j)(g_j) = \int_{G_k} f_k(t_k) u_j(p^k_j(t_k) g_j) m_k(dt_k)$$

for every $f, u \in \mathcal{E}$ and $g \in G = \prod_{\alpha \in J} G_\alpha$, where $k = \phi(j)$, $j \in J$, $g_j \in G_j$ (see [11–13,23]).

Now we take the positive cone

$F := \{f : f \in \mathcal{E}, \forall j \in J\ f_j(g_j) \ge 0$ for $m_j -$ almost all $g_j \in G_j\}$ in $\mathcal{E}$ and put

$f + h = \{(f+h)_j(g_j) = f_j(g_j) + h_j(g_j)\ \forall j \in J\ \forall g_j \in G_j\}$,

$f \times h = f \tilde{*} h$ for each $f, h \in F$ and define

$u \le f$ in F if and only if $u_j(g_j) \le f_j(g_j)$ for each $j \in J$ and m_j-almost all $g_j \in G_j$. Therefore, $w_j(g_j) \ge 0$ for every $f, u \in F$, $j \in J$ and m_j-almost all $g_j \in G_j$, where $w = f \times u$, since m_k is the probability measure, $f_k(t_k) \ge 0$ and $u_j(p^k_j(t_k) g_j) \ge 0$ for m_k-almost all $t_k \in G_k$ and m_j-almost all $g_j \in G_j$ correspondingly. Thus $f \times u \in F$ for each $f, u \in F$.

If $f, h, q, u \in F$ and $f \le q$, $h \le u$, then $f_j(g_j) + h_j(g_j) \le q_j(g_j) + u_j(g_j)$ and

$$(f_k \tilde{*} h_j)(g_j) = \int_{G_k} f_k(t_k) h_j(p^k_j(t_k) g_j) m_k(dt_k)$$

$$\le \int_{G_k} q_k(t_k) u_j(p^k_j(t_k) g_j) m_k(dt_k) = (q_k \tilde{*} u_j)(g_j)$$

for m_j-almost all $g_j \in G_j$ and hence $f + h \leq q + u$ and $f \times h \leq q \times u$. Then we infer that

$$((f_k + h_k)\tilde{*}u_j)(g_j) = \int_{G_k} (f_k(t_k) + h_k(t_k))u_j(p_j^k(t_k)g_j)m_k(dt_k)$$

$$= \int_{G_k} f_k(t_k)u_j(p_j^k(t_k)g_j)m_k(dt_k) + \int_{G_k} h_k(t_k)u_j(p_j^k(t_k)g_j)m_k(dt_k)$$

$$= (f_k\tilde{*}u_j)(g_j) + (h_k\tilde{*}u_j)(g_j)$$

for every $f, h, u \in \mathcal{E}$ and $g \in G = \prod_{i \in J} G_i$, where $k = \phi(j)$, $j \in J$, $\pi_j(g) = g_j \in G_j$, $\pi_j : G \to G_j$ is the projection, consequently, $(f + h) \times u = (f \times u) + (h \times u)$. Analogously it can be verified that $u \times (f + h) = (u \times f) + (u \times h)$ for every $f, h, u \in \mathcal{E}$.

For each $f, h \in F$ there exists an element $u \in F$ so that $f \leq u$ and $h \leq u$, for example, either $u = f + h$ or u given by the formula $u_j(g_j) = \max(f_j(g_j), h_j(g_j))$ for each $j \in J$ and m_j-almost all $g_j \in G_j$.

Take on F the topology τ_n inherited from the norm topology on $\mathcal{E}$. This implies that $(F, +, \times, <, \tau_n)$ is the directed topological non-associative ringoid with the left and right distributivity.

There is the decomposition $f = f^+ - f^-$ for each $f \in \mathcal{E}$, where $f^+ \in F$ and $f^- \in F$, $f_j^+(g_j) := \max(f_j(g_j), 0)$ for each $j \in J$ and $g_j \in G_j$.

If f and h in F are incomparable, there exist $j, l \in J$ (may be either $j = l$ or $j \neq l$) such that $m_j(A_j^+) > 0$ and $m_l(A_l^-) > 0$, where

$A_j^+ = A_j^+(f, h) := \{g_j : g_j \in G_j,\ f_j(g_j) > h_j(g_j)\}$ and

$A_l^- = A_l^-(f, h) := \{g_l : g_l \in G_l,\ f_l(g_l) < h_l(g_l)\}$. Then for

$0 < b < \min(m_j(A_j^+), m_l(A_l^-)) \min(1, \|(f_j - h_j)|_{A_j^+}\|_{L^1(A_j^+)}, \|(f_l - h_l)|_{A_l^-}\|_{L^1(A_l^-)})/4$

each element v in the ball $B(F, h, b) := \{q : q \in F, \|q - h\|_{\mathcal{E}} < b\}$ is incomparable with f, since $\|q_j\|_{L^1(G_j)} \leq \|q_j\|_{L^1_{G_k}(G_j)}$ for each $q \in \mathcal{E}$ and $j \in J$, while m_j is the probability measure for each $j \in J$. On the other hand, if $v < u$ in F, there exists $l \in J$ so that $m_l(A_l^-(v, u)) > 0$ and $v_j(g_j) \leq u_j(g_j)$ for m_j-almost all $g_j \in G_j$ for each $j \in J$. Therefore, for $0 < b$ prescribed by the inequality given above and each $q \in B(F, h, b)$ the inequality $q \leq f$ is impossible, consequently, either q is incomparable with f or $f < q$. Thus each set of the form $S(f) := \{h : h \in F,$ either $f < h$ or f is incomparable with $h\}$ is open in (F, τ_n), where $f \in F$.

19. Note. Certainly relative to the discrete topology the aforementioned ringoids are also topological ringoids. Other examples can be constructed from these using the theorems and the propositions presented above.

3. Skew Morphisms of Ordered Semirings and Ringoids

3.1. Morphisms and Their Properties

1. Notation. Let $\times_2$ denote the mapping on $[K \times C(X, K)] \cup [C(X, K) \times K]$ with values in $C(X, K)$ such that

(1) $c \times_2 f := g^c + f =: g^c \times_2 f$ and $f \times_2 c := f + g^c =: f \times_2 g^c$ for each $c \in K$ and $f \in C(X, K)$, where $g^c(x) := c$ for each $x \in X$, whilst the sum is taken element-wise $(f + g)(x) = f(x) + g(x)$ for every $f, g \in C(X, K)$ and $x \in X$.

2. Definition. We call a mapping ν on $C(X, K)$ (or $C_+(X, K)$ or $C_-(X, K)$) with values in K an idempotent (K-valued) morphism if it satisfies for each $f, g, g^c \in C(X, K)$ (or in $C_+(X, K)$ or $C_-(X, K)$ correspondingly) the following five conditions

(1) $\nu(g^c) = c$;

(2) $\nu(c \times_2 f) = c + \nu(f) =: c \times_2 \nu(f)$ and

(3) $\nu(f \times_2 c) = \nu(f) + c =: \nu(f) \times_2 c$;

(4) $\nu(f \vee g) = \nu(f) \vee \nu(g)$ when f, g satisfy Condition 2.12(3) and

(5) $\nu(f \wedge g) = \nu(f) \wedge \nu(g)$ if f, g satisfy Condition 2.12(3),
where $a \vee b = \max(a, b)$ and $a \wedge b = \min(a, b)$ for each $a, b \in K$ when either $a < b$ or $a = b$ or $b < a$.

A mapping (morphism) ν on $C(X, K)$ (or $C_+(X, K)$ or $C_-(X, K)$) with values in K we call order preserving (non-decreasing), if

(6) $\nu(f) \le \nu(g)$ for each $f \le g$
in $C(X, K)$ (or $C_+(X, K)$ or $C_-(X, K)$ respectively), *i.e.*, when $f(x) \le g(x)$ for each $x \in X$.

A morphism ν will be called K-homogeneous on $C(X, K)$ (or $C_+(X, K)$ or $C_-(X, K)$) if

(7) $\nu(bf) = b\nu(f)$ and

(8) $\nu(fb) = \nu(f)b$

for each f in $C(X, K)$ (or $C_+(X, K)$ or $C_-(X, K)$ correspondingly) and $b \in K$.

3. Remark. If a morphism satisfies Condition 2(4), then it is order preserving.

The evaluation at a point morphism δ_x defined by the formula:

(1) $\delta_x f = f(x)$

is the idempotent K-homogeneous morphism on $C(X, K)$, where x is a marked point in X.

If morphisms $\nu_1, ..., \nu_n$ are idempotent and the multiplication in K is distributive, then for each constants

(2) $c_1 > 0, ..., c_n > 0$ in K with

(3) $c_1 + ... + c_n = 1$ morphisms

(4) $c_1\nu_1 + ... + c_n\nu_n$ and

(5) $\nu_1 c_1 + ... + \nu_n c_n$

are idempotent. Moreover, if the multiplication in K is commutative, associative and distributive and constants satisfy Conditions $(2, 3)$ and morphisms $\nu_1, ..., \nu_n$ are K-homogeneous, then morphisms of the form $(4, 5)$ are also K-homogeneous.

The considered here theory is different from the usual real field **R**, since **R** has neither an infimum nor a supremum, i.e. it is not well-ordered and satisfy neither 2.3(DW) nor 2.4(1).

4. Lemma. *Suppose that either*

(1) *K is well-ordered and satisfies Conditions 2.4$(1-3)$ or*

(2) *X is linearly ordered and K is directed and satisfies Conditions 2.3(DW) and 2.4$(1-3)$. Then there exists an idempotent K-homogeneous morphism ν on $C(X, K)$ in case (1), on $C_+(X, K)$ and $C_-(X, K)$ in case (2). Moreover, if $K \subset On$ and K is infinite, X is not a singleton, $\aleph_0 \le |K|$, $|X| > 1$, then ν has not the form either 3(4) or 3(5) with the evaluation at a point morphisms $\nu_1, ..., \nu_n$ relative to the standard addition in On.*

Proof. Suppose that ν is an order preserving morphism on $C(X, K)$ (or $C_+(X, K)$ or $C_-(X, K)$). If f, g in $C(X, K)$ (or $C_+(X, K)$ or $C_-(X, K)$ respectively) satisfy Condition 2.12(3), then in accordance with Lemma 2.12 there exists $f \vee g$ and $f \wedge g$ in the corresponding $C(X, K)$ (or $C_+(X, K)$ or $C_-(X, K)$). Since $f \vee g \ge f$ and $f \vee g \ge g$ and $f \wedge g \le f$ and $f \wedge g \le g$ and the morphism ν is order preserving, then $\nu(f) \vee \nu(g) \le \nu(f \vee g)$ and $\nu(f \wedge g) \le \nu(f) \wedge \nu(g)$.

Let also E be a subset in X, we put

(3) $\nu(f) = \nu_E(f) = \sup_{x \in E} f(x)$.

This morphism exists due Conditions 2.4(1,3), since in both cases (1) and (2) of this lemma, the image $f(E)$ is linearly ordered and is contained in K.

From the fact that the addition preserves ordering on K (see §2.1) it follows that Properties $(1-3, 7, 8)$ are satisfied for the morphism ν given by Formula (3). If $f \le g$ on X, then for each $a \in f(E)$ there exists $b \in g(E)$ so that $a \le b$, consequently, $\nu(f) \le \nu(g)$, *i.e.*, 2(6) is fulfilled.

We consider any pair of functions f, g in $C(X, K)$ (or $C_+(X, K)$ or $C_-(X, K)$) satisfying Condition 8(3). In case (2) a topological space X is linearly ordered, in case (1) K is well-ordered, hence $f(X)$, $g(X)$, $f(E)$ and $g(E)$ are linearly ordered in K. Then for each $a \in f(E) \cup g(E)$ there exist $b \in (f \vee g)(E)$ so that $a \le b$, while for each $c \in (f \vee g)(E)$ there exists $d \in f(E) \cup g(E)$ so that $c \le d$, hence $\nu(f \vee g) = \nu(f) \vee \nu(g)$. Moreover, for each $a \in f(E) \cup g(E)$ there exists $b \in (f \wedge g)(E)$ so that $b \le a$ and for each

$c \in (f \wedge g)(E)$ there exists $d \in f(E) \cup g(E)$ so that $d \leq c$, consequently, $\nu(f \wedge g) = \nu(f) \wedge \nu(g)$. Thus Properties 2(4, 5) are satisfied as well.

If E is chosen such that there exists $U \in H_X$ with $E \subset U$, then this morphism ν is continuous on $C(X,K)$, $C_+(X,K)$ and $C_-(X,K)$ (see §§2.3, 2.4, 2.9 and 2.10 also).

If a set X is not a singleton, $|X| > 1$, and $K \subset On$ is infinite, $\aleph_0 \leq |K|$, then taking a set E in X different from a singleton, $|E| > 1$, we get that the morphism given by Formula (3) can not be presented with the help of evaluation at a point morphisms $\nu_1 = \delta_{x_1}, ..., \nu_n = \delta_{x_n}$ by Formula either 3(4) or 3(5) relative to the standard addition in On, since functions f in $C(X,K)$ (or $C_+(X,K)$ or $C_-(X,K)$) separate points in X (see Remark 2.5(3)).

5. Remark. Relative to the idempotent addition $x \vee y = \max(x,y)$ the morphism ν_E given by 4(3) has the form $\nu_E(f) = \vee_{x \in E} \delta_x(f)$.

Let $I(X,K)$ denote the set of all idempotent K-valued morphisms, while $I_h(X,K)$ denotes its subset of idempotent homogeneous morphisms.

A set F of all continuous K-valued morphisms on $C(X,K)$ is supplied with the weak* topology having the base consisting of the sets

(1) $< \mu; g_1, ..., g_n; b >_1 := \{\nu : \nu \in F, \ \forall j = 1, ..., n \ \nu(g_j) < \mu(g_j) + b\}$;

$< \mu; g_1, ..., g_n; b >_2 := \{\nu : \nu \in F, \ \forall j = 1, ..., n \ \mu(g_j) < \nu(g_j) + b\}$;

$< \mu; g_1, ..., g_n; b >_3 := \{\nu : \nu \in F, \ \forall j = 1, ..., n \ \nu(g_j) < b + \mu(g_j)\}$;

$< \mu; g_1, ..., g_n; b >_4 := \{\nu : \nu \in F, \ \forall j = 1, ..., n \ \mu(g_j) < b + \nu(g_j)\}$

and their finite intersections, where $0 < b \in K$, $g_1, ..., g_n \in C(X,K)$, $\mu \in F$.

6. Definitions. A morphism $\nu : C(X,K) \to K$ is called weakly additive, if it satisfies Conditions 2(2, 3);

normalized at $c \in K$, if Formula 2(1) is fulfilled;

(1) non-expanding if $\nu(f) \leq \nu(h) + c$ when $f \leq h + g^c$ and $\nu(f) \leq c + \nu(h)$ when $f \leq g^c + h$ for any $f, h \in C(X,K)$ and $c \in K$,

where ν may be non-linear or discontinuous as well.

The family of all order preserving weakly additive morphisms on a Hausdorff topological space X with values in K will be denoted by $\mathcal{O}(X,K)$.

If $E \subset C(X,K)$ satisfies the conditions: $g^0 \in E$, $g + b$ and $b + g \in E$ for each $g \in E$ and $b \in K$, then E is called an A-subset.

7. Lemma. *If $\nu : C(X,K) \to K$ is an order preserving weakly additive morphism, then it is non-expanding.*

Proof. Suppose that $f, h \in C(X,K)$ and $b \in K$ are such that $f(x) \leq (h(x) + c)$ or $f(x) \leq (c + h(x))$ for each $x \in X$, then 2(2, 3, 6) imply that $\nu(f) \leq (\nu(h) + c)$ or $\nu(f) \leq (c + \nu(h))$ respectively. Thus the morphism ν is non-expanding.

8. Corollary. *Suppose that a topological ringoid K is well-ordered, satisfies 1(1) and with the interval topology, $X \in H$, $C(X,K)$ is supplied with the topology of §2.9. Then any order preserving weakly additive morphism $\nu : C(X,K) \to K$ is continuous.*

Proof. This follows from Lemma 7 and §§2.3, 2.4, since each subset $\{f : f \leq g\}$ and $\{f : g \leq f\}$ is closed in $C(X,K)$ in the topology of §2.9, where $g \in C(X,K)$.

9. Lemma. *Suppose that A is an A-subset (a left or right submodule over K) in $C(X,K)$ and $\nu : A \to K$ is an order preserving weakly additive morphism (left or right K-homogeneous with left or right distributive ringoid K correspondingly). Then there exists an order preserving weakly additive morphism $\mu : C(X,K) \to K$ such that its restriction on A coincides with ν.*

Proof. One can consider the set $\mathcal{F}$ of all pairs (B, μ) so that B is an A-subset (a left or right submodule over K respectively), $A \subseteq B \subseteq C(X,K)$, μ is an order preserving weakly additive morphism on B the restriction of which on A coincides with ν. The set $\mathcal{F}$ is partially ordered: $(B_1, \mu_1) \leq (B_2, \mu_2)$ if $B_1 \subseteq B_2$ and μ_2 is an extension of μ_1. In accordance with Zorn's lemma a maximal element (E, μ) in $\mathcal{F}$ exists.

If $E \neq C(X,K)$, there exists $g \in C(X,K) \setminus E$. Let $E_- := \{f : f \in E, f \leq g\}$ and $E_+ := \{f : f \in E, g \leq f\}$, then $\mu(h) \leq \mu(q)$ for each $h \in E_-$ and $q \in E_+$, consequently, an element $b \in K$ exists such that $\mu(E_-) \leq b \leq \mu(E_+)$ due to Conditions 2.3(DW) and 2.4(1) imposed on K. Then we put $F = E \cup \{g + g^c, g^c + g : c \in K\}$ (F is a minimal left or right module over K containing E and g correspondingly). Then one can put $\mu(g + g^c) = b + c$ and $\mu(g^c + g) = c + b$. Moreover, one gets $\mu(d(g + g^c)) = d\mu(g) + dc$ or $\mu((g + g^c)d) = \mu(g)d + cd$ for each $d \in K$ correspondingly for each $c \in K$. Then μ is an order preserving weakly additive morphism (left or right homogeneous correspondingly) on F. This contradicts the maximality of A.

10. Theorem. *If a ringoid K is well-ordered and satisfies $1(1)$, with the interval topology and K is locally compact, $X \in H_X$. Then $\mathcal{O}(X,K)$ is compact relative to the weak* topology.*

Proof. In view of Lemma 8 each $\nu \in \mathcal{O}(X,K)$ is continuous. The set $\mathcal{O}(X,K)$ is supplied with the weak* topology (see §5).

For each $\nu \in \mathcal{O}(X,K)$ one has $\nu(g^c) = \nu(g^c + g^0) = c$, since $g^c + g^0 = g^c$ and $\nu(g^c + g^0) = c + 0$. On the other hand, for each $g \in C(X,K)$ due to Condition 2.4(1) a supremum exists, $\|g\| := \sup_{x \in X} g(x) \in K$. Each segment $[a,b]$ in K is closed, bounded and hence compact relative to the interval topology. Therefore, $\mathcal{O}(X,K)$ is contained in the Tychonoff product $S = \prod\{[0, \|g\|] : g \in C(X,K)\}$, since $g \leq h$ and hence $\nu(g) \leq \nu(h)$ when $h(x) = \|g\|$ for each $x \in X$. This product is compact as the Tychonoff product of compact topological spaces by Theorem 3.2.13 [21]. It remains to prove, that $\mathcal{O}(X,K)$ is closed in S, since a closed subspace of a compact topological space is compact (see Theorem 3.1.2 [21]).

Each compact Hausdorff space has a uniformity compatible with its topology (see Theorems 3.19 and 8.1.20 [21]). To each element $y \in S$ a morphism $y : C(X,K) \to K$ corresponds, since $[0, \|g\|] \subset K$ for each $g \in C(X,K)$. If $\nu_n \in \mathcal{O}(X,K)$ is a net converging to q in S, then Properties $2(2,3,6)$ for each ν_n imply Properties $2(2,3,6)$ for q, since each segment $[a,b]$ in K is compact and hence complete as the uniform space due to Theorem 8.3.15 [21], where $a < b \in K$. Therefore, $\lim_n = q \in \mathcal{O}(X,K)$ according to Lemma 7 and Corollary 8. Thus $\mathcal{O}(X,K)$ is complete as the uniform space by Theorem 8.3.20 [21] and hence closed in S in accordance with Theorem 8.3.6 [21].

11. Proposition. *In the topological space $\mathcal{O}(X,K)$ the subsets $I(X,K)$ and $I_h(X,K)$ are closed.*

Proof. From the definitions above it follows that $I_h(X,K) \subset I(X,K) \subset \mathcal{O}(X,K)$. If ν_k is a net in $I(X,K)$ (or in $I_h(X,K)$) converging to a morphism $\mu \in \mathcal{O}(X,K)$ relative to the weak* topology (see also §1.6 [21]), then μ satisfies Conditions $2(1-5)$ (or to $2(1-5,7,8)$ respectively). Thus $I(X,K)$ and $I_h(X,K)$ are closed in $\mathcal{O}(X,K)$.

12. Corollary. *If the conditions of Theorem 10 are satisfied, then the topological spaces $I(X,K)$ and $I_h(X,K)$ are compact.*

3.2. Categories of Semirings, Ringoids and Morphisms

13. Definition. If topological spaces X and Y are given and $f : X \to Y$ is a continuous mapping, then it induces the mapping $\mathcal{O}(f) : \mathcal{O}(X,K) \to \mathcal{O}(Y,K)$ according to the formula: $(\mathcal{O}(f)(\nu))(g) = \nu(g(f))$ for each $g \in C(Y,K)$ and $\nu \in \mathcal{O}(X,K)$.

By $I(f)$ will be denoted the restriction of $\mathcal{O}(f)$ onto $I(X,K)$.

A T_1 topological space will be called K-completely regular (or K Tychonoff space), if for each closed subset F in X and each point $x \in X \setminus F$ a continuous function $h : X \to K$ exists such that $h(x) = 0$ and $h(F) = \{c\}$, i.e. h is constant on F, where $c \neq 0$.

Let RK denote a category such that a family $Ob(RK)$ of its objects consists of all K-regular topological spaces, a set of morphisms $Mor(X,Y)$ consists of all continuous mappings $f : X \to Y$ for every $X, Y \in Ob(RK)$, i.e. RK is a subcategory in the category of topological spaces. We denote by $\mathcal{O}K$ a category with objects $Ob(\mathcal{O}K) = \{\mathcal{O}(X,K) : X \in Ob(RK)\}$ and families of morphisms $Mor(\mathcal{O}(X,K), \mathcal{O}(Y,K))$.

14. Lemma. (1). *There exists a covariant functor $\mathcal{O}$ in the category RK.* (2). *Moreover, if a topological ringoid K is well-ordered, satisfies 2.4(1) and with the interval topology, when $f \in Mor(X,Y)$, $X \in H_X$, $Y \in H_Y$, $X,Y \in Ob(RK)$, then $\mathcal{O}(f)$ is continuous.*

Proof. (1). If $X,Y \in Ob(RK)$ and $f \in Mor(X,Y)$, $g \le h$ in $C(Y,K)$, then $g \circ f \le h \circ f$ in $C(X,K)$, consequently, $(\mathcal{O}(f)(v))(g) = v(g \circ f) \le v(h \circ f) = (\mathcal{O}(f)(v))(h)$ for each $v \in \mathcal{O}(X,K)$. If $c \in K$, $g^c \in C(Y,K)$, then $g^c \circ f \in C(X,K)$, $(\mathcal{O}(f)(v))(g^c + h) = v(g^c \circ f + h \circ f) = c + v(h \circ f) = c + (\mathcal{O}(f)(v))(h)$ and $(\mathcal{O}(f)(v))(h + g^c) = v(h \circ f + g^c \circ f) = v(h \circ f) + c = (\mathcal{O}(f)(v))(h) + c$ for each $h \in C(Y,K)$. If $1_X \in Mor(X,X)$, $1_X(x) = x$ for each $x \in X$, then $1_X \circ q = q$ for each $q \in Mor(Y,X)$ and $t \circ 1_X = t$ for each $t \in Mor(X,Y)$. On the other hand, $(\mathcal{O}(1_X)(v))(g) = v(g \circ 1_X) = v(g)$ for each $g \in C(X,K)$, *i.e.*, $\mathcal{O}(1_X) = 1_{\mathcal{O}(X)}$. Evidently, $(\mathcal{O}(f \circ s)(v))(g) = v(g \circ f \circ s) = (\mathcal{O}(s)(v)(g \circ f) = ((\mathcal{O}(f) \circ \mathcal{O}(s))(v))(g)$.

(2). If v_j is a net converging to v in $\mathcal{O}(X,K)$ relative to the weak* topology, then $\lim_j(\mathcal{O}(f)(v_j))(g) = \lim_j v_j(g \circ f) = v(g \circ f) = (\mathcal{O}(f)(v))(g)$ for each $f \in Mor(X,Y)$ and $g \in C(Y,K)$, since $\mathcal{O}(X,K)$ and $\mathcal{O}(Y,K)$ are weakly* compact according to Theorem 10, consequently, $\mathcal{O}$ is continuous from $\mathcal{O}(X,K)$ to $\mathcal{O}(Y,K)$.

15. Proposition. *If $f \in Mor(X,Y)$ for $X,Y \in Ob(RK)$, then*
$\mathcal{O}(f)(I(X,K)) \subseteq I(Y,K)$.

Proof. If $g,h \in C(Y,K)$ are such that $g \vee h$ or $g \wedge h$ exists and $f : X \to Y$ is a continuous mapping, then

$(\mathcal{O}(f)(v))(g \vee h) = v(g \circ f \vee h \circ f) = v(g \circ f) \vee v(h \circ f) = (\mathcal{O}(f)(v))(g) \vee (\mathcal{O}(f)(v))(h)$ or

$(\mathcal{O}(f)(v))(g \wedge h) = v(g \circ f \wedge h \circ f) = v(g \circ f) \wedge v(h \circ f) = (\mathcal{O}(f)(v))(g) \wedge (\mathcal{O}(f)(v))(h)$.

Then for each element $c \in K$ one gets

$(\mathcal{O}(f)(v))(g^c \times_2 h) = v(g^c \circ f \times_2 h \circ f) = v(g^c \circ f) \times_2 v(h \circ f) = c \times_2 (\mathcal{O}(f)(v))(h)$ and

$(\mathcal{O}(f)(v))(h \times_2 g^c) = v(h \circ f \times_2 g^c \circ f) = v(h \circ f) \times_2 v(g^c \circ f) = (\mathcal{O}(f)(v))(h) \times_2 c$.

16. Definitions. A covariant functor $F : RK \to RK$ will be called epimorphic (monomorphic) if it preserves epimorphisms (monomorphisms). If $\phi : A \hookrightarrow X$ is an embedding, then $F(A)$ will be identified with $F(\phi)(F(A))$.

If for each $f \in Mor(X,Y)$ and each closed subset A in Y, the equality $(F(f)^{-1})(F(A)) = F(f^{-1}(A))$ is satisfied, then a covariant functor F is called preimage-preserving. When $F(\bigcap_{j \in J} X_j) = \bigcap_{j \in J} F(X_j)$ for each family $\{X_j : j \in J\}$ of closed subsets in $X \in Ob(RK)$ the monomorphic functor F is called intersection-preserving.

If a functor F preserves inverse mapping system limits, it is called continuous.

A functor F is said to be weight-preserving when $w(X) = w(F(X))$ for each $X \in Ob(RK)$, where $w(X)$ denotes the topological weight of $X \in Ob(RK)$.

A functor is said to be semi-normal when it is continuous, monomorphic, epimorphic, preserves weights, intersections, preimages and the empty space.

If a functor is continuous, monomorphic, epimorphic, preserves weights, intersections and the empty space, then it is called weakly semi-normal.

17. Lemma. *Let Y be a normal topological space, let also A and B be nonintersecting closed subsets in Y, where T is a well-ordered set supplied with the interval topology. Suppose also that $c_1 < c_2 \in T$ are such that for each $a,b \in T$ with $c_1 \le a < b \le c_2$ an element $d \in T$ exists such that $a < d < b$ (i.e. a segment $[c_1,c_2]$ is without gaps). Then a continuous function $f : Y \to T$ exists such that $f(A) = \{c_1\}$ and $f(B) = \{c_2\}$.*

Proof. Consider the segment $[c_1,c_2]$ in T. There exists a set E dense in $[c_1,c_2]$ such that

(1) $|E| = d([c_1,c_2])$, $\inf E = c_1$, $\sup E = c_2$,

where $d(X)$ denotes the density of a topological space X, $|E|$ denotes the cardinality of E. There exist open subsets U and V in X such that

(2) $A \subset U$, $B \subset V$, $U \cap V = \emptyset$.

We define open subsets V_t in X such that

(3) $cl_X V_t \subset V_s$ for each $t < s \in E$,

(4) $A \subset V_{c_1}$, $B \subset X \setminus V_{c_2}$,

where $cl_X G$ denotes the closure of a set G in X.

Sets V_t will be defined by the transfinite induction. For this one can put $V_{c_1} = U$ and $V_{c_2} = X \setminus B$. Therefore, $A \subset V_{c_1} \subset X \setminus V = cl_X(X \setminus V) \subset V_{c_2}$, consequently, $cl_X V_{c_1} \subset V_{c_2}$. In view of the Zermelo theorem there exists an ordinal P such that $|P| = |E|$, a bijective surjective mapping $\theta : P \to E$ exists such that $\inf P = 0$, $1 \in P$, $\theta(0) = c_1$ and $\theta(1) = c_2$. Suppose that V_{t_j} satisfying Condition (3) are constructed for $j = 1,...,n$, $j \in P$. There exist elements $a_n = \inf\{t_j : j \le n, t_j < t_{n+1}\}$ and $b_n = \sup\{t_j : j \le n, t_j < t_{n+1}\}$. Therefore, $cl_X V_{a_n} \subset V_{b_n}$. From the normality of X it follows that open sets U and V exist such that $cl_X V_{a_n} \subset U$, $X \setminus V_{b_n} \subset V$ and $U \cap V = \emptyset$, consequently, $U \subset X \setminus V \subset V_{b_n}$ and hence $cl_X U \subset cl_X(X \setminus V) = X \setminus V \subset V_{b_n}$. Then one puts $V_{t_{n+1}} = U$. This means that there exists a countable infinite sequence V_{t_j} for $j \in \omega_0 \subseteq P$ satisfying Conditions $(3,4)$. If $\{t_j : j \in \omega_0\}$ is not dense in $[c_1, c_2]$ the process continues. Suppose that α is an ordinal such that $\omega_0 \subseteq \alpha \subset P$, V_{t_j} is defined for each $j \in \alpha$. If the set $\{t_j : j \in \alpha\}$ is not dense in $[c_1, c_2]$, there exists a segment

(5) $[a,b] \subset [c_1, c_2]$ such that $[a,b] \cap \{t_j : j \in \alpha\} = \emptyset$. We put $L = \bigcup_{t_j < a; j \in \alpha} V_{t_j}$ and $M = \bigcap_{b < t_j; j \in \alpha} V_{t_j}$. From $(3,4)$ it follows that the set L is open in X and $L \subset M$. On the other hand,

(6) $V_{t_l} \subset L \subset cl_X L \subset cl_X M \subset cl_X V_{t_j} \subset V_{t_k}$ for every $l, j, k \in \alpha$ such that $t_l < a$ and $b < t_j < t_k$. If

(7) $cl_X L$ is not contained in $Int_X M$ this segment $[a,b]$ is skipped, where $Int_X M$ is an interior of M in X. If $cl_X L \subset Int_X M$ one can put $V_a = L$ and $V_b = Int_X M$. Then the process continues for $[a,b]$.

The family $\mathcal{F} = \{(V_j : j \in \alpha) : \alpha \subset P\}$ is ordered by inclusion: $(V_j : j \in \alpha) \le (W_k : k \in \beta)$ if and only if a bijective monotonously increasing mapping $\theta : \alpha \to \beta$ exists such that $V_j = W_{\theta(j)}$ for each $j \in \alpha$. If a subfamily $\{(V_j : j \in \alpha) : \alpha_k \subset P, k \in \Lambda\}$ is linearly ordered, then its union is in $\mathcal{F}$. In view of the Kuratowski-Zorn lemma there exists a maximal element $(V_j : j \in \alpha_1)$ in $\mathcal{F}$ for some ordinal $\alpha_1 \subset P$ such that conditions $(3,4)$ are satisfied.

Put $f(x) = \inf\{t : x \in V_t\}$ for $x \in V_{c_2}$ and $f(x) = c_2$ when $x \in X \setminus V_{c_2}$. Therefore, $f(x) \in [c_1, c_2]$ for each $x \in X$, $f(A) \subset \{c_1\}$ and $f(B) \subset \{c_2\}$. Since $[c_1, c_2]$ is supplied with the interval topology it is sufficient to prove that $f^{-1}([c_1, a))$ and $f^{-1}((b, c_2])$ are open in $[c_1, c_2]$ for each $c_1 < a \le c_2$ and $c_1 \le b < c_2$. From $(3,4)$, also from $(6,7)$ when (5) is fulfilled, and the definition of f it follows that $f^{-1}([c_1, a)) = \bigcup\{V_{t_j} : t_j < a, j \in \alpha_1\}$ and $f^{-1}((b, c_2]) = \bigcup\{X \setminus cl_X V_{t_j} : b < t_j, j \in \alpha_1\}$ are open in $[c_1, c_2]$.

18. Lemma. *If X is well-ordered and E is a segment $[a,b]$ in X, while K satisfies Condition 2.3(DW), then each $f \in C_+(E,K)$ has a continuous extension $g \in C_+(X,\mathbf{K})$.*

Proof. Since $f(E) =: A$ is linearly ordered in K, then by 2.3(DW) there exists a well ordered subset B in K such that $A \subset B$. So putting $g(x) = \inf A$ for each $x < a$ in X, whilst $g(x) = \sup A$ for each $b < x$ in X one gets the continuous extension $g \in C_+(X,K)$ of f, that is $g|_E(y) = f(y)$ for each $y \in E$, since $\inf A$ and $\sup A$ exist in K due to 2.3(DW) and 2.4(1).

19. Definition. It will be said that a pair (X,K) of a topological space X and a ringoid K has property (CE) if for each closed subset E in X and each continuous function $f : E \to K$, *i.e.*, $f \in C(E,K)$ or $f \in C_+(E,K)$ or $f \in C_-(E,K)$, there exists a continuous extension $g : X \to K$, *i.e.*, $g|_E = f$ so that $g \in C(X,K)$ or $g \in C_+(X,K)$ or $g \in C_-(X,K)$ respectively.

Henceforward, it will be supposed that a pair (X,K) has property (CE).

20. Definitions. If Hausdorff topological spaces X and Y are given and $f : X \to Y$ is a continuous mapping, K_1, K_2 are ordered topological ringoids (or may be particularly semirings) with an order-preserving continuous algebraic homomorphism $u : K_1 \to K_2$ then it induces the mapping $\mathcal{O}(f,u) : \mathcal{O}(X,K_1) \to \mathcal{O}(Y,K_2)$ according to the formula:

(1) $(\mathcal{O}(f,u)(v))(g) = u[v(g_1(f))]$ for each $g_1 \in C(Y,K_1)$ and $v \in \mathcal{O}(X,K_1)$, where $u \circ g_1 = g \in C(Y,K_2)$, $g_1 \in C(Y,K_1)$, $(\mathcal{O}(f,u)(v))$ is defined on $(\hat{f},\hat{u})(C(X,K_1)) = \{t : t \in C(Y,K_2); \forall x \in X\ t(x) = u(h \circ f(x)), h \in C(Y,K_1)\}$.

By $I(f,u)$ will be denoted the restriction of $\mathcal{O}(f,u)$ onto $I(X,K)$. The shorter notations $\mathcal{O}(f)$ and $I(f)$ are used when K is fixed, i.e. $u = id$. When $X = Y$ and $f = id$ we write simply $\mathcal{O}_2(u)$ and $I_2(u)$ respectively omitting $f = id$.

Let $\mathcal{S}$ denote a category such that a family $Ob(\mathcal{S})$ of its objects consists of all topological spaces, a family of morphisms $Mor(X,Y)$ consists of all continuous mappings $f : X \to Y$ for every X, $Y \in Ob(\mathcal{S})$.

Let $\mathcal{K}$ be the category objects of which $Ob(\mathcal{K})$ are all ordered topological ringoids satisfying Conditions 2.3 and 2.4, $Mor(A,B)$ consists of all order-preserving continuous algebraic homomorphisms for each $A, B \in \mathcal{K}$. Then by $\mathcal{K}_w$ we denote its subcategory of well-ordered ringoids and their order-preserving algebraic continuous homomorphisms.

We denote by $\mathcal{OK}$ a category with the families of objects $Ob(\mathcal{OK}) = \{\mathcal{O}(X,K) : X \in Ob(\mathcal{S}), K \in Ob(\mathcal{K}_w)\}$ and morphisms $Mor(\mathcal{O}(X,K_1),\mathcal{O}(Y,K_2))$ for every $X, Y \in Ob(\mathcal{S})$ and $K_1, K_2 \in Ob(\mathcal{K}_w)$. Furthermore, $\mathcal{IK}$ stands for a category with families of objects $Ob(\mathcal{IK}) = \{I(X,K) : X \in Ob(\mathcal{S}), K \in Ob(\mathcal{K}_w)\}$ and morphisms $Mor(I(X,K_1), I(Y,K_2))$ for every $X, Y \in Ob(\mathcal{S})$ and $K_1, K_2 \in Ob(\mathcal{K})$.

By $\mathcal{S}_l$ will be denoted a category objects of which are linearly ordered topological spaces, while $Mor(X,Y)$ consists of all monotone nondecreasing continuous mappings $f : X \to Y$, that is $f(x) \le f(y)$ for each $x \le y \in X$, where $X, Y \in Ob(\mathcal{S}_l)$. Then we put $\mathcal{O}_l(f,u) : \mathcal{O}_l(X,K_1) \to \mathcal{O}_l(Y,K_2)$ for each $X, Y \in Ob(\mathcal{S}_l)$ and $f \in Mor(X,Y)$, $K_1, K_2 \in Ob(\mathcal{K})$, $u \in Mor(K_1,K_2)$ according to the formula:

(2) $(\mathcal{O}_l(f,u)(v))(g) = u[v(g_1(f))]$ for each $g_1 \in C_+(Y,K_1)$ and $u \circ g_1 = g \in C(Y,K_2)$ and $v \in \mathcal{O}_l(X,K_1)$, where $(\mathcal{O}_l(f,u)(v))$ is defined on $(\hat{f},\hat{u})(C_+(X,K_1)) := \{t : t \in C_+(Y,K_2); \forall x \in X\ t(x) = u(h \circ f(x)), h \in C_+(Y,K_1)\}$. Then the category $\mathcal{O}_l\mathcal{K}$ with families of objects $Ob(\mathcal{O}_l\mathcal{K}) = \{\mathcal{O}_l(X,K) : X \in Ob(\mathcal{S}_l), K \in Ob(\mathcal{K})\}$ and morphisms $Mor(\mathcal{O}_l(X,K_1),\mathcal{O}_l(Y,K_2))$ and the category $\mathcal{I}_l\mathcal{K}$ with $Ob(\mathcal{I}_l\mathcal{K}) = \{I_l(X,K) : X \in Ob(\mathcal{S}_l), K \in Ob(\mathcal{K})\}$ and $Mor(I_l(X,K_1), I_l(Y,K_2))$ are defined.

Subcategories of left homogeneous continuous morphisms we denote by $\mathcal{O}_h\mathcal{K}, \mathcal{O}_{l,h}\mathcal{K}, \mathcal{I}_h\mathcal{K}, \mathcal{I}_{l,h}\mathcal{K}$ correspondingly. These morphisms are taken on subcategories $\mathcal{K}_{w,l}$ in $\mathcal{K}$ or $\mathcal{K}_l$ in $\mathcal{K}$ of left distributive topological ringoids.

21. Lemma. *There exist covariant functors $\mathcal{O}$, $\mathcal{O}_h$ and $\mathcal{O}_l$, $\mathcal{O}_{l,h}$ in the categories $\mathcal{S}$ and $\mathcal{S}_l$ respectively.*

Proof. Suppose that $X, Y \in Ob(\mathcal{S})$ and $f \in Mor(X,Y)$, while $g \le h$ in $C(Y,K)$, where $K \in Ob(\mathcal{K}_w)$ (or in $\mathcal{K}_{w,l}$) is marked, then $g \circ f \le h \circ f$ in $C(X,Y)$. Therefore one gets $(\mathcal{O}(f)(v))(g) = v(g \circ f) \le v(h \circ f) = (\mathcal{O}(f)(v))(h)$ for each $v \in \mathcal{O}(X,K)$. Now if $c \in K$, $g^c \in C(Y,K)$, then $g^c \circ f \in C(X,K)$, but also the equalities are fulfilled $(\mathcal{O}(f)(v))(g^c + h) = v(g^c \circ f + h \circ f) = c + v(h \circ f) = c + (\mathcal{O}(f)(v))(h)$ and $(\mathcal{O}(f)(v))(h + g^c) = v(h \circ f + g^c \circ f) = v(h \circ f) + c = (\mathcal{O}(f)(v))(h) + c$ for each $h \in C(Y,K)$. Then for $1_X \in Mor(X,X)$, that is $1_X(x) = x$ for each $x \in X$, one deduces $1_X \circ q = q$ for each $q \in Mor(Y,X)$ and $t \circ 1_X = t$ for each $t \in Mor(X,Y)$. On the other hand, $(\mathcal{O}(1_X)(v))(g) = v(g \circ 1_X) = v(g)$ for each $g \in C(X,K)$, i.e. $\mathcal{O}(1_X) = 1_{\mathcal{O}(X)}$. But at the same time, the equalities are valid: $(\mathcal{O}(f \circ s)(v))(g) = v(g \circ f \circ s) = (\mathcal{O}(s)(v)(g \circ f) = ((\mathcal{O}(f) \circ \mathcal{O}(s))(v))(g)$, since the composition of continuous mappings is continuous.

Moreover, if $v \in \mathcal{O}_h(X,K)$, then $(\mathcal{O}(f)(v))(bg) = v(bg \circ f) = bv(g \circ f) = (b(\mathcal{O}(f)(v))(g)$. Furthermore, for the categories $\mathcal{O}_l$ (or $\mathcal{O}_{l,h}$) the proof is analogous with $X, Y \in Ob(\mathcal{S}_l)$, $C_+(X,K)$ and $C_+(Y,K)$, where $K \in Ob(\mathcal{K})$ (or $K \in Ob(\mathcal{K}_l)$) is marked.

22. Proposition. *Suppose that $f \in Mor(X,Y)$ for $X, Y \in Ob(\mathcal{S})$ or in $Ob(\mathcal{S}_l)$. Then*

$\mathcal{O}(f)(I(X,K)) \subseteq I(Y,K)$ and $\mathcal{O}_h(f)(I_h(X,K)) \subseteq I_h(Y,K)$ for $K \in Ob(\mathcal{K}_{w,l})$ or $\mathcal{O}_l(f)(I_h(X,K)) \subseteq I_l(Y,K)$ or $\mathcal{O}_{l,h}(f)(I_{l,h}(X,K)) \subseteq I_{l,h}(Y,K)$ for $K \in Ob(\mathcal{K})$ or $K \in Ob(\mathcal{K}_l)$ correspondingly.

Proof. If $g, h \in C(Y,K)$ are such that $g \vee h$ or $g \wedge h$ exists (see Condition (3) in Lemma 2.12) and $f : X \to Y$ is a continuous mapping, $v \in I(X,K)$ (or $I_l(X,K)$), then we infer that

$(\mathcal{O}(f)(v))(g \vee h) = v(g \circ f \vee h \circ f) = v(g \circ f) \vee v(h \circ f) = (\mathcal{O}(f)(v))(g) \vee (\mathcal{O}(f)(v))(h)$ or

$(\mathcal{O}(f)(v))(g \wedge h) = v(g \circ f \wedge h \circ f) = v(g \circ f) \wedge v(h \circ f) = (\mathcal{O}(f)(v))(g) \wedge (\mathcal{O}(f)(v))(h)$.

Furthermore, for each $c \in K$ we deduce that

$(\mathcal{O}(f)(v))(g^c \times_2 h) = v(g^c \circ f \times_2 h \circ f) = v(g^c \circ f) \times_2 v(h \circ f) = c \times_2 (\mathcal{O}(f)(v))(h)$ and

$(\mathcal{O}(f)(v))(h \times_2 g^c) = v(h \circ f \times_2 g^c \circ f) = v(h \circ f) \times_2 v(g^c \circ f) = (\mathcal{O}(f)(v))(h) \times_2 c$.

Then for $\nu \in I_h(X,K)$ (or $I_{l,h}(X,K)$) one gets $(\mathcal{O}(f)(\nu))(bg) = \nu(bg \circ f) = b\nu(g \circ f) = (b(\mathcal{O}(f)(\nu)))(g)$.

23. Definitions. A covariant functor $F : \mathcal{S} \to \mathcal{S}$ will be called epimorphic (monomorphic) if it preserves continuous epimorphisms (monomorphisms). If $\phi : A \hookrightarrow X$ is a continuous embedding, then $F(A)$ will be identified with $F(\phi)(F(A))$.

If for each $f \in Mor(X,Y)$ and each closed subset A in Y, the equality $(F(f)^{-1})(F(A)) = F(f^{-1}(A))$ is satisfied, then a covariant functor F is called preimage-preserving. In the case $F(\bigcap_{j\in J} X_j) = \bigcap_{j\in J} F(X_j)$ for each family $\{X_j : j \in J\}$ of closed subsets in $X \in Ob(\mathcal{S})$ (or in $Ob(\mathcal{S}_l)$), the monomorphic functor F is called intersection-preserving.

If a functor F preserves inverse mapping system limits, it is called continuous.

A functor is said to be semi-normal when it is monomorphic, epimorphic, also preserves intersections, preimages and the empty space.

If a functor is monomorphic, epimorphic, also preserves intersections and the empty space, then it is called weakly semi-normal.

24. Proposition. *The functor $\mathcal{O}$ (or $\mathcal{O}_h$, $\mathcal{O}_l$, $\mathcal{O}_{l,h}$) is monomorphic.*

Proof. Let $X, Y \in Ob(\mathcal{S})$ (or in $Ob(\mathcal{S}_l)$ respectively) with a continuous embedding $s : X \hookrightarrow Y$ (order-preserving respectively). Then we suppose that $\nu_1 \neq \nu_2 \in \mathcal{O}(X,K)$ (or in $\mathcal{O}_h(X,K)$, $\mathcal{O}_l(X,K)$, $\mathcal{O}_{l,h}(X,K)$ correspondingly). This means that a mapping $g \in C(X,K)$ (or in $C_+(X,K)$ correspondingly) exists such that $\nu_1(g) \neq \nu_2(g)$. A function $u \in C(Y,K)$ (or in $C_+(Y,K)$ respectively) exists such that $u \circ s = g$, hence $(\mathcal{O}(s)(\nu_k)))(u) = \nu_k(u \circ s) = \nu_k(g)$. Thus $\mathcal{O}(s)(\nu_1) \neq \mathcal{O}(s)(\nu_2)$ (or $\mathcal{O}_h(\nu_1) \neq \mathcal{O}_h(\nu_2)$, $\mathcal{O}_l(\nu_1) \neq \mathcal{O}_l(\nu_2)$, $\mathcal{O}_{l,h}(\nu_1) \neq \mathcal{O}_{l,h}(\nu_2)$ correspondingly).

25. Corollary. *The functors I, I_h, I_l and $I_{l,h}$ are monomorphic.*

Proof. This follows from Proposition 24 and Definitions 20.

26. Proposition. *The functors $\mathcal{O}$, $\mathcal{O}_h$, $\mathcal{O}_l$ and $\mathcal{O}_{l,h}$ are epimorphic, when $X \in H_X$ (see §14 also).*

Proof. Let $f : X \to Y$ be a continuous surjective mapping, $\nu \in \mathcal{O}(Y,K)$ (or in $\mathcal{O}_h(Y,K)$, $\mathcal{O}_l(Y,K)$, $\mathcal{O}_{l,h}(Y,K)$ respectively). The set L of all continuous mappings $g \circ f : X \to K$ with $g \in C(Y,K)$ (or in $C_+(Y,K)$ correspondingly) is the A-subset according to Definitions 6 or the left module over K in $C(X,K)$ (or in $C_+(X,K)$). Then we put $\mu(g \circ f) = \nu(g)$. This continuous morphism has an extension from L to a continuous morphism $\mu \in \mathcal{O}(X,K)$ (or in $\mathcal{O}_h(X,K)$, $\mathcal{O}_l(X,K)$, $\mathcal{O}_{l,h}(X,K)$ correspondingly) due to Lemmas 9, 14 and Corollary 8.

27. Lemma. *Let L be a submodule over K of $C(X,K)$ or $C_+(X,K)$ relative to the operations $\vee$, $\wedge$, $\times_2$ and containing all constant mappings $g^c : X \to K$, where $c \in K$. Let also $\nu : L \to K$ be an idempotent (left homogeneous) continuous morphism. For each $f \in C(X,K) \setminus L$ or $C_+(X,K) \setminus L$ there exists an idempotent (left homogeneous) continuous extension μ_M of ν on a minimal closed submodule M containing L and f.*

Proof. For each $g \in M$ we put

(1) $\mu_M(g) = \nu(g) = \inf\{\nu(h) : g \leq h, h \in L\}$.

This implies that $\nu(g_1) \leq \nu(g_2)$ for each $g_1 \leq g_2 \in M$. Then

$\nu(g^c \times_2 g) = \inf\{\nu(h) : h \in L, g^c \times_2 g \leq h\} =$

$\inf\{\nu(g^c \times_2 q) : q \in L, g^c \times_2 g \leq g^c \times_2 q\} = c \times_2 \inf\{\nu(q) : q \in L, g \leq q\} = c \times_2 \nu(g)$ and

$\nu(g \times_2 g^c) = \inf\{\nu(h) : h \in L, g \times_2 g^c \leq h\} = \inf\{\nu(q \times_2 g^c) : q \in L, q \times_2 g^c \geq g \times_2 g^c\}$

$= \inf\{\nu(q) : q \in L, q \geq g\} \times_2 c = \nu(g) \times_2 c$.

On the other hand for each $g_1, g_2 \in M$ one gets

$\nu(g_1) \vee \nu(g_2) = \inf\{\nu(g) : g \in L, g_1 \leq g\} \vee \inf\{\nu(q) : q \in L, g_2 \leq q\}$

$= \inf\{\nu(g) \vee \nu(q) : g, q \in L, g_1 \leq g, g_2 \leq q\} \geq \inf\{\nu(g \vee q) : g, q \in L, g_1 \vee g_2 \leq g \vee q\} = \nu(g_1 \vee g_2)$.

From the inequalities $g_k \leq g_1 \vee g_2$ for $k = 1$ and $k = 2$ it follows, that $\nu(g_k) \leq \nu(g_1 \vee g_2)$, consequently, $\nu(g_1) \vee \nu(g_2) = \nu(g_1 \vee g_2)$. Then

$\nu(g_1) \wedge \nu(g_2) = \inf\{\nu(g) : g \in L, g_1 \leq g\} \wedge \inf\{\nu(q) : q \in L, g_2 \leq q\}$

$= \inf\{\nu(g) \wedge \nu(q) : g, q \in L, g_1 \leq g, g_2 \leq q\} \leq \inf\{\nu(g \wedge q) : g, q \in L, g_1 \wedge g_2 \leq g \wedge q\} = \nu(g_1 \wedge g_2)$.

But $\nu(g_k) \geq \nu(g_1 \wedge g_2)$, since $g_k \geq g_1 \wedge g_2$ for $k = 1$ and $k = 2$, consequently, $\nu(g_1) \wedge \nu(g_2) = \nu(g_1 \wedge g_2)$. If ν is left homogeneous, then $\inf\{\nu(bh) : bh \geq bg, h \in L\} = \inf\{\nu(bh) : h \geq g, h \in$

$L\} = b\inf\{\nu(h) : h \geq g, h \in L\}$ for each $b \in K$, consequently, ν is left homogeneous on M. If ν is continuous and g_k is a net in M converging to $g \in M$ (see §2.9), then $\nu(g) = \inf\{\nu(h) : g \leq h, h \in L\} = \lim_k \inf\{\nu(h) : g_k \leq h, h \in L\} = \lim_k \nu(g_k)$.

28. Lemma. *If suppositions of Lemma 27 are satisfied, then there exists an idempotent (left homogeneous) continuous morphism λ on $C(X,K)$ or $C_+(X,K)$ respectively such that $\lambda|_L = \nu$.*

Proof. The family of all extensions (M, μ_M) of ν on closed submodules M of $C(X,K)$ or $C_+(X,K)$ respectively is partially ordered by inclusion: $(M, \mu_M) \leq (N, \mu_N)$ if and only if $M \subset N$ and $\nu_N|_M = \nu_M$. In view of the Kuratowski-Zorn lemma [20] there exists the maximal closed submodule P in $C(X,K)$ or $C_+(X,K)$ correspondingly and an idempotent extension ν_P of ν on P. If $P \neq C(X,K)$ or $C_+(X,K)$ correspondingly by Lemma 27 this morphism ν_P could be extended on a module L containing P and some $g \in C(X,K) \setminus P$ or in $C(X,K)_+ \setminus P$ respectively. This contradicts the maximality of (P, ν_P). Thus $P = C(X,K)$ or $C_+(X,K)$ correspondingly.

29. Proposition. *The functors I, I_l and I_h, $I_{l,h}$ are epimorphic.*

Proof. Let a continuous mapping $f : X \to Y$ be epimorphic. We consider the set L of all continuous mappings $g \circ f : X \to K$ such that $g \in C(Y,K)$ or $C_+(Y,K)$. Then L is a submodule of $C(X,K)$ or $C_+(X,K)$ relative to the operations $\vee$, $\wedge$, $\times_2$ and L contains all constant mappings $g^c : X \to K$, where $c \in K$. Then we put $\mu(g \circ f) = \nu(g)$ for $\nu \in I(X,K)$ or in $I_l(X,K)$, $I_h(X,K)$ or $I_{l,h}(X,K)$. In view of Lemma 28 there is a continuous extension of μ from L onto $C(Y,K)$ or $C_+(Y,K)$ such that $\mu \in I(Y,K)$ or in $I_l(Y,K)$, $I_h(Y,K)$ or $I_{l,h}(Y,K)$ correspondingly.

30. Definition. It is said that $\nu \in \mathcal{O}(X,K)$ (or $\nu \in \mathcal{O}_l(X,K)$) is supported on a closed subset E in X, if $\nu(f) = 0$ for each $f \in C(X,K)$ or in $C_+(X,K)$ such that $f|_E \equiv 0$. A support of ν is the intersection of all closed subsets in X on which ν is supported.

31. Proposition. *Let $\nu \in \mathcal{O}(X,K)$ or in $\mathcal{O}_l(X,K)$. Then ν is supported on $E \subset X$ if and only if $\nu(f) = \nu(g)$ for each $f, g \in C(X,K)$ or in $C_+(X,K)$ correspondingly such that $f|_E \equiv g|_E$. Moreover, E is a support of ν if and only if ν is supported on E and for each proper closed subset F in E, i.e. $F \subset E$ with $F \neq E$, there are $f, h \in C(X,K)$ or in $C_+(X,K)$ respectively with $f|_F \equiv h|_F$ such that $\nu(f) \neq \nu(h)$.*

Proof. Consider $\nu \in \mathcal{O}(X,K)$ such that $\nu(f) = \nu(g)$ for each functions $f, g : X \to K$ with $f|_E = g|_E$. A continuous morphism ν induces a continuous morphism $\lambda \in \mathcal{O}(E,K)$ such that $\lambda(h) = \nu(h)$ for each $h \in C(X,K)$ with $h|_{X \setminus E} = 0$. Denote by id the identity embedding of a closed subset E into X. Each function $t : E \to K$ has an extension on X with values in K by Condition 19(CE). Then $\mathcal{O}(id)(\lambda) = \nu$, since $\nu(g^0) = 0$ and hence $\nu(s) = 0$ for each $s \in C(X,K)$ such that $s|_E \equiv 0$.

If $\nu \in \mathcal{O}(X,K)$ and ν is supported on E, then by Definition 30 there exists a morphism $\lambda \in \mathcal{O}(E,K)$ such that $\mathcal{O}(id)(\lambda) = \nu$. Therefore the equalities are valid: $\nu(f) = \lambda(f|_E) = \lambda(g|_E) = \nu(g)$ for each functions $f, g \in C(X,K)$ such that $f|_E = g|_E$.

If E is a support of ν, then by the definition this implies that ν is supported on E. Suppose that $F \subset E$, $F \neq E$ and for each $f, g \in C(X,K)$ with $f|_F \equiv g|_F$ the equality $\nu(f) = \nu(g)$ is satisfied, then a support of ν is contained in F, hence E is not a support of ν. This is the contradiction, hence there are $f, g \in C(X,K)$ with $f|_F \equiv g|_F$ such that $\nu(f) \neq \nu(g)$.

If ν is supported on E and for each proper closed subset F in E there are $f, h \in C(X,K)$ with $f|_F \equiv h|_F$ such that $\nu(f) \neq \nu(h)$, then ν is not supported on any such proper closed subset F, consequently, each closed subset G in X on which ν is supported contains E, i.e. $E \subset G$. Thus E is the support of ν.

32. Proposition. *The functors $\mathcal{O}$, I, $\mathcal{O}_l$, I_l, $\mathcal{O}_{l,h}$, $I_{l,h}$ preserve intersections of closed subsets.*

Proof. If E is a closed subset in X, then there is the natural embedding $C(E,K) \hookrightarrow C(X,K)$ (or $C_+(E,K) \hookrightarrow C_+(X,K)$, when $X \in Ob(\mathcal{S}_l)$) due to Condition 19(CE). Therefore, $\mathcal{O}(E \cap F, K) \subset \mathcal{O}(E,K) \cap \mathcal{O}(F,K)$ (or $\mathcal{O}_l(E \cap F, K) \subset \mathcal{O}_l(E,K) \cap \mathcal{O}_l(F,K)$ respectively). For any closed subsets E and F in X and each functions $f, g \in C(X,K)$ (or $C_+(X,K)$) with $f|_{E \cap F} \equiv g|_{E \cap F}$ there exists a function $h \in C(X,K)$ (or $C_+(X,K)$) such that $h|_E = f$ and $h|_F = g$ due to 19(CE). Therefore $\nu(f) = \nu(h)$ and $\nu(g) = \nu(h)$ for each $\nu \in \mathcal{O}(E,K) \cap \mathcal{O}(F,K)$ (or in $\mathcal{O}_l(E,K) \cap \mathcal{O}_l(F,K)$). In view of Proposition 31 the functors $\mathcal{O}$ and $\mathcal{O}_l$ preserve intersections of closed subsets. This implies that the functors I, I_l, $\mathcal{O}_{l,h}$ and $I_{l,h}$ also have this property.

33. Proposition. *Let $\{X_b; p^b_a; V\} =: P$ be an inverse system of topological spaces X_b, where V is a directed set, $p^b_a : X_b \to X_a$ is a continuous mapping for each $a \le b \in V$, $p_b : X = \lim P \to X_b$ is a continuous projection. Then the mappings*

(1) $s = (\mathcal{O}(p_b) : b \in V) : \mathcal{O}(X,K) \to \mathcal{O}(P,K)$ and $s_h = (\mathcal{O}_h(p_b) : b \in V) : \mathcal{O}_h(X,K) \to \mathcal{O}_h(P,K)$

(2) $t = (I(p_b) : b \in V) : I(X,K) \to I(P,K)$ and $t_h = (I_h(p_b) : b \in V) : I_h(X,K) \to I_h(P,K)$

are bijective and surjective continuous algebraic homomorphisms. Moreover, if $X_b \in Ob(\mathcal{S}_l)$ and p^b_a is order-preserving for each $a < b \in V$, then the mappings

(3) $s_l = (\mathcal{O}_l(p_b) : b \in V) : \mathcal{O}_l(X,K) \to \mathcal{O}_l(P,K)$ and $s_{l,h} = (\mathcal{O}_{l,h}(p_b) : b \in V) : \mathcal{O}_{l,h}(X,K) \to \mathcal{O}_{l,h}(P,K)$

(4) $t_l = (I_l(p_b) : b \in V) : I_l(X,K) \to I_l(P,K)$ and $t_{l,h} = (I_{l,h}(p_b) : b \in V) : I_{l,h}(X,K) \to I_{l,h}(P,K)$

also are bijective and surjective continuous algebraic homomorphisms.

Proof. We consider the inverse system $\mathcal{O}(P) = (\mathcal{O}(X_a); \mathcal{O}(p^a_b); V\}$ and its limit space $Y = \lim \mathcal{O}(P)$. Then $\mathcal{O}(p^b_a)\mathcal{O}(p_b) = \mathcal{O}(p_a)$ for each $a \le b \in V$, since $p^b_a \circ p_b = p_a$. Let $q : \mathcal{O}(X,K) \to Y$ denote the limit map of the inverse mapping system $q = \lim\{\mathcal{O}(p_a); \mathcal{O}(p^a_b); V\}$ (see also §2.5 [21]).

A continuous morphism ν is in $\mathcal{O}(X,K)$ if and only if $\mathcal{O}(p_a)(\nu) \in \mathcal{O}(X_a, K)$ for each $a \in V$, since

(5) $f \in C(X,K)$ if and only if $f = \lim\{f_b; p^b_a; V\}$ and

(6) $\mathcal{O}(p_a)(\nu)(f_a) = \nu(f_a \circ p_a) = \nu_a(f_a)$, where $\nu_a \in \mathcal{O}(X_a, K)$, $f_b \in C(X_b, K)$, $f_b = f_a \circ p^b_a$ for each $a \le b \in V$, $p^b_b = id$, $f(x) = \{f_a \circ p_a(x) :\ a \in V\} \in \theta(K)$ for each $x = \{x_a : a \in V\} \in X$, where $\{x_a : a \in V\}$ is a thread of P such that $x_a \in X_a$, $p^b_a(x_b) = x_a$ for each $a \le b \in V$, $\theta : K \to K^X$ is an order-preserving continuous algebraic embedding, $\theta(K)$ is isomorphic with K.

If $\nu, \lambda \in \mathcal{O}(X,K)$ are two different continuous morphisms, then this means that a continuous function $f \in C(X,K)$ exists such that $\nu_1(f) \ne \nu_2(f)$. This is equivalent to the following: there exists $a \in V$ such that $(\mathcal{O}(p_a)(\nu))(f) \ne (\mathcal{O}(p_a)(\lambda))(f)$. Thus the mappings s and analogously t are surjective and bijective.

On the other hand,

(7) $\nu_b(f_b \vee g_b) = \nu_b(f_b) \vee \nu_b(g_b)$ and

(8) $\nu_b(f_b \wedge g_b) = \nu_b(f_b) \wedge \nu_b(g_b)$ for each $b \in V$ and each $\nu_b \in I(X_b, K)$ and every $f_b, g_b \in C(X_b, K)$ such that either $f_b(x) < g_b(x)$ or $f_b(x) = g_b(x)$ or $g_b(x) < f_b(x)$ for each $x \in X_b$, also

(9) $\nu_b(g^c \times_2 f_b) = c \times_2 \nu_b(f_b)$ and

(10) $\nu_b(f_b \times_2 g^c) = \nu_b(f_b) \times_2 c$ for each $c \in K$ and $f_b \in C(X_b, K)$. Taking the inverse limit in Equalities $(5-10)$ gives the corresponding equalities for $\nu \in I(X,K)$, where $\nu = \lim\{\nu_a; I(p^b_a); V\}$, hence t is the continuous algebraic homomorphism due to Theorem 2.5.8 [21].

Analogously s preserves Properties $(9,10)$, that is $\lambda = \lim\{\lambda_a; \mathcal{O}(p^b_a); V\}$ is weakly additive, where $\lambda_b \in \mathcal{O}(X_b, K)$ for each $b \in V$. Suppose that $f \le g \in C(X,K)$, then $f_b \le g_b$ for each $b \in V$ due to (5). From $\lambda_b(f_b) \le \lambda_b(g_b)$ for each $b \in V$, the inverse limit decomposition $\lambda = \lim\{\lambda_b; \mathcal{O}(p^b_a); V\}$ and Formula (6) it follows that λ is order-preserving.

If $X_b \in Ob(\mathcal{S}_l)$ for each $b \in V$, then a topological space X is linearly ordered: $x = \{x_b : b \in V\} \le y = \{y_b : b \in V\}$ if and only if $x_b \le y_b$ for each $b \in V$, where $x, y \in X$ are threads of the inverse system P such that $p^b_a(x_b) = x_a$ for each $a \le b \in V$. Since p^b_a is order-preserving for each $a \le b \in V$ and each f_b is non-decreasing, then f is nondecreasing and hence $f \in C_+(X,K)$ for each $f = \lim\{f_b; p^b_a; V\}$, where $f_b \in C_+(X_b, K)$ and $f_b = f_a \circ p^b_a$ for each $a \le b \in V$ and $x \in X$, $f(x) = \{f_a \circ p_a(x) : a \in V\}$.

Moreover, $\nu \in \mathcal{O}_h(X,K)$ is left homogeneous if and only if $\theta(p_a)(\nu)$ is left homogeneous for each $b \in V$, since $(\mathcal{O}_h(p_a)(\nu))(f_a) = \nu(f_a \circ p_a) = \nu_a(f_a)$. Applying Lemma 2.5.9 [21] one gets properties of mappings in Formulas $(3,4)$.

34. Lemma. *There exist covariant functors $\mathcal{O}_2$, I_2, and $\mathcal{O}_{l,2}$, $I_{l,2}$ and $\mathcal{O}_{h,2}$, $I_{h,2}$ and $\mathcal{O}_{l,h,2}$, $I_{l,h,2}$ in the categories $\mathcal{K}_w$ and $\mathcal{K}$ and $\mathcal{K}_{w,l}$ and $\mathcal{K}_l$ respectively.*

Proof. If $K_1, K_2, K_3 \in Ob(\mathcal{K}_w)$, $u \in Mor(K_1, K_2)$, $v \in Mor(K_2, K_3)$, $\nu \in I(X, K_1)$, then $(I_2(vu)(\nu))(f) = v \circ u \circ \nu(f_1) = [I_2(v)(I_2(u)(\nu))](f)$ for each $f_1 \in C(X, K_1)$ such that $f(x) = v \circ u \circ f_1(x)$ for each $x \in X$, where $X \in Ob(\mathcal{S})$. That is $I_2(vu) = I_2(v)I_2(u)$. On the other hand, the equality $I_2(id) = 1$ is fulfilled.

If $f(x) \le g(x)$, then $u(f(x)) \le u(g(x))$, where $x \in X$, $f, g \in C(X, K_1)$. Therefore, if a mapping either $f \vee g$ or $f \wedge g$ exists in $C(X, K_1)$, then $u(f \vee g) = u(f) \vee u(g)$ or $u(f \wedge g) = u(f) \wedge u(g)$ in $C(X, K_2)$ respectively. If $f, g \in C(X, K_1)$, then $u(f(x) + g(x)) = u(f(x)) + u(g(x))$ for each $x \in X$, particularly, this is valid for $f = g^c$ or $g = g^c$, where $c \in K_1$. Therefore, $u(g^c \times_2 g) = g^{u(c)} \times_2 u(g)$ and $u(g \times_2 g^c) = u(g) \times_2 g^{u(c)}$. To each $\nu_n \in \mathcal{O}(X, K_n)$ and $u \in Mor(K_n, K_{n+1})$ there corresponds a morphism $u \circ \nu_n$ on $(\hat{id}, \hat{u})(C(X, K_n))$, $(\hat{id}, \hat{u})(C(X, K_n)) \hookrightarrow C(X, K_{n+1})$ (see §20). If $u : K_n \to K_{n+1}$ is not an epimorphism, the image $(\hat{id}, \hat{u})(C(X, K_n))$ is a proper submodule over $u(K_n)$ in $C(X, K_{n+1})$.

If $K_n, K_{n+1} \in Ob(\mathcal{K})$ and $X \in Ob(\mathcal{S}_l)$, $u \in Mor(K_n, K_{n+1})$, then $\hat{u} : C_+(X, K_n) \to C_+(X, K_{n+1})$ is a continuous homomorphism. If $K_n, K_{n+1} \in Ob(\mathcal{K}_l)$ and $X \in Ob(\mathcal{S}_l)$ (or $K_n, K_{n+1} \in Ob(\mathcal{K}_{w,l})$ and $X \in Ob(\mathcal{S})$) and $\nu \in \mathcal{O}_h(X, K_n)$ or in $I_h(X, K_n)$, $u \in Mor(K_n, K_{n+1})$, then $u \circ \nu \in \mathcal{O}_h(X, K_{n+1})$ or in $I_h(X, K_{n+1})$ respectively.

This and the definitions above imply that $\mathcal{O}_2(u) : \mathcal{O}(X, K_1) \to \mathcal{O}(X, K_2)$, $I_2(u) : I(X, K_1) \to I(X, K_2)$ and $\mathcal{O}_{l,2}(u)$, $I_{l,2}(u)$ and $\mathcal{O}_{h,2}(u)$, $I_{h,2}(u)$ and $\mathcal{O}_{l,h,2}(u)$, $I_{l,h,2}(u)$ are the homomorphisms. Thus we deduce that $\mathcal{O}_2 : \mathcal{K}_w \to \mathcal{OK}$ and $\mathcal{O}_{l,2} : \mathcal{K} \to \mathcal{O}_l\mathcal{K}$, $I_2 : \mathcal{K}_w \to \mathcal{IK}$ and $I_{l,2} : \mathcal{K} \to \mathcal{I}_l\mathcal{K}$, $\mathcal{O}_{h,2} : \mathcal{K}_{w,l} \to \mathcal{O}_h\mathcal{K}$, $I_{h,2} : \mathcal{K}_{w,l} \to \mathcal{I}_h\mathcal{K}$, $\mathcal{O}_{l,h,2} : \mathcal{K}_l \to \mathcal{O}_{l,h}\mathcal{K}$ and $\mathcal{I}_{l,h,2} : \mathcal{K}_l \to \mathcal{I}_{l,h}\mathcal{K}$ are the covariant functors on the categories $\mathcal{K}_w$, $\mathcal{K}$, $\mathcal{K}_{w,l}$ and $\mathcal{K}_l$ correspondingly with values in the categories of skew idempotent continuous morphisms, when a set $X \in Ob(\mathcal{S})$ or in $Ob(\mathcal{S}_l)$ correspondingly is marked.

35. Proposition. *The bi-functors I on $\mathcal{S} \times \mathcal{K}_w$, I_l on $\mathcal{S}_l \times \mathcal{K}$, I_h on $\mathcal{S} \times \mathcal{K}_{w,l}$ and $I_{l,h}$ on $\mathcal{S}_l \times \mathcal{K}_l$ preserve pre-images.*

Proof. In view of Proposition 24 and Lemma 34 I, I_l, I_h and $I_{l,h}$ are the covariant bi-functors, *i.e.*, the functors in $\mathcal{S}$ or $\mathcal{S}_l$ and the functors in $\mathcal{K}_w$ or $\mathcal{K}$ or $\mathcal{K}_{w,l}$ or $\mathcal{K}_l$ correspondingly as well. For any functor F the inclusion $F(f^{-1}(B)) \subset (F(f))^{-1}(F(B))$ is satisfied, where, for example, B is closed in $Y \in Ob(\mathcal{S})$.

Suppose the contrary that I does not preserve pre-images. This means that there exist $X, Y \in Ob(\mathcal{S})$ and $K_1, K_2 \in Ob(\mathcal{K}_w)$ or $X, Y \in Ob(\mathcal{S}_l)$ and $K_1, K_2 \in Ob(\mathcal{K})$, $f \in Mor(X, Y)$, $u \in Mor(K_1, K_2)$, $A \subset X$ and $B \subset Y$, where B is closed and hence A is closed when $A = F^{-1}(B)$, $\nu \in I(X, K_1)$ such that $I(f, u)(\nu) \in I(B, K_2)$ but $\nu \notin I(f^{-1}(B), u^{-1}(K_2))$ (or $\nu \in I_l(X, K_1)$, $I_l(f, u)(\nu) \in I_l(B, K_2)$ and $\nu \notin I_l(f^{-1}(B), u^{-1}(K_2))$ respectively). One can choose two functions $g, h \in C(X, K_1)$ such that

(1) $g|_A = h|_A$,
(2) $0 < c_1 = u[\inf_{x \in X} g(x)]$, $0 < c_2 = u[\inf_{x \in X} h(x)]$ and
(3) $u[\nu(g)] \ne u[\nu(h)]$.

There exist functions $s, t \in C(X, K_1)$ such that

(4) $s|_A = g|_A$ and $t|_A = h|_A$, while
(5) $s|_{X \setminus A} = t|_{X \setminus A}$ and
(6) $s(x) \le g(x)$ and $s(x) \le h(x)$ for each $x \in X \setminus A$, where g, h satisfy Conditions $(1-3)$ due to property 19(CE). There are also functions $q, r \in C(X, K_1)$ such that
(7) $q|_{X \setminus A} = g|_{X \setminus A}$ and $r|_{X \setminus A} = h|_{X \setminus A}$ with
(8) $q(x) = r(x)$ and $q(x) \le c$ for each $x \in A$, where
(9) $c \in K_1$, $c < \inf_{x \in X} g(x)$, $c < \inf_{x \in X} h(x)$ such that $u(c) < c_1$ and $u(c) < c_2$.
Evidently, $c_1 \le u[\nu(g)]$ and $c_2 \le u[\nu(h)]$. Then
(10) $\nu(g) = \nu(s \vee q) = \nu(s) \vee \nu(q)$ and
(11) $\nu(h) = \nu(t \vee r) = \nu(t) \vee \nu(r)$ and $u[\nu(q)] \ne u[\nu(r)]$.

On the other hand, there are functions $q_1, r_1 \in C(Y, K_2)$, $q_2, r_2 \in C(Y, K_1)$ such that $q_2 \circ f = q$, $r_2 \circ f = r$, $u \circ q_2 = q_1$, $u \circ r_2 = r_1$ and $q_2|_B = r_2|_B$. Therefore, from Properties $(7-10)$ it follows that

(12) $(I(f,u)(v))(q_1) = u[v(q)] \leq u(c)$ and $(I(f,u)(v))(r_1) = u[v(r)] \leq u(c)$. The condition $s = t$ on A and on $X \setminus A$ imply that

(13) $v(s) = v(t)$. Therefore,

(14) $u(v(g)) = u(v(s)) \vee u(v(q))$ and $u(v(h)) = u(v(t)) \vee u(v(r))$, which follows from $(10, 11)$. But Formulas $(4-6, 12-14)$ contradict the inequality $u[v(g)] \neq u[v(h)]$, since u is the order-preserving continuous algebraic homomorphism from K_1 into K_2. Thus the bi-functors I and I_l preserve pre-images. The proof in other cases is analogous.

36. Corollary. *If $v \in I(X,K)$ or $v \in I_l(X,K)$, $f \in Mor(X,Y)$, $u \in Mor(K_1,K_2)$, where $X,Y \in Ob(\mathcal{S})$ and $K_1,K_2 \in Ob(\mathcal{K}_w)$ or $X,Y \in Ob(\mathcal{S}_l)$ and $K_1,K_2 \in Ob(\mathcal{K})$, then $supp(I(f,u)(v)) = f(supp(u[v]))$ or*

$supp(I_l(f,u)(v)) = f(supp(u[v]))$ correspondingly.

37. Definitions. Suppose that Q is a category and F,G are two functors in Q. Suppose also that a transformation $p : F \to G$ is defined for each $X \in Q$, that is a continuous mapping $p_X : F(X) \to G(X)$ is given. If $p_Y \circ F(f) = G(f) \circ p_X$ for each mapping $f \in Mor(X,Y)$ and every objects $X,Y \in Ob(Q)$, then the transformation $p = \{p_X : X \in Ob(Q)\}$ is called natural.

If $T : Q \to Q$ is an endofunctor in a category Q and there are natural transformations the identity $\eta : 1_Q \to T$ and the multiplication $\psi : T^2 \to T$ satisfying the relations $\psi \circ T\eta = \psi \circ \eta T = 1_T$ and $\psi \circ \psi T = \psi \circ T\psi$, then one says that the triple $\mathbf{T} := (T, \eta, \psi)$ is a monad.

38. Theorem. *There are monads in the categories $\mathcal{S} \times \mathcal{K}_w$, $\mathcal{S}_l \times \mathcal{K}$, $\mathcal{S} \times \mathcal{K}_{w,l}$ and $\mathcal{S}_l \times \mathcal{K}_l$.*

Proof. Let $\bar{g}(v) := v(g)$ for $g \in C(X,K)$ and $v \in I(X,K)$, where $X \in Ob(\mathcal{S})$ and $K \in Ob(\mathcal{K}_w)$. Therefore, this induces the morphism $\bar{g} : I(X,K) \to K$. Then

$$\overline{g^b \times_2 g}(v) = v(g^b \times_2 g) = b \times_2 v(g) = b \times_2 \bar{g}(v) \text{ and}$$

$$\overline{g \times_2 g^b}(v) = v(g \times_2 g^b) = v(g) \times_2 b = \bar{g}(v) \times_2 b,$$

where $g^b(x) = b$ for each $x \in X$, that is

(1) $\overline{g \times_2 g^b} = \bar{g} \times_2 \overline{g^b}$ and (1') $\overline{g^b \times_2 g} = \overline{g^b} \times_2 \bar{g}$

for each $g \in C(X,K)$ and $b \in K$.

Then we get $\overline{g \vee h}(v) = v(g \vee h) = v(g) \vee v(h) = \bar{g}(v) \vee \bar{h}(v) = (\bar{g} \vee \bar{h})(v)$. Moreover, we deduce that $\overline{g \wedge h}(v) = v(g \wedge h) = v(g) \wedge v(h) = \bar{g}(v) \wedge \bar{h}(v) = (\bar{g} \wedge \bar{h})(v)$. Thus we get the equalities

(2) $\overline{g \vee h} = \bar{g} \vee \bar{h}$ and (2') $\overline{g \wedge h} = \bar{g} \wedge \bar{h}$.

If additionally v is left homogeneous and $K \in Ob(\mathcal{K}_{w,l})$, then $\overline{bg} = v(bg) = bv(g) = b\bar{g}(v)$. Therefore, we infer that $\overline{bg} = b\bar{g}$ for every $b \in K$ and $g \in C(X,K)$.

For $\lambda \in I(I(X,K),K)$ we put $\xi_{X,K}(\lambda)(g) = \lambda(\bar{g})$ for each $g \in C(X,K)$. Then $\xi_{X,K}(\lambda)(g^b) = \lambda(\overline{g^b}) = \lambda(q^b) = b$, where $q^b : I(X,K) \to K$ denotes the constant mapping $q^b(y) = b$ for each $y \in I(X,K)$. From Formulas $(1,1')$ it follows that

$$\xi_{X,K}(\lambda)(g^b \times_2 g) = \lambda(\overline{g^b \times_2 g}) = \lambda(b \times_2 \bar{g}) = b \times_2 \lambda(\bar{g}) = b \times_2 \xi_{X,K}(\lambda)(g) \text{ and}$$

$$\xi_{X,K}(\lambda)(g \times_2 g^b) = \lambda(\overline{g \times_2 g^b}) = \lambda(\bar{g} \times_2 \overline{g^b}) = \lambda(\bar{g}) \times_2 b = \xi_{X,K}(\lambda)(g) \times_2 b.$$

On the other hand, from Formulas $(2,2')$ we get that

$$\xi_{X,K}(v)(g \vee h) = v(\overline{g \vee h}) = v(\bar{g} \vee \bar{h}) = v(\bar{g}) \vee v(\bar{h}) = \xi_{X,K}(v)(g) \vee \xi_{X,K}(v)(h) \text{ and}$$

$$\xi_{X,K}(v)(g \wedge h) = v(\overline{g \wedge h}) = v(\bar{g} \wedge \bar{h}) = v(\bar{g}) \wedge v(\bar{h}) = \xi_{X,K}(v)(g) \wedge \xi_{X,K}(v)(h)$$

for each $b \in K$, $g,h \in C(X,K)$. Thus $\xi_{X,K} : I(I(X,K),K) \to I(X,K)$.

If $\lambda \in I_h(I_h(X,K),K)$ for some $K \in Ob(\mathcal{K}_{w,l})$, then $\xi_{X,K}(\lambda)(bg) = \lambda(\overline{bg}) = \lambda(b\bar{g}) = b\lambda(\bar{g})$, hence $\xi_{X,K} : I_h(I_h(X,K),K) \to I_h(X,K)$. Analogously the mapping $\xi_{X,K} : \mathcal{O}(\mathcal{O}(X,K),K) \to \mathcal{O}(X,K)$ is defined for each $X \in Ob(\mathcal{S})$ and $K \in \mathcal{K}_w$, also $\xi_{X,K} : \mathcal{O}_l(\mathcal{O}_l(X,K),K) \to \mathcal{O}_l(X,K)$, $\xi_{X,K} : I_l(I_l(X,K),K) \to I_l(X,K)$ for each $X \in Ob(\mathcal{S}_l)$ and $K \in \mathcal{K}$, $\xi_{X,K} : I_h(I_h(X,K),K) \to I_h(X,K)$ for $X \in Ob(\mathcal{S})$ and $K \in \mathcal{K}_{w,l}$, $\xi_{X,K} : I_{l,h}(I_{l,h}(X,K),K) \to I_{l,h}(X,K)$ for $X \in Ob(\mathcal{S}_l)$ and $K \in \mathcal{K}_l$. One also puts $\eta : Id_Q \to \mathcal{O}$ or $\eta : Id_Q \to I$ for $Q = \mathcal{S} \times \mathcal{K}_w$, also $\eta : Id_Q \to \mathcal{O}_l$ or $\eta : Id_Q \to I_l$ for $Q = \mathcal{S}_l \times \mathcal{K}$ correspondingly.

Next we verify that the transformations η and ξ are natural for each $f \in Mor(X \times K_1, Y \times K_2)$, i.e. $f = (s,u), s \in Mor(X,Y), u \in Mor(K_1,K_2)$:

$$\eta_{(Y,K_2)} \circ \mathcal{O}((s,u)) = \mathcal{O}(id_Y, id_{K_2})) \circ \mathcal{O}((s,u))$$

$$= \mathcal{O}((s,u)) = \mathcal{O}((s,u)) \circ \mathcal{O}(id_X, id_{K_1})) = \mathcal{O}((s,u)) \circ \eta_{(X,K_1)},$$

$$\xi_{(Y,K_2)} \circ \mathcal{O}((s,u))[\mathcal{O}^2(X,K_1)] = \xi_{(Y,K_2)}(\mathcal{O}(\bar{s},\bar{u})[\mathcal{O}(X,K_1)])$$

$$= \mathcal{O}((s,u)) \circ \eta_{(X,K_1)}[\mathcal{O}^2(X,K_1)]),$$

where $\mathcal{O}^{m+1}(X,K) := \mathcal{O}(\mathcal{O}^m(X,K),K)$ for each natural number m (see also §20 and Proposition 35).

For each $\nu \in \mathcal{O}(X,K)$ and $g \in C(X,K)$ one gets

$$\xi_{X,K} \circ \eta_{(\mathcal{O}(X,K),K)}(\nu)(g) = \eta_{(\mathcal{O}(X,K),K)}(\nu)(\bar{g}) = \bar{g}(\nu) = \nu(g) \text{ and}$$

$$\xi_{X,K} \circ \mathcal{O}(\eta_{(X,K)})(\nu)(g) = (\mathcal{O}(\eta_{(X,K)}(\nu))(\bar{g}) = \nu(\bar{g} \circ \eta_{(X,K)}) = \nu(g).$$

Let now $\tau \in \mathcal{O}^3(X,K)$ and $g \in C(X,K)$, then

$$\xi_{(X,K)} \circ \xi_{\mathcal{O}(X,K)}(\tau)(g) = (\xi_{\mathcal{O}(X,K)}(\tau))(\bar{g}) = \tau(\bar{\bar{g}}) \text{ and}$$

$$\xi_{(X,K)} \circ \mathcal{O}(\xi_{(X,K)})(\tau)(g) = (\mathcal{O}(\xi_{(X,K)})(\tau))(\bar{g}) = \tau(\bar{g} \circ \xi_{(X,K)}) = \tau(\bar{\bar{g}}),$$

where $\bar{\bar{g}} \in C(\mathcal{O}^2(X,K),K)$ is prescribed by the formula $(\bar{\bar{g}})(\nu) = \nu(\bar{g})$ for each $\nu \in \mathcal{O}^2(X,K)$. Thus $\mathbf{O} := (\mathcal{O},\eta,\xi)$ is the monad. Since I is the restriction of the functor $\mathcal{O}$, the triple $\mathbf{I} := (I,\eta,\xi)$ is the monad in the category $\mathcal{S} \times \mathcal{K}_w$ as well. Analogously $\mathbf{O}_l := (\mathcal{O}_l,\eta,\xi)$ and $\mathbf{I}_l := (I_l,\eta,\xi)$ form the monads in the category $\mathcal{S}_l \times \mathcal{K}$; $\mathbf{O}_h = (\mathcal{O}_h,\eta,\xi)$ and $\mathbf{I}_h = (I_h,\eta,\xi)$ are the monads in $\mathcal{S} \times \mathcal{K}_{w,l}$; $\mathbf{O}_{l,h} = (\mathcal{O}_{l,h},\eta,\xi)$ and $\mathbf{I}_{l,h} = (I_{l,h},\eta,\xi)$ are the monads in $\mathcal{S}_l \times \mathcal{K}_l$.

39. Proposition. *If a sequence*

(1) *... $\to K_n \to K_{n+1} \to K_{n+2} \to$ ... in $\mathcal{K}_w$ (or in $\mathcal{K}$) is exact, then sequences*
(2) *... $\to \mathcal{O}_2(X,K_n) \to \mathcal{O}_2(X,K_{n+1}) \to \mathcal{O}_2(X,K_{n+2}) \to$ ... and*
(3) *... $\to I_2(X,K_n) \to I_2(X,K_{n+1}) \to I_2(X,K_{n+2}) \to$ ... are exact (analogously for $\mathcal{O}_{l,2}$ and $I_{l,2}$ correspondingly).*

Proof. A sequence

... $\to K_n \to K_{n+1} \to K_{n+2} \to$ ... is exact means that $s_n(K_n) = ker(s_{n+1})$ for each n, where $s_n : K_n \to K_{n+1}$ is an order-preserving continuous algebraic homomorphism, $ker(s_{n+1}) = s_{n+1}^{-1}(0)$. Each continuous homomorphism s_n induces the continuous homomorphism $\mathbf{s}_n : C(X,K_n) \to C(X,K_{n+1})$ point-wise $(\mathbf{s}_n(f))(x) = s_n(f(x))$ for each $x \in X$. Therefore, we get that $\mathbf{s}_n(f \vee g) = \mathbf{s}_n(f) \vee \mathbf{s}_n(g)$ or $\mathbf{s}_n(f \wedge g) = \mathbf{s}_n(f) \wedge \mathbf{s}_n(g)$, when $f \vee g$ or $f \wedge g$ exists, where $f,g \in C(X,K_n)$. Moreover, the equalities $(\mathbf{s}_n(f+g))(x) = \mathbf{s}_n(f(x)+g(x)) = s_n(f(x)) + s_n(g(x)) = [\mathbf{s}_n(f) + \mathbf{s}_n(g)](x)$ and $[\mathbf{s}_n(fg)](x) = s_n(f(x)g(x)) = s_n(f(x))s_n(g(x)) = [(\mathbf{s}_n(f))(\mathbf{s}_n(g))](x)$ are fulfilled, consequently, $\mathbf{s}_n(C(X,K_n)) = \mathbf{s}_{n+1}^{-1}(0)$, since $f_{n+2} \in C(X,K_{n+2})$ is zero if and only if $f_{n+2}(x) = 0$ for each $x \in X$. Thus the sequence

... $\to C(X,K_n) \to C(X,K_{n+1}) \to C(X,K_{n+2}) \to$ ... is exact.

Then a continuous morphism $\lambda_{n+2} \in \mathcal{O}(X, K_{n+2})$ is zero on $\mathbf{s}_{n+1}(C(X, K_{n+1}))$ if and only if $\lambda_{n+2}(f_{n+2}) = 0$ for each $f_{n+2} \in \mathbf{s}_{n+1}(C(X, K_{n+1}))$. Therefore, $\mathbf{s}_{n+1}(\lambda_{n+1}) = 0 = \lambda_{n+2}$ on $\mathbf{s}_{n+1}[\mathbf{s}_n(C(X, K_n))]$ if and only if $\lambda_{n+1}(f_{n+1}) \in s_n(K_n)$ for each $f_{n+1} \in \mathbf{s}_n(C(X, K_n))$. At the same time we have that $\mathbf{s}_{n+1}[\mathbf{s}_n(C(X, K_n))] \subset \mathbf{s}_{n+1}(C(X, K_{n+1}))$, consequently, $\mathcal{O}_2(s_n) = ker\mathcal{O}_2(s_{n+1})$. Thus the sequences $(2, 3)$ are exact, analogously for other functors I_2, $\mathcal{O}_{l,2}$ and $I_{2,l}$.

3.3. Lattices Associated with Actions of Groupoids on Topological Spaces

40. Lemma. *Let G be a topological groupoid with a unit acting on a topological space X such that to each element $g \in G$ a continuous mapping $v_g : X \to X$ corresponds having the properties*

(1) $v_g v_h = v_{gh}$ for each $g, h \in G$ and

(2) $v_e = id$, where $e \in G$ is the unit element, $id(x) = x$ for each $x \in X$. If K is a topological ringoid with the associative sub-ringoid L, $L \supset \{0, 1\}$, such that

(3) $a(bc) = (ab)c$ for each $a, b \in L$ and $c \in K$, a continuous mapping $\rho : G^2 \to L \setminus \{0\}$ satisfies the cocycle condition

(4) $\rho(g, x)\rho(h, v_g x) = \rho(gh, x)$ and

(5) $\rho(e, x) = 1 \in K$ for each $g, h \in G$ and $x \in X$, then

(6) $T_g f(x) := \rho(g, x)\hat{v}_g f(x)$ is a representation of G by continuous in the $g \in G$ variable mappings T_g of $C(X, K)$ into $C(X, K)$, when f is marked, where $f \in C(X, K)$, $\hat{v}_g f(x) := f(v_g(x))$ for each $g \in G$ and $x \in X$.

Proof. For each $g, h \in G$ one has $T_g(T_h f(x)) = \rho(g, x)\hat{v}_g[\rho(h, x)\hat{v}_h f(x)] = \rho(gh, x)\hat{v}_{gh} f(x) = T_{gh} f(x)$, hence $T_g T_h = T_{gh}$. Moreover, $T_e f = f$, since $v_e = id$ and $\rho(e, x) = 1$, i.e., $T_e = I$ is the unit operator on $C(X, K)$. Mappings $T_g f(x)$ are continuous in the $g \in G$ variable as compositions and products of continuous mappings.

The continuous mappings T_g are (may be) generally non-linear relative to K. If K is commutative, distributive and associative, then T_g are K-linear on $C(X, K)$.

41. Definition. A continuous morphism ν on $C(X, K)$ or $C_+(X, K)$ we call semi-idempotent, if it satisfies the property:

(1) $\nu(g + f) = \nu(g) + \nu(f)$ for each $f, g \in C(X, K)$ or $C_+(X, K)$ respectively, where $(g + f)(x) = g(x) + f(x)$ for each $x \in X$.

Suppose that G is a topological groupoid with the unit continuously acting on a topological space X and satisfying Conditions 40(1,2). A continuous morphism λ on $C(X, K)$ or $C_+(X, K)$ we call (T, G)-invariant if

(2) $\hat{T}_g\lambda = \lambda$, where $(\hat{T}_g\lambda)(f) := \lambda(T_g f)$ for each $g \in G$ and f in $C(X, K)$ or $C_+(X, K)$ correspondingly.

Let $S_+(G, K)$ denote the family of all semi-idempotent continuous morphisms, when K is commutative and associative relative to the addition for (G, K), let also $S_\vee(G, K)$ (or $S_\wedge(G, K)$) denote the family of all continuous morphisms satisfying Conditions 2(4) (or 2(5) correspondingly) for general K. Denote by $H_+(G, K)$ (or $H_\vee(G, K)$ or $H_\wedge(G, K)$) the family of all G-invariant semi-idempotent (or in $S_\vee(G, K)$ or in $S_\wedge(G, K)$ correspondingly) continuous morphisms for (X, K), when $X = G$ as a topological space. We supply these families with the operations of the addition

(3) $\nu(f) +_i \lambda(f) =: (\nu +_i \lambda)(f)$ in $S_j(G, K)$ for $i = 1, 2, 3$ and $j = +, \vee, \wedge$ respectively and the multiplication being the convolution of continuous morphisms

(4) $(\nu * \lambda)(f) = \nu(\lambda(T_g f))$ in $S_j(G, K)$, where $g \in G$, $j \in \{+, \vee, \wedge\}$.

Then we put $H_h(G, K)$, $S_h(G, K)$, $H_{\vee,h}(G, K)$, $S_{\vee,h}(G, K)$, $H_{\wedge,h}(G, K)$ and $S_{\wedge,h}(G, K)$ for the subsets of all left homogeneous morphisms in $H_+(G, K)$, $S_+(G, K)$, $H_\vee(G, K)$, $S_\vee(G, K)$, $H_\wedge(G, K)$, $S_\wedge(G, K)$ correspondingly.

42. Proposition. *If ν is a (T,G)-invariant semi-idempotent continuous morphism, then its support is contained in $\bigcap_{n=1}^{\infty} T^n(X)$, where*

$$T(A) := \bigcup_{g\in G} supp(\rho(g,x)\hat{v}_g(\chi_A(x)))$$

for a closed subset A in X. Moreover, if K has not divisors of zero a support of ν is G-invariant and contained in $\bigcap_{n=1}^{\infty} P^n(X)$, where

$$P(X) = \bigcup_{g\in G} v_g(X).$$

Proof. If $\nu(f) \neq 0$, then $\nu(T_g f) \neq 0$ for each $g \in G$, when a continuous morphism ν is (T,G)-invariant. On the other hand, if $supp(f) \subset supp(\nu)$, then $supp(\rho(g,x)\hat{v}_g f(x)) \subset supp(\nu)$ for each $g \in G$. At the same time, $\bigcup_{g\in G} supp(T_g f) \subset \bigcup_{g\in G} supp(\hat{v}_g f)$, since $\rho(g,x) \in L\setminus\{0\}$ for each $g \in G$ and $x \in X$. If $f = \chi_{supp(\nu)}$, then $supp(\nu) \subset T(supp(\nu)) \subset T(X)$, hence by induction we deduce that $supp(\nu) \subset T^n(X)$ for each natural number n, where χ_A is the characteristic function of a set A, so that $\chi_A(x) = 1$ for each $x \in A$ while $\chi_A(x) = 0$ for each $x \notin A$.

If K has not divisors of zero, then $supp(\hat{T}_g\nu) = \hat{v}_g supp(\nu) \subset supp(\nu)$ for each element $g \in G$, hence $\bigcup_{g\in G} \hat{v}_g supp(\nu) = supp(\nu)$, since $e \in G$ and $v_e = id$. That is $supp(\nu)$ is G-invariant. Since $supp(\nu) \subset X$, then $supp(\nu) \subset P(X)$ and by induction $supp(\nu) \subset P^n(X)$ for each natural number n.

43. Proposition. *If G is a topological groupoid with a unit or a topological monoid, then $S_+(G,K)$, $S_\vee(G,K)$ and $S_\wedge(G,K)$ for general T_g and K (or $S_h(G,K)$, $S_{h,\vee}(G,K)$ and $S_{h,\wedge}(G,K)$ for $T_g \equiv \hat{v}_g$ or when K is commutative and associative relative to the multiplication) supplied with the convolution 41(4) as the multiplication operation are topological groupoids with a unit or monoids correspondingly.*

Proof. Certainly, the definitions above imply the inclusion $S_h(G,K) \subset S_+(G,K)$. If $\nu,\lambda \in I_h(G,K)$, then $(\nu*\lambda)(bf) = \nu(\lambda(T_g(bf))) = \nu(b\lambda(T_g f)) = b((\nu*\lambda)(f))$, when either $T_g \equiv \hat{v}_g$ for each $g \in G$ or K is commutative and associative relative to the multiplication. We mention that the evaluation morphism δ_e at e belongs to $S_h(G,K)$ and has the property $\nu*\delta_e = \delta_e*\nu = \nu$ for each $\nu \in S(G,K)$, where e is a unit element in G, $\delta_x f = f(x)$ for each $f \in C(X,K)$ and $x \in X$. Thus δ_e is the neutral element in $S(G,K)$.

For a topological monoid G one has $\hat{v}_s(\hat{v}_u f(x)) = f(s(ux)) = f((su)x) = \hat{v}_{su} f(x)$ for each $f \in C(G,K)$ and $s,u,x \in G$ so that $f((su)x)$ is a function continuous in the variables s, u and x in G. Since ν and λ are continuous on $C(G,K)$, then $\nu*\lambda$ is continuous on $C(G,K)$.

If G is a topological monoid, then $(\nu*(\lambda*\phi))(f) = \nu^u((\lambda*\phi)(T_u f)) = \nu^u(\lambda^s(\phi(T_s T_u f)) = \nu^u(\lambda^s(\phi(T_{su} f))) = (\nu*\lambda)^{su}(\phi(T_{su} f)) = [(\nu*\lambda)*\phi](f)$ for every $f \in C(G,K)$ and $u,s \in G$ and $\nu,\lambda,\phi \in S_j(G,K)$, where $\nu^u(h)$ means that a continuous morphism ν on a function h acts by the variable $u \in G$, consequently, $\nu*(\lambda*\phi) = (\nu*\lambda)*\phi$. Thus the family $S_j(G,K)$ is associative, when G is associative, where $j \in \{+,\vee,\wedge,h,(h,\vee),(h,\wedge)\}$ for the corresponding T_g and K.

From §§2.3, 2.4, 2.9 and 5 it follows that the mapping $(\nu,\lambda) \mapsto \nu*\lambda$ is continuous.

44. Theorem. *If G is a topological groupoid with a unit or a topological monoid, then $S_+(G,K)$ (for K commutative and associative relative to $+$), $S_\vee(G,K)$ and $S_\wedge(G,K)$ for general T_g (or $S_{\vee,h}(G,K)$ and $S_{\wedge,h}(G,K)$ for either $T_g \equiv \hat{v}_g$ or when K is commutative and associative relative to the multiplication) are topological ringoids or semirings correspondingly.*

Proof. If $f,g \in C(X,K)$ or in $C_+(X,K)$ and $f\vee g$ or $f\wedge g$ exists (see Condition (3) in Lemma 2.12), ν,λ are continuous morphisms satisfying Condition either 2(4) or 2(5) respectively, then

(1) $(\nu +_i \lambda)(f +_i g) = \nu(f +_i g) +_i \lambda(f +_i g) = (\nu(f) +_i \nu(g)) +_i (\lambda(f) +_i \lambda(g)) = (\nu(f) +_i \lambda(f)) +_i (\nu(g) +_i \lambda(g)) = (\nu +_i \lambda)(f) +_i (\nu +_i \lambda)(g)$
for $i = 1,2,3$, where $+_1 = +$, $+_2 = \vee$, $+_3 = \wedge$. That is, the continuous morphism $\nu +_i \lambda$ satisfies Property 41(1) for $i = 1$ or 2(4) for $i = 2$ or 2(5) when $i = 3$ correspondingly. If additionally ν and λ are left homogeneous, then

(2) $(\nu +_i \lambda)(bf) = \nu(bf) +_i \lambda(bf) = b\nu(f) +_i b\lambda(f) = b(\nu +_i \lambda)(f)$ for each $b \in K$.

On the other hand, we deduce that
$((\nu_1 +_i \nu_2) * \lambda)(f) = (\nu_1 +_i \nu_2)(\lambda(T_g f)) = \nu_1(\lambda(T_g f)) +_i \nu_2(\lambda(T_g f))$
$= (\nu_1 * \lambda)(f) +_i (\nu_2 * \lambda)(f)$ and
$(\lambda * (\nu_1 +_i \nu_2))(f) = \lambda((\nu_1 +_i \nu_2)(T_g f)) = \lambda(\nu_1(T_g f)) +_i \lambda(\nu_2(T_g f))$
$= (\lambda * \nu_1)(f) +_i (\lambda * \nu_2)(f)$
for each $\nu_1, \nu_2, \lambda \in S_j(G,K)$ and $f \in C(G,K)$ or in $C_+(G,K)$ correspondingly, for $i = 1,2,3$ and $i = i(j)$ respectively, where $+_1 = +$, $+_2 = \vee$ and $+_3 = \wedge$. Thus, the right and left distributive rules are satisfied:

(3) $(\nu_1 +_i \nu_2) * \lambda = \nu_1 * \lambda +_i \nu_2 * \lambda$ and
(4) $\lambda * (\nu_1 +_i \nu_2) = \lambda * \nu_1 +_i \lambda * \nu_2$
for $i = 1,2,3$ respectively. From the definitions of these operations and Proposition 43 their continuity follows.

Therefore, Formulas $(1-4)$ and Proposition 43 imply that $S_+(G,K)$, $S_\vee(G,K)$, $S_\wedge(G,K)$, $S_{\vee,h}(G,K)$ and $S_{\wedge,h}(G,K)$ are left and right distributive topological ringoids or semirings correspondingly.

45. Theorem. *If G is a topological groupoid with a unit, $X = G$ as a topological space (see §41), then $H_j(G,K)$ is a closed ideal in $S_j(G,K)$, where $j = +$ (for K commutative and associative relative to $+$) or $j = \vee$ or $j = \wedge$ or $j = (\vee, h)$ or $j = (\wedge, h)$ with $\rho(u,x) \equiv 1$; $j = (\vee, h)$ or $j = (\wedge, h)$ for commutative and associative K relative to the multiplication with general T_u.*

Proof. We mention that $\hat{T}_g(b_1\lambda_1 +_i b_2\lambda_2)(f) = b_1\lambda_1(T_g f) +_i b_2\lambda_2(T_g f)$, where the operation denoted by the addition $+_i$ is either $+$ or $\vee$ or $\wedge$ for $i = 1$ or $i = 2$ or $i = 3$ correspondingly (and also below in this section), consequently, $b_1\lambda_1 +_i b_2\lambda_2 \in H_j(G,K)$ for each $\lambda_1, \lambda_2 \in H_j(G,K)$ and $b_1, b_2 \in K$, $i = i(j)$.

In Formula 41(4) after the action of a morphism λ on a continuous function $T_g f(x)$ in the variable x one gets that $\lambda(T_g f) =: h(g)$ is a continuous function in the variable g and ν is acting on this function, i.e. $(\nu * \lambda)(f) = \nu(h(x))$, where $x, g \in G$. This implies that
$\nu * (\lambda(f +_i t)) = \nu * (\lambda(f) +_i \lambda(t)) = \nu(\lambda(T_g f) +_i \lambda(T_g t))$
$= \nu(\lambda(T_g f)) +_i \nu(\lambda(T_g t)) = (\nu * \lambda)(f) +_i (\nu * \lambda)(t)$ for $i = 1,2,3$,
consequently, the convolution operation maps from $S_j(G,K)^2$ into $S_j(G,K)$.

The property being G-invariant provides closed subsets in $S_j(G,K)$, since if a net of continuous mappings g_k converges to a continuous mapping g an each g_k is G-invariant, then $g = \lim_k g_k$ is G-invariant as well.

If $\lambda \in H_j(G,K)$ and $\nu \in S_j(G,K)$, then
$(\hat{T}_s(\nu * \lambda))(f) = \hat{T}_s(\nu^u(\lambda^x(T_u f(x)))) = \nu^u(\lambda^x(T_s(T_u f(x)))$
$= \nu^u(\lambda^x(T_u f(x)))) = (\nu * \lambda)(f)$ and
$(\hat{T}_s(\lambda * \nu))(f) = \hat{T}_s(\lambda^u(\nu^x(T_u f(x)))) = \lambda^u(\nu^x(T_s(T_u f(x))))$
$= \lambda^u(T_s(\nu^x(T_u f(x)))) = (\lambda * \nu)(f)$,
since $\lambda^u(T_s g(u)) = \lambda^u(g(u)) = \lambda(g)$, particularly with $g(x) = T_u f(x)$ or $g(u) = \nu^x(T_u f(x))$ correspondingly, whilst $T_s \equiv \hat{v}_s$ in the cases $j = +$ or $j = \vee$ or $j = \wedge$ with $\rho \equiv 1$, or for general $T_u f(x) = \rho(u,x)\hat{v}_s f(x)$ in the cases of homogeneous continuous morphisms $j = (\vee, h)$ or $j = (\wedge, h)$ (see §43 also), hence $\nu * \lambda$, $\lambda * \nu \in H_j(G,K)$. Therefore, the latter formula and Theorem 44 imply that
$(\nu +_i H_j(G,K)) * H_j(G,K) \subset (\nu * H_j(G,K)) +_i (H_j(G,K) * H_j(G,K))$
$\subset H_j(G,K) +_i H_j(G,K) \subset H_j(G,K)$ and
$H_j(G,K) * (\nu +_i H_j(G,K)) \subset (H_j(G,K) * \nu) +_i (H_j(G,K) * H_j(G,K))$
$\subset H_j(G,K) +_i H_j(G,K) \subset H_j(G,K)$
for each $\nu \in S_j(G,K)$ and $+_i$ corresponding to j, that is $H_j(G,K)$ is the right and left closed ideal in $S_j(G,K)$.

4. Conclusions

Skew continuous morphisms of ordered ringoids, semirings, algebroids and non-associative algebras can be used for studies of their structures and representations.

Conflicts of Interest: The author declares no conflict of interest.

References

1. Baez, J.C. The octonions. *Bull. Am. Mathem. Soc.* **2002**, *39*, 145–205.
2. Birkhoff, G. *Lattice Theory*; Mathematical Society: Providence, RI, USA, 1967.
3. Bourbaki, N. *Algebra*; Springer: Berlin, German, 1989.
4. Dickson, L.E. The Collected Mathematical Papers; Chelsea Publishing Co.: New York, NY, USA, 1975; Volumes 1–5.
5. Grätzer, G. *General Lattice Theory*; Akademie-Verlag: Berlin, German, 1978.
6. Kasch, F. *Moduln und Ringe*; Teubner: Stuttgart, German, 1977.
7. Kurosh, A.G. *Lectures on General Algebra*; Nauka: Moscow, Russian, 1973.
8. Schafer, R.D. *An Introduction to nonassociative Algebras*; Academic Press: New York, NY, USA, 1966.
9. Litvinov, G.L.; Maslov, V.P.; Shpiz. G.B. Idempotent functional analysis: an algebraic approach. *Math. Notes* **2001**, *65*, 696–729.
10. Ludkovsky, S.V. Topological transformation groups of manifolds over non-Archimedean fields, representations and quasi-invariant measures. *J. Mathem. Sci. NY Springer* **2008**, *147*, 6703–6846.
11. Ludkovsky, S.V. Topological transformation groups of manifolds over non-Archimedean fields, representations and quasi-invariant measures, II. *J. Mathem. Sci., N. Springer* **2008**, *150*, 2123–2223.
12. Ludkovsky, S.V. Operators on a non locally compact group algebra. *Bull. Sci. Math. Paris Ser. 2* **2013**, *137*, 557–573, doi:10.1016/j.bulsci.2012.11.008.
13. Ludkovsky, S.V. Meta-centralizers of non locally compact group algebras. *Glasg. Mathem. J.* **2015**, *57*, 349–364, doi:10.1017/S0017089514000330.
14. Weil, A. *L'intégration Dans Les Groupes Topologiques et Ses Applications*; Hermann: Paris, France, 1940.
15. Bucur, I.; Deleanu, A. *Introduction to the Theory of Categories and Functors*; Wiley: London, UK, 1968.
16. Fedorchuk, V.V. Covariant functors in the category of compacta, absolute retracts and *Q*-manifolds. *Rissian Math. Surv.* **1981**, *36*, 211–233.
17. Mitchell, B. *Theory of Categories*; Academic Press, Inc.: London, UK, 1965.
18. Shchepin, E.V. Functors and uncountable powers of compacta. *Russ. Math. Surv.* **1981**, *36*, 1–71.
19. Mendelson, E. *Introduction to Mathematical Logic*; D. van Nostrand Co., Inc.: Princeton, NJ, USA, 1964.
20. Kunen, K. *Set Theory*; North-Holland Publishing Co.: Amsterdam, Dutch, 1980.
21. Engelking, R. *General Topology*; Heldermann: Berlin, German, 1989.
22. Barwise, J. Ed. *Handbook of Mathematical Logic*; North-Holland Publishing Co.: Amsterdam, Dutch, 1977.
23. Ludkovsky, S.V. Properties of quasi-invariant measures on topological groups and associated algebras. *Ann. Math. Blaise Pascal* **1999**, *6*, 33–45.

Permissions

All chapters in this book were first published in Mathematics, by MDPI; hereby published with permission under the Creative Commons Attribution License or equivalent. Every chapter published in this book has been scrutinized by our experts. Their significance has been extensively debated. The topics covered herein carry significant findings which will fuel the growth of the discipline. They may even be implemented as practical applications or may be referred to as a beginning point for another development.

The contributors of this book come from diverse backgrounds, making this book a truly international effort. This book will bring forth new frontiers with its revolutionizing research information and detailed analysis of the nascent developments around the world.

We would like to thank all the contributing authors for lending their expertise to make the book truly unique. They have played a crucial role in the development of this book. Without their invaluable contributions this book wouldn't have been possible. They have made vital efforts to compile up to date information on the varied aspects of this subject to make this book a valuable addition to the collection of many professionals and students.

This book was conceptualized with the vision of imparting up-to-date information and advanced data in this field. To ensure the same, a matchless editorial board was set up. Every individual on the board went through rigorous rounds of assessment to prove their worth. After which they invested a large part of their time researching and compiling the most relevant data for our readers.

The editorial board has been involved in producing this book since its inception. They have spent rigorous hours researching and exploring the diverse topics which have resulted in the successful publishing of this book. They have passed on their knowledge of decades through this book. To expedite this challenging task, the publisher supported the team at every step. A small team of assistant editors was also appointed to further simplify the editing procedure and attain best results for the readers.

Apart from the editorial board, the designing team has also invested a significant amount of their time in understanding the subject and creating the most relevant covers. They scrutinized every image to scout for the most suitable representation of the subject and create an appropriate cover for the book.

The publishing team has been an ardent support to the editorial, designing and production team. Their endless efforts to recruit the best for this project, has resulted in the accomplishment of this book. They are a veteran in the field of academics and their pool of knowledge is as vast as their experience in printing. Their expertise and guidance has proved useful at every step. Their uncompromising quality standards have made this book an exceptional effort. Their encouragement from time to time has been an inspiration for everyone.

The publisher and the editorial board hope that this book will prove to be a valuable piece of knowledge for researchers, students, practitioners and scholars across the globe.

List of Contributors

Sudip Ratan Chandra
Department of Mathematics, Jadavpur University, West Bengal 700032, India

Diganta Mukherjee
Indian Statistical Institute, Kolkata, West Bengal 700108, India

André Liemert and Alwin Kienle
Institut für Lasertechnologien in der Medizin und Meßtechnik an der Universität Ulm, Helmholtzstr. 12, D-89081 Ulm, Germany

Alastair A. Abbott and Cyril Branciard
Institut Néel, CNRS and Université Grenoble Alpes, 38042 Grenoble Cedex 9, France

Pierre-Louis Alzieu
Institut Néel, CNRS and Université Grenoble Alpes, 38042 Grenoble Cedex 9, France
Ecole Normale Supérieure de Lyon, 69342 Lyon, France

Michael J. W. Hall
Centre for Quantum Computation and Communication Technology (Australian Research Council), Centre for Quantum Dynamics, Griffith University, Brisbane 4111, Australia

Carlos Meniño Cotón
Instituto de Matemática, Universidade Federal do Rio de Janeiro, Rio de Janeiro 21941-909, Brazil

Ali Karcı
Department of Computer Engineering, Faculty of Engineering, İnönü University, 44280 Malatya, Turkey

Nur Alam
Department of Mathematics, Pabna University of Science & Technology, Pabna 6600, Bangladesh

Fethi Bin Muhammad Belgacem
Department of Mathematics, Faculty of Basic Education, PAAET, Al-Ardhiya 92400, Kuwait

Mohammad Shahzad
Nizwa College of Applied Sciences, Ministry of Higher Education, Nizwa 611, Oman

Erin Denette and Mustafa R. S. Kulenović
Department of Mathematics, University of Rhode Island, Kingston, RI 02881-0816, USA

Esmir Pilav
Department of Mathematics, University of Sarajevo, 71000 Sarajevo, Bosnia and Herzegovina

Rayaprolu Bharavi Sharma
Department of Mathematics, Kakatiya University, Warangal, Telangana-506009, India

Kalikota Rajya Laxmi
Department of Mathematics, SRIIT, Hyderabad, Telangana-501301, India

Amritansu Ray
Department of Mathematics, Rajyadharpur Deshbandhu Vidyapith, Serampore, Hooghly 712203, West Bengal, India

S. K. Majumder
Department of Mathematics, Indian Institute of Engineering Science and Technology (IIEST), Shibpur, Howrah 711103, West Bengal, India

Igor Korepanov
Moscow Technological University, 20 Stromynka Street, Moscow 107076, Russia

Abdul Khaliq
Department of Mathematics, Faculty of Science, King Abdulaziz University, P.O. Box 80203, Jeddah 21589, Saudi Arabia

E.M. Elsayed
Department of Mathematics, Faculty of Science, King Abdulaziz University, P.O. Box 80203, Jeddah 21589, Saudi Arabia
Department of Mathematics, Faculty of Science, Mansoura University, Mansoura 35516, Egypt

Sergey Victor Ludkowski
Department of Applied Mathematics, Moscow State Technical University MIREA, avenue Vernadsky 78, Moscow 119454, Russia

Index

A

Abelian Groups, 50, 53

Algebraic Realizations, 149

Analytic Functions, 121-123, 131

Analytic Properties, 70, 73, 77, 79, 83

Aspherical Optical Cavities, 19

Associative Semirings, 179

B

Barrier Option, 1-5, 9, 11, 17

Berezin Integrals, 149, 151, 163

Bi-starlike Functions, 122

Bi-univalent Functions, 121-123, 130

Birkhoff Normal Forms, 109, 111, 113, 115, 117, 120

Boundedness, 51, 109, 165-166, 169, 176

Burg's Modified Entropy, 132-133, 135, 137, 139, 143, 146-147

C

Cancer Progression, 98

Cancerous Abnormal Cells, 98

Caputo Fractional Derivative, 19

Chaos Control, 98-99, 101, 105, 107-108

Chaotic Dynamics, 98-99, 107

Coefficient Inequalities, 121, 123-124, 129, 131

Cohomology, 50-51, 53-57, 59-63, 65-69

D

Derivative Operator, 70, 83

Difference Equation, 109-110, 120, 165-167, 169, 175, 177-178

E

Entropic Uncertainty Relations, 33-36, 44, 47-48

Entropy Optimization, 132, 147-148

Ergodic Group Action, 50

Exact Solutions, 85-86, 91-92, 95-97

Expansion Method, 85-86, 91, 94-97

F

Fermionic Relations, 149

First-passage Distributions, 1

Foliated Cocycles, 50, 66

Foliations, 50-52, 59, 61-62, 66-69

Föllmer-schweizer Minimal Measure, 3

Fractional Calculus, 32, 70, 84

Fractional Order Derivative, 70, 84

Fractional Order Derivatives, 70, 73, 77, 79, 83

Fractional Schrödinger Equation, 19-20, 23-25, 27, 31-32

G

Global Attractor, 165, 167, 169

Grassmann-gaussian Exponent, 149

H

Harmonic Potential, 19

Heisenberg's Discussion, 33

Holonomy Grupoids, 50

I

Inflection Points, 70, 83

Interconnection, 50

Invariant Measures, 50-52, 61, 66-67, 179, 202

K

Kam(kolmogorov-arnold-moser), 109

L

L'hôpital's Rule, 70

Lattice Ringoids, 179, 181, 183, 185, 187, 189, 191, 193, 195, 197, 199, 201

Lévy Model, 1, 3, 5, 9, 11, 17

Lie Algebra, 98-99, 108

Linear Potential, 19-20, 23, 25, 27, 31-32

M

Maximum Entropy Probability Density (mepd), 132

Mellin Transform, 1-3, 5, 9, 11, 16-18

Microtubules Nonlinear Models Dynamics, 85, 87, 91, 95, 97

Mittag-leffler Matrix Function, 19, 31

Multiplicative Expression, 149, 151, 153, 155, 157, 159, 161, 163

N

Non-associative Algebras, 179, 181, 202

Non-associative Ringoids, 179

P

Pachner Moves, 149-151, 164

Partial Differential Equations(pdes), 85

Partial Integro-differential Equation (pide), 1-2, 5

Periodic Solutions, 85, 92, 96, 109, 111, 177

Position-momentum Uncertainty Relation, 33-34

Pricing Expression, 1

Probability Distribution, 47, 132, 134-137, 142, 146-147

Propagation Dynamics, 19, 32

Proportionality Coefficient, 149

Q

Qualitative Properties, 165, 167, 169, 175, 177

Quantum Measurement, 33

Quantum Mechanics, 19, 24-25, 31-33, 35, 48, 179

R

Rational Difference Equations, 110, 165-166, 176-178

Rational Map, 109

Riesz Fractional Derivative, 19

Robertson-schrödinger Uncertainty Relations, 33

S

Second Hankel Determinants, 121, 123, 129, 131

Shannon's Entropy, 132, 135-136, 146-147

Skew Continuous Morphisms, 179, 181, 183, 185, 187, 189, 191, 193, 195, 197, 199, 201-202

Skew Morphisms, 179, 186

Solitary Solutions, 85

State Space Exact Linearization, 98-99, 101

Statistical Mechanics, 132-133, 135-137, 139, 142-143, 145, 147-148

Stochastic Model, 1

T

Taylor Maclaurin's Series, 122

Three Dimensional Cancer Model, 98-99, 101, 105, 107

Time Reversal, 109, 111, 113, 115, 117

Time Reversal Symmetry, 109, 111, 113, 115, 117

Trigonometric Solutions, 85

U

Uncertainty Relations, 33-37, 40, 43-49

Univalent Functions, 121-123, 130-131

Printed in the USA
CPSIA information can be obtained
at www.ICGtesting.com
JSHW051324221024
72173JS00006B/1288

9 781632 385543